NEW RESEARCH ON SUPERCONDUCTIVITY

New Research on Superconductivity

Barry P. Martins
Editor

Nova Science Publishers, Inc.
New York

For permission to use material from this book please contact us:
Telephone 631-231-7269; Fax 631-231-8175
Web Site: http://www.novapublishers.com

Library of Congress Cataloging-in-Publication Data
Available Upon Request

ISBN 978-1-59454-197-1

Published by Nova Science Publishers, Inc. ✢ New York

CONTENTS

PREFACE

Superconductivity is the ability of certain materials to conduct electrical current with no resistance and extremely low losses. High temperature superconductors, such as La2-xSrxCuOx (Tc=40K) and YBa2Cu3O7-x (Tc=90K), were discovered in 1987 and have been actively studied since. In spite of an intense, world-wide, research effort during this time, a complete understanding of the copper oxide (cuprate) materials is still lacking. Many fundamental questions are unanswered, particularly the mechanism by which high-Tc superconductivity occurs. More broadly, the cuprates are in a class of solids with strong electron-electron interactions. An understanding of such "strongly correlated" solids is perhaps the major unsolved problem of condensed matter physics with over ten thousand researchers working on this topic. High-Tc superconductors also have significant potential for applications in technologies ranging from electric power generation and transmission to digital electronics. This ability to carry large amounts of current can be applied to electric power devices such as motors and generators, and to electricity transmission in power lines. For example, superconductors can carry as much as 100 times the amount of electricity of ordinary copper or aluminum wires of the same size. Many universities, research institutes and companies are working to develop high-Tc superconductivity applications and considerable progress has been made. This new volume brings together new leading-edge research in the field.

Chapter 1 - In this article, the authors provide electrical, magnetic, tunneling, and Raman spectroscopic evidence for room-temperature superconductivity in carbon nanotubes. Electrical measurements indicate that the superconducting transition temperatures in carbon nanotubes can vary from 0.44 K to 750~K. The temperature dependencies of the resistive transitions agree quantitatively with the well-established Langer-Ambegaokar-McCumber-Halperin (LAMH) theory for the resistive transition of quasi-one-dimensional superconductors. Because of a finite number of transverse conduction channels, the four-probe resistance of an individual tube will not go to zero in the superconducting state. This would lead to a false impression that these tubes are not superconducting. Nonetheless, the on-tube resistance at room temperature is found to be indistinguishable from zero for many individual multi-walled nanotubes. The very small but finite room-temperature on-tube resistance is consistent with quantum phase slips due to the finite number of transverse channels. Magnetic measurements show a small Meissner effect in aligned and physically separated multi-walled nanotubes up to room temperature. The small Meissner effect is due to the fact that the diameter of the tubes is much smaller than the magnetic penetration depth.

At low temperatures, the directionally averaged diamagnetic susceptibility for physically coupled multi-walled nanotubes increases by a factor of 4.3 compared with that for physically separated tubes, which is consistent with the enhancement of the Meissner effect due to the Josephson coupling among the superconducting tubes. Further, Raman data and single-particle tunneling spectra consistently show the existence of single particle excitation gaps of larger than 100 meV. The authors suggest microscopic pairing mechanism for high-temperature superconductivity in carbon nanotubes.

Chapter 2 - Since the discovery of superconductivity in the hydrous cobaltate $Na_xCoO_2{\cdot}yH_2O$, intensive studies, particularly for the synthesis of superconductors with high transition temperature, have been performed. As a result, various phase diagrams have been reported, in which the Na content and the period of time after intercalation of water have been employed as tuning parameters. Recently, the valence of Co was investigated and the presence of oxonium H_3O^+ was recognized. It has been accepted that the chemical formula for the superconducting cobaltate can be expressed as $Na_x(H_3O)_zCoO_2{\cdot}yH_2O$, which includes at least three parameters x, y, and z. Magnetic ordering was discovered in a sample with a large z content, but with the same crystal structure as the superconducting sample. Much attention has been paid to the relationship between superconductivity and magnetism. In order to investigate the relationship, the development of the phase diagram is desirable. It is necessary to find the crucial physical parameters that are responsible for the superconductivity, and give consideration to the superconducting mechanisms.

The authors succeeded in controlling the chemical composition by a modified soft-chemical method, and established a phase diagram using samples with various chemical compositions. It was found that two compositional parameters, Na content (x) and Co valence ($s = 4\text{-}x\text{-}z$), are necessary to determine the physical properties of the compound. In the case of $s = +3.48$, the superconducting region has a dome-shape in the range of $0.33 < x < 0.38$, while in the case of $s = +3.40$, the magnetically ordered phase appears in the range of $0.33 < x < 0.35$, sandwiched by two superconducting regions at $x < 0.33$ and $x > 0.35$. The previously reported phase diagrams may be certain projections of the x-s-T phase diagram onto the x-T or s-T plane.

Microscopic experiments are powerful tools used for the study of superconductivity. They found that the nuclear quadrupole resonance frequency of Co, which is strongly sample-dependent, is a good parameter to use for representation of the phase diagram. With the increasing frequency of Co, the superconducting transition temperature increases, and the magnetically ordered phase appears. Our analysis strongly suggests that the change in the frequency corresponds to pressing of the CoO_2 layer along the c axis, and therefore the microscopic phase diagram indicates that the superconductivity and magnetism of this system are governed by the thickness of the CoO_2 layer. This scenario is consistent with the chemical phase diagram, because the ion exchange between Na and H_3O ions would change the thickness of the CoO_2 layer. In addition, the spin-lattice relaxation rate indicates that the superconductivity is closely related to quantum critical magnetic fluctuations, which is a key factor for understanding the superconductivity.

In this article, the authors summarize the sample dependence of various physical parameters and present the phase diagrams, which are based on chemical analyses and nuclear quadrupole measurements. The superconducting mechanisms are also discussed on the basis of our experimental results.

Chapter 3 - The tunneling studies by the authors are reviewed on the following matters: Determination of the spectral function $\alpha^2F(\omega)$ for slightly overdoped $Bi_2Sr_2CaCu_2O_8$ (Bi2212), good correspondence of $\alpha^2F(\omega)$ with the phonon structures, corresponding phonon modes calculated by Paltzer and De Wette, energy gap $2\Delta(0)$ and critical temperature Tc estimated from $\alpha^2F(\omega)$, layer number dependence of Tc, and unusually large $2\Delta(0)/k_BT_c$ for underdoped Bi2212. It is emphasized that the large $2\Delta(0)/k_BT_c$ for underdoped Bi2212 cannot be explained by a usual electron-phonon (e-ph) interaction alone and that this is as substantial as a high value of Tc for high-Tc superconductivity.

$\alpha^2F(\omega)$ was applied for calculating dc resistivity, spin-lattice relaxation rate, isotope effect on Tc, dispersion relation and damping rate of a quasi-particle, and optical damping rate. They are compared with corresponding experimental ones. The comparison determines what is explained and what is not explained by the e-ph interaction.

Furthermore, the authors discuss a role of spin in high-Tc superconductivity. For explaining the large $2\Delta(0)/k_BT_c$ for underdoped Bi2212, a pair-break model by Cu magnetic moments is proposed. Finally, the authors propose an experiment for clarifying whether a spin enhancement of the e-ph interaction really occurs or not.

Chapter 4 - The authors describe the behavior of the Bose-Einstein condensation (BEC) critical temperature in a relativistic ideal Bose gas with and without the antibosons expected to be pair-produced abundantly at sufficiently hot temperatures. Based on the respective similarities between bosons/antibosons and two-electron (2e)/two-hole (2h) Cooper pairs (CPs) in an interacting nonrelativistic many-fermion system as described by a generalized BEC theory of superconductivity, the authors hope to elucidate the empirically exceptionally efficient role of hole charge carriers in elevating the critical temperature of superconductors (including room-temperatures *with electron-phonon interactions* as compared with electron charge carriers. The latter is done within a so-called *ideal* complete boson-fermion (BF) model of superconductivity which is a non-interacting ternary BF mixture consisting of electrons and *both* 2e- and 2h-CPs. It is the BE condensate of the 2h-CPs that emerges with the highest T_cs. This is consistent with empirical trends for either ``exotic" or conventional superconductors.

Chapter 5 - The resistivity, Hall effect and Nernst effect of overdoped $Bi_2Sr_2CaCu_2O_{8+\delta}$ single crystals were studied to investigate superconducting fluctuations above T_c. The superconducting transition temperature (T_{c0}) for the three measured samples with various oxygen content is 86.5, 83 and 77 K, and their doping level (hole concentration, p) is estimated to be 0.180, 0.193 and 0.204 respectively. The pseudogap opening temperature (T^*) determined from the temperature dependence of resistivity decreases slightly from 126 K to 100 K with increasing doping level p. The temperature scale (T_1) determined from the cotangent of Hall angle ($\cot\theta$) is consistent with the onset temperature (T_v) of vortex-like excitation detected by Nernst effect, and both decrease with increasing p. Moreover, none of the three temperature scales (T^*, T_1 and T_v) drops to zero as p increases across the proposed quantum critical point (QCP) of p ~0.19. This result indicates that both pseudogap opening and superconducting fluctuations exist above T_c in the overdoped region of electronic phase diagram of cuprates and there is no evidence for QCP at $p \sim 0.19$. The authors suggest that the temperature range between T_c and T_v (or T_1) is characterized by finite pairing amplitude with strong phase fluctuations. As T is above T_v, the phase correlation disappears completely and only the pairing amplitude fluctuations remain. The strong superconducting phase fluctuation in the pseudogap regime implies that the pseuodgap is a precursor to superconductivity.

Chapter 6 - Superconductivity has been recently found in two Li containing compounds, Li_2Pd_3B and Li_2Pt_3B. They show superconducting transition at the temperatures, 7.2 K and 2.6 K respectively. Those materials have the same cubic structure ($P4_332$) composed of distorted octahedrons, whose corners are occupied by Pd/Pt. This structure does not have mirror or inversion symmetry along any directions. It is a very interesting feature of those materials in relation to the symmetry of superconductivity. Their superconductivity is induced by Li addition. Controlling the Li stoichiometry is a key to synthesize high quality samples in the preparation process of arc-melting because Li is volatile at high temperatures. Specific heat of these compounds was measured for confirming their bulk superconductivity and investigating superconducting properties. The values of dC/T_C were evaluated to be around 17 mJ/mol K^2 for Li_2Pd_3B and 7.2 mJ/mol K^2 for Li_2Pt_3B. Electronic specific heat coefficient (γ) and Debye temperature (θ_D) were derived from the normal state data in a magnetic field over H_{C2} and lead to 8.3 mJ/mol K^2 and 202 K for Li_2Pd_3B, 9.6 mJ/mol K^2 and 240 K for Li_2Pt_3B, respectively. Pressure effect was also measured with a SQUID magnetometer using a low temperature hydrostatic micro pressure cell at pressures up to around 1 GPa. The superconducting transition temperatures of both Li_2Pd_3B and Li_2Pt_3B decreased linearly with pressure. The pressure dependence, dT_C/dP, were –0.091 and –0.090 K/GPa, respectively. These values are comparable to those of conventional metallic superconductors, but much smaller than the value, –1.03 K/GPa, of MgB_2.

Chapter 7 - The authors consider the $\lambda\phi^4$ theory in the D-dimensional Euclidean space with d ($d \leq D$) compactified dimensions. Parameterizing the mass term in the form $m^2(T)= \alpha (T-T_0)$, the authors retrieve Ginzburg-Landau (GL) models with the order parameter confined in d spatial dimensions. They use methods of Quantum Field Theory to discuss second-order phase transitions in such framework. Specifically, the authors consider the 3-dimensional GL model compactified in d=1,2 and 3 space coordinates describing a film, a wire and a grain, respectively. The authors determine the dependence of the transition temperature with the relevant size of the system for these cases. In particular, they address the question of the existence of a minimal size of the system for sustaining the superconducting phase. The authors also consider the critical behavior of type-II superconducting films in a magnetic field. The results are in qualitative agreement with the experimental measures for the case of superconducting transitions.

EXPERT COMMENTARY

In: New Research on Superconductivity
Editor: Barry P. Martins, pp. 3-10
ISBN: 978-1-59454-197-1

T_C'S OF HIGH-TEMPERATURE SUPERCONDUCTORS VIA THE BCS PARAMETERS OF THEIR CONSTITUENTS: THE ROAD NOT TAKEN

*G. P. Malik**
(Formerly) Theory Group, School of Environmental Sciences, Jawaharlal Nehru University, New Delhi 110067, India

In about twenty years since Bednorz and Müller [1] reported the discovery of superconductivity in La-Ba-Cu-O system at about 30 K, which marks the advent of the era of high-T_c superconductivity, an unprecedented amount of activity has taken place in the field. While a large number of new high-T_c superconductors (HTSCs) have thus been reported, and a host of experimental data about them recorded, it seems fair to say that theory has lagged behind experiment in this field of activity. It is natural to ask: Why? Might one have taken an approach based on premises other than the ones followed? I wish to suggest here one such approach.

First, the basic requirements of a theory of high-T_c superconductivity. These have been succinctly listed by Rice [2] as follows (one may add to them; e.g., require explanation of the d-wave feature of pairing):

(1) The theory should describe the full complex phase diagram of the HTSCs.
(2) It should reveal the special conditions in the cuprates that lead to this very special behavior; from this should follow suggestions for other materials that would show similar behavior.
(3) It should give the correct order of magnitude of the T_c's, and
(4) It should explain the increase in T_c as the number of layers is increased to two and three.

* E-mail addresses: *gp7malik@bol.net.in*, malik@mail.jnu.ac.in; Present address: B-208 Sushant Lok I, Gurgaon 122002, India

To meet these requirements, the thrust of theoretical activity seems to have been concerned with inventing non-phononic mechanisms of forming Cooper pairs (CPs), such as: electronic mechanisms (including plasmons that invoke "demons", which are akin to phonons), magnetic mechanisms (where magnons play the role of phonons as in the Hubbard and the RVB models), and mechanisms based on the proximity effect. Detailed accounts of these have been given in [3]. Let us for a moment assume that it is established that one or more of these mechanisms are responsible for the formation of CPs. The question then is: what next? Are we going to undertake to "reduce" the Hamiltonian and the wave function – as in the BCS theory - for the highly anisotropic HTSC as a whole, or are we going to do so for each layer of it and then do some sort of "patching"? Note that all theoretical attempts, past and present, are based on the belief that superconductivity occurs, ultimately, due to the formation of CPs. That these then go on to be coherently "linked" with all the other similar pairs is the essence of the BCS theory. What is questioned is the applicability of this theory to HTSCs. The reason is: for acceptable values of the Debye temperature (θ) and the interaction parameter ([N(0)V]), one of the basic equations of this theory viz.,

$$1 = [N(0)V] \int_0^{\theta/2T_c} \frac{\tanh x}{x} dx, \tag{1}$$

or

$$T_c \cong 1.14\theta \exp(-1/[N(0)V]) \tag{2}$$

cannot lead to a T_c greater than about 30 K.

Ideally, one would imagine that properties of the sub-systems that comprise a complex system should play an active role in determining the properties of the latter. In such a scenario, the BCS theory would provide a natural starting point for explanation of the properties of HTSCs.; after all, it is the only successful theory for superconducting elements that one has. The game would then become well defined: figure out how the BCS parameters of the constituents of a HTSC change when they form the latter. These considerations suggest the need to have a framework that: (a) manifestly takes into account the fact that HTSCs are multi-component compounds, and (b) does not have the limitation of Eq. (1). I delineate below how such a framework may be created via a synthesis of four theoretical strands, all of which originated in the 1950s; and refer to it as the framework of the Extended Cooper Problem (ECP).

Recall that analysis of the Cooper problem [4] - a landmark work of the 1950s - was carried out in the framework of non-relativistic quantum mechanics. Interestingly enough, there took place three important developments in quantum field theory in the same decade, two of which are well known: the Bethe-Salpeter equation (BSE) and Matsubara's finite-temperature propagator. The third development, not so well known, was introduction by Okubo [5] of the concept of a superpropagator (christened so by Salam) as the vacuum expectation value of the time-ordered product of a superfield:

$$\langle 0|T[U\{\phi(x)\}U\{\phi(0)\}]|0\rangle, \tag{3}$$

where the superfield, $U\{\phi(x)\}$, is a non-polynomial function of the field $\phi(x)$. Examples:

$$U\{\phi(x)\} = \exp[g\phi(x)], \quad \frac{1}{1+g\phi(x)}, \quad \frac{1}{1+g^2\phi^2(x)}, \ldots \tag{4}$$

where g is a coupling constant. The virtues of non-polynomial theories, in a different context, were extolled by Salam and others [6]. What concerns us here is the physical significance of the superpropagator. For the second superfield in Eq. (4), for example, this construct has the form

$$\Delta_F^{SP}(k) \approx (k^2 g^2)^{-1} \exp(k^2 g^2 / 2)\, W_{-2,1/2}(k^2 g^2), \tag{5}$$

where $W_{\lambda,\mu}$ is the Whitatker function. By using an appropriate representation of this function, one has [7]

$$\Delta_F^{SP}(k) \approx \frac{1}{k^2 g^2} - \int_0^{\infty} \frac{\sigma(t)dt}{(k^2 g^2 + t)}, \tag{6}$$

where $\sigma(t) = e^{-t}(2-t)$. Thus, a superpropagator resembles a weighted superposition of propagators.

Let us now reformulate the Cooper problem in a relativistic framework viz., the framework of BSE. In such a formulation, the CPs are bound via exchanges of phonon propagators in the ladder approximation. Choosing

$$\Delta_F^{Cooper}(\mathbf{q}-\mathbf{p}) = -\frac{V}{(2\pi)^3} \quad for\ E_F \langle \frac{\hbar^2\mathbf{q}^2}{m}, \frac{\hbar^2\mathbf{p}^2}{m} \langle E_F + \hbar\omega_D \ (V>0) \tag{7}$$
$$= 0, \qquad \text{(otherwise)}$$

and subjecting the resulting BS equation to the usual process of "reduction" [8], one obtains the standard expression for binding energy of a CP. There is as yet no gain: one has simply obtained a known result via a more complex theoretical framework. An advantage of this approach emerges when one temperature-generalizes the BSE for the Cooper problem via the Matsubara method. This results in the BCS expression for the T_c of a superconductor [9] with somewhat more ease than in the BCS theory. However, this again is a known result. The real advantage of the relativistic approach emerges when one temperature-generalizes the BSE for the Cooper problem by replacing the propagator in it by a superpropagator. In the spirit of the Cooper problem, choose:

$$\Delta_F^{Cooper,\,SP}(\mathbf{q}-\mathbf{p})$$

$$= (-)\sum_{i=1}^{n}\frac{V_i}{(2\pi)^3} \quad for\ E_F-(\omega_D)_i<\mathbf{p}^2/2m,\ \mathbf{q}^2/2m<E_F+(\omega_D)_i \quad (V_i>0)$$

$$= 0 \qquad otherwise, \tag{8}$$

where n denotes the number of *species* of phonons responsible for binding a CP, and $(\omega_D)_i$ is the Debye frequency of lattice vibrations of the ith species of ions in the sample. We are thus led to the following basic equation of the ECP [9] (zero external magnetic field):

$$1=[N(0)V]_1^c\int_0^{\theta_1^c/2T_c}\frac{\tanh x}{x}dx+[N(0)V]_2^c\int_0^{\theta_2^c/2T_c}\frac{\tanh x}{x}dx+..... \tag{9}$$

Here $[N(0)V]_i$ and $[N(0)V]_i^c$ denote the interaction parameters of the ith species in its free and composite states, respectively; so also for θ_i and θ_i^c.

Since the individual terms on the rhs of Eq. (9) are identical in form with the rhs of Eq. (1), they embody the same one-phonon exchange mechanism as the latter equation. Therefore in Tl-2212 (Tl-2 hereafter), for example, while the phonons responsible for the formation of a CP could belong to any one of the three positively charged sub-lattices in it viz., the Tl, Ba, or Ca sub-lattices, the consistency condition which determines the T_c takes into cognizance all these possibilities. This is in essence the content of Eq. (9). A rather pleasant insight is obtained by recognizing that the structure of Eq. (9) is formally identical with the mathematical expression for the probability of a number of mutually exclusive events as a sum of the probabilities of these events individually.

That the basic equation of the ECP for the T_c of a multi-component superconductor is a generalization of the BCS equation is rather pleasing because it enables one to calculate the T_c of a multi-component HTSC beyond the limit of the latter equation, while remaining within the realm of low coupling. We now have in place the framework that enables one to tackle the problem of HTSCs on the basis of the BCS parameters of their constituents. Let us see how it works by taking the example of Tl-2, as dealt with in [10].

(1) The free-state BCS parameters of the constituents of Tl-2 in the order (T_c, θ_D, [N(0)V]) are:

Tl (2.4, 90, 0.2668),
Ba (5, 110, 0.311),
Ca (15, 229, 0.3507), (12)
Cu (-, 348, -), and
O (0.6, 91, 0.1943),

where [N(0)V] for any constituent is determined by substituting its experimentally determined values of T_c and θ_D in Eq. (1).

(2) Appeal to the experimentally determined charge structure of the HTSC to figure out the number of *species* of phonons that may be responsible for the formation of CP's of *electrons*. For Tl-2, this number is 3; for this HTSC therefore we have three terms on the rhs of Eq. (9). Our task now is reduced to determining $[N(0)V]_i^c$ and θ_i^c, where i=1, 2, 3.

(3) As a first approximation, assume that $[N(0)V]_i^c = [N(0)V]_i$ i.e., the constituents of a HTSC retain the values of their free-state interaction parameters..

(4) In order to determine θ_i^c, recall that oscillators with intrinsic frequencies ω_1 and ω_2, when connected by a weak spring, vibrate in two different modes whose frequencies can be expressed in terms of ω_1 and ω_2 - as in the polariton effect. Assume now that constituents in each layer of a HTSC are similarly coupled. In terms of Debye temperatures, for the layer A_xB_{1-x} whose Debye temperature is $\theta(x)$, we then have [11]:

$$x\,\theta_A^c + (1-x)\,\theta_B^c = \theta(x), \text{ and}$$

$$\frac{\theta_A^c}{\theta_B^c} = \left[\frac{1+\sqrt{a_2/(a_1+a_2)}}{1-\sqrt{a_2/(a_1+a_2)}}\right]^{1/2}, \tag{13}$$

where a_1 and a_2 are the atomic mass numbers of elements A and B respectively. Note that coupling between the layers is neglected. Applying Eqs. (13) to the TlO, BaO, and CuO_2 layers in Tl-2, taking $\theta(x)$ =254 [12] in each case, we obtain

$$\begin{aligned} \theta_{Tl}^c &= 289, \quad \theta_O^c = 219 \\ \theta_{Ba}^c &= 296, \quad \theta_O^c = 212 \\ \theta_{Cu}^c &= 341, \quad \theta_O^c = 210 \end{aligned} \tag{14}$$

We thus have three values for θ_O^c; if required, one may use the average of these values (we will not require it).

(5) Solution of Eq. (9) – with appropriate substitutions from Eqs. (12) and (14) - yields (within the parentheses is given the experimental value):

$$T_c\,(\text{Tl-2212}) = 112.7 \text{ K } (110 \text{ K}).$$

Similarly,

$$T_c\,(\text{Tl-2223}) = 128.7 \text{ K } (125 \text{ K})$$

T_c (Tl-2201)=55.2 K (0 –90 K).

Note that for Tl-2201, we must invoke a 2-phonon exchange mechanism because it does not contain Ca.

Applying the above procedure to Hg-based cuprates, which are structurally similar to the Tl-based cuprates, we obtain [10]:

T_c ($HgBa_2CuO_{4.18}$)=99.2 K (97 K)

T_c ($HgBa_2CaCu_2O_{6.26}$)=120.5 K (123 K)

T_c ($HgBa_2Ca_2Cu_3O_{8.44}$)=133.4 K (135 K).

As a final test of our approach here, we present in Table 1 the results corresponding to some rather fine-tuned experiments; for details see [10].

Table 1. Variation of the T_c of $Tl_2(Ba_{2-x}Sr_x)CaCu_2O_8$ with increasing x, where the experimental data have been taken from Ref. [13].

x	Expt.T_c (onset) (K)	Expt. T_c(0) (K)	T_c (K)via the ECP
0.0	116	106	115.0
0.2	113	94	109.2
0.6	108	94	97.7
1.0	90	82	86.3
1.4	72	42	75.0
2.0	61	28	58.3

Having used the framework of the ECP for both the conventional superconducting binary alloys [11,14] and HTSCs such as MgB_2 [15] and the Tl- and Hg-based cuprates [10], I conclude this note with the following remarks in order not only to draw attention to some of its salient features, but also to suggest and facilitate further work.

(1) By using the experimentally determined values of T_c and θ of the constituents of a HTSC, we have obtained their [N(0)V]-values via Eq. (1). Thus, the detailed mechanism responsible for these values is a non-issue.

(2) Note that while the Debye temperatures of the Tl-based cuprates have been found from the specific heat data, those of the Hg-based family have been determined via the elastic constants. That the two approaches give significantly different results is well known: for Hg, for example, $\theta_D^{sp.ht} = 75K$, while $\theta_D^{el.cons.} = 167K$ [16]! Consistency therefore demands that the [N(0)V]-values of the constituents of the Hg-based cuprates should be calculated by using their θ_D-values as determined via the elastic constants.

(3) Because the [N(0)V]'s calculated on the basis of $\theta_D^{el.cons.}$'s are higher than those based on $\theta_D^{sp.ht}$'s, it turns out that we require a 2-phonon exchange mechanism for all the three Hg-based cuprates. This is in contrast with the Tl-based cuprates. Further, among the three candidates for the 2-phonon exchange mechanism, as expected, the ones

containing Hg give much better fits. However, it is not possible to decide whether it is the (Hg+Ba)- or the (Hg+Ca)-scenario that is in operation [10]. $\theta_D^{sp.ht}$'s of the Hg-based cuprates are needed to shed more light on this issue.

(4) Even though rather pleasant, it is somewhat surprising that the constituents of a HTSC should retain their free-state [N(0)V]-values. This may not always be the case. More extensive use of (9), for conventional as well HTSCs, should yield further insights. Calculation of T_c's of $Tl_2(Ba_{2-x}Sr_x)CaCu_2O_8$ provides one such insight [10].

(5) In contrast with the BCS-McMillan theory, but in accord with experiment, the basic equation of the ECP has naturally led to an explanation of the increase in T_c's of the cuprates with their θ_D's. I draw attention to Ledbetter's [17] experiments concerned with such dependence for five cuprate classes.

(6) For MgB_2, one has to invoke the proximity effect [15]. Can we develop criteria about when this effect is to be invoked?

(7) An immediate task: tackle the Bi-based cuprates.

(8) A clean test of the BCS theory is provided by combining Eq. (1) with the gap equation at T=0. This eliminates the "black box", [N(0)V]. A challenging problem: explore if such a universal relation exists in the present approach.

(9) Calculation of H_c via the present approach will be tackled elsewhere.

(10) Agreement between theory and experiment achieved here, for such structures as the cuprates, ought not to be taken lightly. Since two of Rice's requirements stated above have been met via the ECP, the need for further exploration with it is obvious.

To end, a quote by Prof. Salam (one of my mentors; ICTP; 1975): "it is not that we do not take numbers led to by our theory seriously, but that we do not take them seriously enough".

Acknowledgements

I thank Profs. L.N. Cooper and A.N. Mitra for their interest in my work.

References

[1] Bednorz, J.G.; Müller, K.A. *Z. Phys.* 1986, B 64, 189-193.
[2] Rice, M *Physics World* 1999 (Dec).
[3] Kresin V.Z.; Morawitz, H.; Wolf, S.A. *Mechanisms of Conventional and High T_c Superconductivity*; Oxford University Press: NY, 1993.
[4] Cooper, L.N. *Phys. Rev.* 1956, 104, 1189-1190.
[5] Okubo, S. *Progr. Theor. Phys.* 1954, 11, 80-94.
[6] Salam, A. (Ed.) *Nonpolynomial Lagrangians, Renormalization and Gravity*; Gordon and Breach: NY, 1971.
[7] Biswas, S.N.; Malik, G.P.; Sudarshan, E.C.G. *Phys. Rev.* 1973, D 7, 2884-2886.
[8] Malik, G.P.; Singh, S; Varma, V.S. *Pramana-J. Phys.* 1992, 38, 319-327.
[9] Malik, G.P.; Malik, U. *Physica B* 2003, 336, 349-352.

[10] Malik, G.P.; Kumar, A.; Malik, U. (submitted).
[11] Malik, G.P.; Malik, U. *Physica B* 2004, 348, 341-346.
[12] Junod, A.; Eckert, D.; Triscone G.; Lee, V.Y.; Muller, J *Physica C* 1989, 159, 215-225.
[13] Hayri, E. A;. Greenblatt, M. *Physica C*, 1988, 156 , 775-780.
[14] Malik, G.P.; Kumar, A. *Physica B* 2004, 351, 191-196.
[15] Malik, G.P.; Kumar, A. *Physica C* 2005, 433, 70-74.
[16] Gschneidner; K.A. Jr. *Solid State Physics* 1964, 16, pages 371, 381.
[17] Ledbetter, H. *Physica C* 1994, 235-240, 1325-1326.

SHORT COMMUNICATION

In: New Research on Superconductivity
Editor: Barry P. Martins, pp. 13-19

ISBN: 978-1-59454-197-1

Plasmon Exchange Model for Superconductivity in Carbon Nanotubes

S. M. Bose and S. Gayen*
Department of Physics, Drexel University, Philadelphia, PA 19104, USA

Abstract

Recent experiments have shown that a single-wall zigzag carbon nanotube can become superconducting at around 15 K. Calculations using the traditional phonon exchange model, however, give a superconducting transition temperature T_C less than 1K. To explain the observed higher critical temperature the plasmon exchange mechanism for superconductivity in nanotubes is explored in this paper. Calculation of the effective interaction between electrons in a nanotube mediated by plasmon exchange shows that this interaction can become attractive. This attractive interaction is then used in the modified Eliashberg theory for strong coupling superconductors to calculate the critical temperature T_C in a nanotube. T_C is found to be sensitively dependent on the dielectric constant of the medium, the effective mass of the electrons and the radius of the nanotube. These results can explain the observed T_C in a nanotube.

PACS: 74.20.Mn, 71.45.Gm, 61.46.+w
Keywords: Superconductivity, Carbon nanotubes, Plasmons

Carbon nanotubes, first dicovered by Iijima in 1991 [1], are a new form of carbon with exotic physical properties [2]. The electronic and transport properties of the carbon nanotubes have been studied extensively [3,4,5,6,7]. It has been shown both theoretically and experimentally that nanotubes of zero helicity are predominantly metallic in character, whereas, nanotubes of nonzero helicity are mostly semiconducting in character [4,8,9]. Using different methods of synthesis, bundles of nanotubes have now been produced [10,11]. The question whether a carbon nanotube or a group of carbon nanotubes can exhibit superconductivity has also been addressed in several recent studies. Carbon nanotubes are

* E-mail address: bose@drexel.edu Telephone: +1-215-895-2718

observed to pass supercurrent between superconducting leads due to proximity effect [12]. Recent experiments by Tang *et al.* [13] have shown the presence of superconductivity in single-wall zigzag nanotubes of radius 2.1 A^o at about 15 K. On the other hand Kochiak *et al.* [14] have reported superconductivity in a bundle of arm-chair nanotubes of radius 7 A^o at about 0.55 K. There have been some theoretical studies of superconductivity in carbon nanotubes [15,16]. Gonzalez [17] has recently pointed out that an electron-phonon mechanism of superconductivity in ropes of carbon nanotubes can give a superconducting transition temperature T_C less than 1K. From these studies it appears that the phonon exchange mechanism can not account for superconductivity in a single-wall carbon nanotube (SWNT) obsereved at finite temperatures. In this paper we introduce the plasmon exchange mechanism for superconductivity in a metallic carbon nanotube with the expectation that a plasmon with its frequency higher than the phonon frequency would give a higher critical temperature in a carbon nanotube. We first calculate the effective interaction between electrons in a nanotube mediated by plasmon exchange and show that this interaction can be attractive. We then use this effective interaction in the Eliasberg theory [18] of strong coupling superconductors as modified by McMillan [19] to calculate the superconducting transition temperature, T_C This model was previously used by Longe and Bose [20] to calculate the critical temperature in high- T_C superconductors.

We first review briefly the excitation of a plasmon in a metallic nanotube. In our model we consider that the length of a carbon nanotube is very large (several microns) compared to its radius a (several angstroms). We assume that the electrons can move parallel to the axis of the tube described by the quantum number q as well as around the tube axis described by the azimuthal quantum number μ. The dielectric function $\varepsilon(Q,\omega)$ of the nanotube is calculated in the random phase approximation (RPA) [21] using

$$\varepsilon(Q,\omega) = \varepsilon + v_o(Q)\Pi(Q,\omega), \qquad (1)$$

where $Q=[q,\mu/a]$, q and μ/a are the components of the wave vector for motions parallel and azimuthal directions, respectively. The polarization propagator in the frequency region of plasmon excitation has been shown [21] to be

$$\Pi(Q,\omega) \approx -\frac{n_s Q^2}{m\omega^2}, \qquad (2)$$

where n_s and m are surface number density and effective mass of the electron, respectively. In Eq. (1) ε is the dielectric constant of the medium and $v_o(Q)$ is the bare Coulomb interaction between two electrons on a nanotube and is given by [21]

$$v_o(Q) = 4\pi e^2 a I_\mu(aq) K_\mu(aq) \qquad (3)$$

where $I_\mu(aq)$ and $K_\mu(aq)$ are modified Bessel functions, e is the electronic charge and the azimuthal quantum number μ runs through all integral values. The plasmon frequencies are obtained from the zeros of the dielectric function as

$$[\omega_\mu(q)]^2 = \frac{4\pi n_s e^2 a}{m\varepsilon} Q^2 I_\mu(aq) K_\mu(aq) \tag{4}$$

In reference 21 it has been shown that the plasmon frequency corresponding to $\mu = 0$ is semi-acoustic in nature whereas the frequencies for $\mu \neq 0$ are optical. Once the dielectric function of the nanotube $\varepsilon(Q,\omega)$ has been determined by Eq. (1), we can write the effective interaction between two electrons on a nanotube due to plasmon exchange as

$$V(Q,\omega) = \frac{v_o(Q)}{\varepsilon(Q,\omega)} = \frac{v_o(Q)}{\varepsilon + v_o(Q)\Pi(Q,\omega)} \tag{5}$$

Substituting for $\Pi(Q,\omega)$ from Eq.(2) we can rewrite Eq. (5) as

$$V(Q,\omega) = \frac{v_o(Q)}{\varepsilon} + \frac{v_o^2(Q) n_s Q^2}{\varepsilon^2 m\omega^2 - v_o(Q)\varepsilon n_s Q^2} \tag{6}$$

The first term on the right hand side of Eq. (6) is the statically screened Coulomb repulsion part and the second term represents the effect of plasmon excitation. We notice that the second term can become attractive and can thus lead to superconductivity in a nanotube.

To calculate the critical temperature using this effective interaction, we use the Eliashberg model [18] of superconductivity in a strong-coupling superconductor as modified by McMillan and others [19,22]. In this paper we use the McMillan model which gives the critical temperature for superconductivity in a strong coupling superconductor as

$$T_c = \frac{<\omega>}{1.45} exp[-\frac{1.04(1+\lambda)}{\lambda - \mu^*(1+0.62\lambda)}] \tag{7}$$

In this equation $<\omega>$ is the average value of the frequency of the boson, the exchange of which is responsible for superconductivity, λ is the coupling strength due to attractive part of the effective interaction and μ^* is the Coulomb repulsion parameter. It has been shown by Allen and Dynes [23] that if the effective interaction between electrons in a superconductor can be written as

$$V(Q,\omega) = v_o(Q) + \frac{2\omega(Q)|M(Q)|^2}{\omega^2 - \omega^2(Q)}, \tag{8}$$

then the above parameters can be obtained from

$$\lambda = \lambda(0) = N(0) < 2\frac{|M(Q)|^2}{\omega(Q)} >_{FS} \tag{9}$$

and

$$\lambda < \omega^2 >= N(0) < 2|M(Q)|^2 \,\omega(Q) >_{FS} \tag{10}$$

where $N(0)$ is the density of states of the electrons at the Fermi surface and $< .... >_{FS}$ indicates that an average of the expression is taken over the Fermi surface. Combining Eqs. (9) and (10) one can obtain $< \omega >$ from

$$< \omega >= \sqrt{\frac{\lambda < \omega^2 >}{\lambda}} \tag{11}$$

We can express our effective interaction (Eq. (6)) in the form of Eq. (8) if we identify

$$\omega^2(Q) = \frac{n_s Q^2 v_o(Q)}{m\varepsilon} \tag{12}$$

and

$$|M(Q)|^2 = \frac{1}{2}\sqrt{\frac{n_s Q^2 v_o^3(Q)}{m\varepsilon^3}} \tag{13}$$

Also for the carbon nanotubes where the electrons have axial and azimuthal motions, the Fermi surface will be cylindrical and the density of states at the Fermi surface will be given by

$$N(0) = \frac{m}{2\pi^2 a}\sum_{\mu}\frac{1}{\sqrt{k_F^2 - (\frac{\mu}{a})^2}} \tag{14}$$

Substituting the values of $\omega^2(Q), |M(Q)|^2$ and $N(0)$ in Eqs. (9) and (10) and carrying out the average over the Fermi surface we have calculated the value of $\lambda < \omega^2 >$ and λ and then $< \omega >$ from Eq. (12). These parameters obviously depend on the dielectric constant ε, the effective mass m, the surface number density n_s of the electron and the radius a of the nanotube. The Coulomb repulsion parameter μ^* depends on other boson frequencies and like many other investigators [19,24] we take its numerical value to be 0.1.

Substituting these values of $<\omega>$, λ and μ^* in the McMillan's expression [Eq. (7)] for T_C, we obtain the critical temperature as a function of the parameters ε, m, a and n_s.

It turns out that in a metallic nanotube the number density of electrons n_s is fixed and is independent of whether it is an arm-chair or zigzag nanotube. Assuming that each carbon atom in such a nanotube contributes one electron to the conduction band, n_s can be shown to be $3.73\times10^{15}/cm^2$. The radii a of a nanotube is known to vary from 0.2 to 1.0 nm. The effective dielectric constant ε has been reported to be of the order of 1.4 [25]. The effective mass Z $(=m/m_e)$ has been reported to be of the order of that in a graphite sheet which is known to be 0.24 in a zigzag nanotube and speculated to be one order larger in an arm-chair nanotube [13]. Since these parameters are not known exactly, we thought it would be interesting to study numerically how T_C varies as a function of their reasonable (measured or speculated) values. In Figure 1 we present a contour plot of T_C as a function of ε and Z for $a = 0.21$ nm corresponding to a zigzag nanotube. The figure clearly shows that T_C decreases with increasing ε and decreasing Z. In Figure 2 we have plotted T_C versus ε for Z = 0.24, 0.26, 0.28 and 0.30 for a nanotube of radius $a = 0.21$ nm and in Figure 3 we have plotted T_C versus Z for ε = 1.3, 1.35, 1.40 and 1.45 for the same nanotube.

These figures make it abundantly clear that the plasmon exchange model for superconductivity shows that the critical temperature in a nanotube is indeed sensitively dependent on the parameters ε and Z for a fixed a and depending on their actual values T_C can lie within a wide range. For example we find that for ε = 1.3 and Z =0.24, T_C = 17 K which is close to what Tang *et al.* [13] have measured in a single-wall zigzag nanotube.

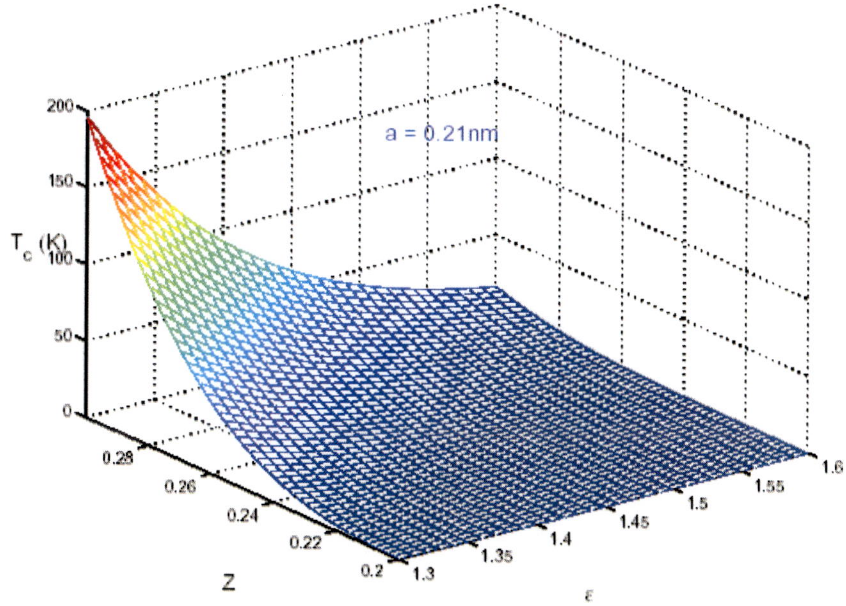

Figure 1. Variation of T_C as a function of ε and Z.

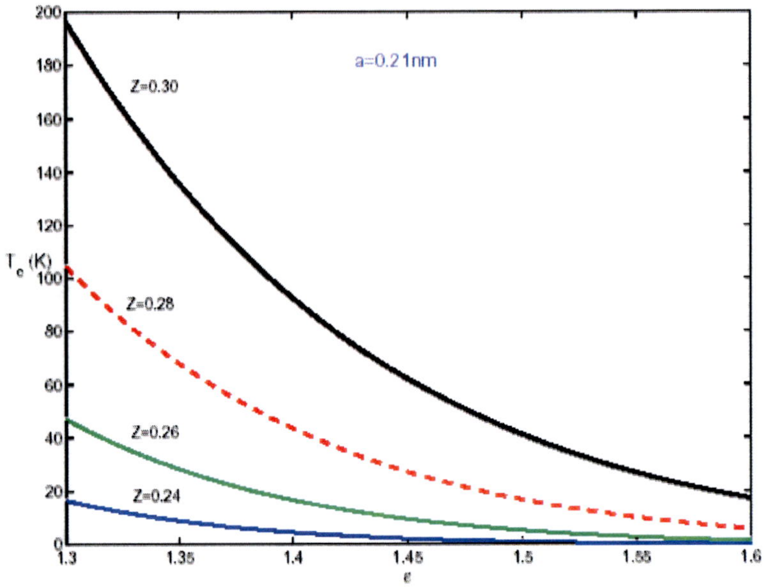

Figure 2. T_C as a function of ε for $Z = 0.24, 0.26, 0.28$ and 0.30.

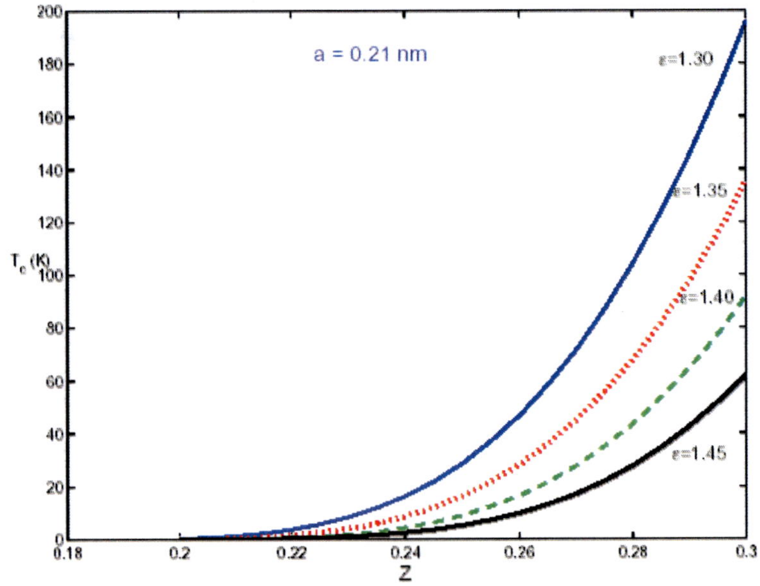

Figure 3. T_C as a function of Z for $\varepsilon = 1.30, 1.35, 1.40$ and 1.45.

In summary, we have studied the plasmon exchange model for superconductivity in a single-wall carbon nanotube. We have first shown that the effective interaction between two electrons mediated by plasmon exchange can become attractive which in its turn can lead to superconductivity. The superconducting critical temperature, calculated by using Eliashberg theory for strong coupling superconductors as modified by McMillan, is found to be sensitively dependent on ε, the dielectric constant of the medium; m, the effective mass of the electron; and a, the radius of the nanotube. For reasonable values of these parameters the calculated value of T_C is found to be in reasonable agreement with the experimental values of a zigzag nanotube [13].

References

[1] S. Iijima, *Nature* 1991, 56.

[2] R. Satio, G. Dresselhaus and M.S. Dresselhaus, *Physical Properties of Carbon Nanotubes* (London: Imperial College Press), 1999.

[3] C. Kane, L. Balents and M.P.A. Fisher, *Phys. Rev. Lett.* 1997, 5086

[4] N. Hamada, S.I. Sawada and A. Oshiyama, *Phys. Rev. Lett.* 1992, 1579

[5] M.S. Dresselhaus, G. Dresselhaus and P.C. Eklund, 1996, *Science of Fullerenes and Carbon Nanotubes* (Sandiego: Academic)

[6] T.W. Odom, J.W. Huang, P. Kim and C.M. Lieber, *Nature,* 1998, 62.

[7] E.L. Ivchenko and B. Spivak, *Phys. Rev. B*, 2002, 155404.

[8] J.W. Mintmire, B.I. Dunlap and C.T. White, *Phys. Rev. Lett.* 1992, 631.

[9] J.W.G. Wildoer *et al., Nature,* 1998, 59.

[10] S. Iijima and T. Ichbashi, *Nature,* 1993, 603; D.S. Bethune, C.H. Kiang, M.S. DeVries, G. Gorman, S. Savoy, and R. Beyers, *Nature,* 1993, 605; T.W. Ebbesen and P.M. Ajayan, *Nature,* 1992, 220

[11] Thess *et al., Science,* 1996, 483; Z.F. Ren *et al., Science,* 1998, 1105.

[12] Yu Kasumov *et al., Science,* 1999, 1508.

[13] Z.K. Tang *et al., Science,* 2001, 2462.

[14] M. Kociak *et al., Phys. Rev. Lett.* 2001, 2416; A. Yu Kasumov *et al., Physica* B, 2003, 333, 1321.

[15] L.X. Benedict *et al., Phys. Rev. B,* 1995, 14935.

[16] Sedeki, L.G. Caron and C. Bourbonnais, *Phys. Rev. B,* 2002, 140515(R).

[17] J. Gonzalez, *Phys. Rev. Lett.* 2002, 076403-1.

[18] G.M. Eliashberg, *Zh. Eksp. Teor. Fiz.* 1960, 966 [1960 *Sov. Phys. JETP* **696**]; G.M. Eliashberg, *Zh. Eksp. Teor. Fiz.* 1960, 1437 [1961 *Sov. Phys. JETP* **1000**].

[19] W.L. McMillan, *Phys. Rev.* 1968, 331

[20] P. Longe and S.M. Bose, *J. Phys.: Condens. Matter,* 1992, 1811.

[21] P. Longe and S.M. Bose, *Phys. Rev. B,* 1993, 18239.

[22] V.Z. Kresin, *Phys. Lett. A*, 1987, 434; *Phys. Rev. B.* 1987, 8716.

[23] P.B. Allen and R.C. Dynes, *Phys. Rev. B,* 1975, 905.

[24] P. Morel and P.W. Anderson, *Phys. Rev.* 1962, 1263.

[25] R. Egger *et al. Eur. Phys. J. B,* 1998, 281.

In: New Research on Superconductivity
Editor: Barry P. Martins, pp. 21-57

ISBN 978-1-59454-197-1

Chapter 1

EVIDENCE FOR ROOM-TEMPERATURE SUPERCONDUCTIVITY IN CARBON NANOTUBES

Guo-meng Zhao*
Department of Physics and Astronomy,
California State University at Los Angeles, Los Angeles, CA 90032, USA

Abstract

In this article, we provide electrical, magnetic, tunneling, and Raman spectroscopic evidence for room-temperature superconductivity in carbon nanotubes. Electrical measurements indicate that the superconducting transition temperatures in carbon nanotubes can vary from 0.44 K to 750 K. The temperature dependencies of the resistive transitions agree quantitatively with the well-established Langer-Ambegaokar-McCumber-Halperin (LAMH) theory for the resistive transition of quasi-one-dimensional superconductors. Because of a finite number of transverse conduction channels, the four-probe resistance of an individual tube will not go to zero in the superconducting state. This would lead to a false impression that these tubes are not superconducting. Nonetheless, the on-tube resistance at room temperature is found to be indistinguishable from zero for many individual multi-walled nanotubes. The very small but finite room-temperature on-tube resistance is consistent with quantum phase slips due to the finite number of transverse channels. Magnetic measurements show a small Meissner effect in aligned and physically separated multi-walled nanotubes up to room temperature. The small Meissner effect is due to the fact that the diameter of the tubes is much smaller than the magnetic penetration depth. At low temperatures, the directionally averaged diamagnetic susceptibility for physically coupled multi-walled nanotubes increases by a factor of 4.3 compared with that for physically separated tubes, which is consistent with the enhancement of the Meissner effect due to the Josephson coupling among the superconducting tubes. Further, Raman data and single-particle tunneling spectra consistently show the existence of single particle excitation gaps of larger than 100 meV. We suggest microscopic pairing mechanism for high-temperature superconductivity in carbon nanotubes.

Keywords: Room-Temperature Superconductivity, Carbon Nanotubes

*E-mail address: gzhao2@calstatela.edu

1. Introduction

It is generally believed that the superconducting transition temperature T_c cannot be higher than 30 K within the conventional phonon-mediated mechanism. Alexandrov and Mott [1] demonstrate that strong electron-phonon coupling can lead to the formation of intersite bipolarons and that the Bose-Einstein condensation of the bipolarons can explain high-temperature superconductivity in cuprates. Ginzburg [2] and Little [3] propose that high-temperature or room-temperature superconductivity could be realized by exchanging high-energy electronic excitations such as excitons and plasmons. Lee and Mendoza show that superconductivity as high as 500 K can be reached through a pairing interaction mediated by undamped acoustic plasmon modes in a quasi-one-dimensional (1D) electronic system [4]. Moreover, high-temperature superconductivity can occur in a multi-layer electronic system due to an attraction of charge carriers in the same conducting layer via exchange of virtual plasmons in neighboring layers [5]. If the plasmon-mediated pairing mechanisms are relevant, one should be able to find high-temperature superconductivity in quasi-one-dimensional and/or multi-layer systems such as cuprates, carbon nanotubes (CNTs), and graphites.

Carbon nanotubes constitute a novel class of quasi-one-dimensional materials which would offer the potential for high-temperature superconductivity. The simplest single-walled nanotube (SWNT) consists of a single graphite sheet which is curved into a long cylinder with a diameter of about 1 nm. Band-structure calculations predict that carbon nanotubes have two types of electronic structures depending on the chirality [6, 7], which is indexed by a chiral vector (n, m): $n - m = 3N + \nu$, where N, n, m are the integers, and $\nu = 0, \pm 1$. The tubes with $\nu = 0$ are metallic while the tubes with $\nu = \pm 1$ are semiconductive. For metallic chirality SWNTs, there are two and six transverse conduction channels when the Fermi level is crossing the first and second subbands, respectively. Multi-walled nanotubes (MWNTs) consist of at least two concentric shells which could have different chiralities. The outer diameter of MWNTs prepared by arc discharge is normally centered around 10-15 nm. The MWNTs possess both quasi-one-dimensional and multi-layer electronic structures. This unique quasi-one-dimensional electronic structure in both SWNTs and MWNTs make them ideal for plasmon-mediated high-temperature superconductivity.

In order to confirm the existence of superconductivity, it is essential to provide two important signatures: the Meissner effect and the sharp resistive transition to the zero resistance state. However, these two essential signatures are less obvious in a quasi-one-dimensional superconducting wire that has a finite number of transverse conduction channels and a very thin transverse dimension. Due to large superconducting fluctuations, the resistive transition in quasi-1D superconductors could be very broad. Because of the finite number of transverse channels, the four-probe resistance never goes to zero even though the on-wire resistance approaches zero. This would lead to a false impression that these thin wires are not superconducting. Because the magnetic penetration depth may be far larger than the transverse dimension, the diamagentic Meissner effect might be negligible, which makes it difficult to distinguish between the Meissner effect and orbital diamagnetism. Therefore, it is difficult to unambiguously identify quasi-1D superconductivity in ultra-thin wires such as carbon nanotubes.

In spite of these difficulties to show the conventional signatures of superconductivity in ultra-thin superconducting wires, Zhao *et al.* [8] reported in 2001 magnetic and electrical evidence for possible superconductivity above room temperature in MWNT mat samples. Since then, Zhao [9, 10, 11, 12, 13, 14] has analyzed the published magnetic, electrical, and optical data for both single-walled and multi-walled carbon nanotubes, and provided over twenty arguments for the existence of superconductivity above room temperature. In this article, we will present electrical, magnetic, tunneling, and Raman spectroscopic evidence for room-temperature superconductivity in carbon nanotubes. In section 2. we make quantitative data analyses on the observed resistive transitions in carbon nanotubes. We find that the resistive transitions in carbon nanotubes agree quantitatively with the Langer-Ambegaokar-McCumber-Halperin (LAMH) theory [15] for quasi-1D superconductors although the superconducting transition temperatures vary from 0.4 K to 750 K for different samples. In section 3., we identify the Meissner effect in the field parallel to the tube axis up to room temperature for aligned MWNTs that are physically separated. The magnitude of the Meissner effect is in quantitative agreement with the predicted penetration depth from the measured carrier density. Furthermore, the diamagnetic susceptibility in closely packed MWNT bundles increases by a factor of over 4 at low temperatures compared with that for physically separated tubes. This is the hallmark of the Josephson coupling among the tubes in bundles. In section 4., we provide Raman spectroscopic evidence for superconductivity at about 665 K in a SWNT mat sample. We quantitatively deduce the superconducting gap $\Delta(0)$ at zero temperature, the mean-field superconducting transition temperature T_{c0}, and the electron-phonon coupling constant for the Raman-active graphitic mode. We find that $\Delta(0)$ = 106 meV and $2\Delta(0)/k_BT_{c0}$ = 3.7. In section 5., we present the tunneling evidence for the existence of large single-particle excitation gaps ($>$ 100 meV) in both SWNTs and MWNTs. In section 6., we provide evidence for negligible on-tube resistances at room temperature in many individual MWNTs, and rule out the possibility of ballistic transport at room temperature. In section 7., we discuss microscopic mechanism for high-temperature superconductivity in carbon nanotubes. We show that SWNT bundles and individual MWNTs are ideal systems for high-temperature superconductivity within the plasmon-mediated mechanism. In section 8., we discuss the pairing symmetry. It appears that s-wave gap symmetry is relevant. In the last section, we will give concluding remarks.

2. The Resistive Transitions in Carbon Nanotubes

2.1. Theoretical Description for the Resistive Transition in Quasi-1D Superconductors

The phenomenon of superconductivity depends on the coherence of the phase of the superconducting order parameter. The phase coherence of the superconducting order parameter leads to the zero-resistance state. For three-dimensional (3D) bulk systems, the transition to the zero-resistance state occurs right below the mean-field superconducting transition temperature T_{c0} such that the resistive transition is very sharp and the transition width is negligibly small. In contrast, the resistive transition in quasi-1D superconductors is broad because of large superconducting fluctuations. A quantum theory to describe the resistive transition in quasi-1D superconductors was developed by Langer, Ambegaokar, McCumber and Halperin [15] over 30 years ago. The theory is based on thermally activated phase slips (TAPS), which cause the resistance to decrease to zero exponentially. In addition to the thermally activated phase slips, there also exist quantum phase slips due to a finite number of transverse channels [16], which prevent ultra-thin wires or tubes from being true superconductors with absolutely zero resistance.

In a theory developed by Langer, Ambegaokar, McCumber and Halperin [15], phase slips occur via thermal activation, leading to a finite width for the resistive transition. The resistance due to the TAPS is given by [17]

$$R_{TA} = \frac{h}{4e^2}\frac{\hbar\Omega}{k_B T}\exp[-\Delta F_\circ(T)/k_B T], \tag{1}$$

where the attempt frequency Ω is [15]

$$\Omega = \frac{\sqrt{3}}{2\pi^{3/2}}\frac{L}{\xi}\sqrt{\frac{\Delta F_\circ(T)}{k_B T}}\frac{1}{\tau_{GL}}. \tag{2}$$

Here L is the length of the wire, $\xi(T)$ is the coherence length, and $\hbar/\tau_{GL} = (8/\pi)k_B(T_{c0} - T)$. The barrier energy $\Delta F_\circ(T)$ is

$$\Delta F_\circ(T) = \frac{8\sqrt{2}}{3}\frac{H_c^2(T)}{8\pi}A\xi, \tag{3}$$

where $H_c^2(T)/8\pi$ is the condensation energy, A is the cross-section area of the wire, and the critical field near T_{c0} is given by $H_c(T) = 1.73H_c(0)(1 - T/T_{c0})$ within the BCS theory [17]. Using $\xi(T) = \xi(0)(1 - T/T_{c0})^{-1/2}$ (Ref. [17]), we then have

$$\frac{\Delta F_\circ(T)}{k_B T} = \frac{3cT_{c0}}{T}(1 - \frac{T}{T_{c0}})^{3/2}, \tag{4}$$

where

$$c = \frac{\Delta F_\circ(0)}{k_B T_{c0}} = \frac{8\sqrt{2}}{3}\frac{H_c^2(0)}{8\pi k_B T_{c0}}A\xi(0). \tag{5}$$

Combining the above equations, we finally get

$$R_{TA} = \frac{m}{T^{1.5}}(1 - \frac{T}{T_{c0}})^{9/4}\exp[-\frac{3cT_{c0}}{T}(1 - \frac{T}{T_{c0}})^{3/2}], \tag{6}$$

with

$$m = 2.55T_{c0}(3cT_{c0})^{1/2}\frac{L}{\xi(0)}, \tag{7}$$

where m is in the unit of kΩK$^{3/2}$. We would like to mention that the exponent in Eq. 6 is factor of 3 larger than that in a similar formula deduced in Ref. [18]. It is possible that the authors in Ref. [18] missed the prefactor of 1.73 in $H_c(T)$.

The condensation energy at zero temperature $H_c^2(0)/8\pi$ is equal to $N(0)\Delta^2(0)/2$ within the BCS theory, where $N(0)$ is the density of states near the Fermi level and $\Delta(0)$ is the superconducting gap at zero temperature. For a single metallic SWNT with two transverse channels, $N(0)A = 4/3\pi a_{C-C}\gamma_\circ$ (Ref. [19]), $\hbar v_F = 1.5a_{C-C}\gamma_\circ$ (Ref. [20]), where

$\gamma_\circ$ is the hopping integral and a_{C-C} (0.142 nm) is the bonding length. Using the BCS relations: $\xi_{BCS} = \hbar v_F/\pi\Delta(0)$ and $\Delta(0)/k_B T_{c0}$ =1.76, and the above relations, one can readily show that

$$c = 0.68\frac{\xi(0)}{\xi_{BCS}}. \tag{8}$$

If a bundle of single-walled nanotubes or a multi-walled nanotube consists of N_{ch} transverse channels, then

$$c = 0.34N_{ch}\frac{\xi(0)}{\xi_{BCS}}. \tag{9}$$

In the clean limit, $\xi(0) = 0.74\xi_{BCS}$ and thus $c = 0.25N_{ch}$.

For two-probe or four-probe measurements on carbon nanotubes with finite transverse channels, the total resistance is $R = R_0 + R_{tube}$, where R_{tube} is the on-tube resistance and $R_0 = R_t = R_Q/tN_{ch}$ for four-probe measurements, or $R_0 = R_Q/tN_{ch} + R_c$ for two probe measurements [21]. Here t is the transmission coefficient ($t \leq 1$), $R_Q = h/2e^2 =$ 12.9 kΩ is the resistance quantum, R_c is the contact resistance, and R_t is the tunneling resistance. Both R_c and R_t should be temperature independent. For ideal contacts, $R_c = 0$ and $t = 1$, so $R_0 = R_t = 12.9$ kΩ/N_{ch} for a bundle comprising N_{ch} transverse channels. For quasi-1D systems, N_{ch} is always finite such that both four-probe and two-probe resistances never go to zero even if the on-tube resistance is zero. Only if N_{ch} goes to infinity, as in the bulk 3D systems, R_t becomes zero such that four probe resistance can go to zero below the superconducting transition temperature. Therefore, the non-zero four-probe resistance in ultra-thin wires does not rule out superconductivity in the wires.

2.2. The Resistive Transition in a SWNT Bundle with T_{c0} = 0.44 K

In 2001, Kociak *et al.* provided the first experimental evidence for superconductivity in single-walled carbon nanotube bundles from the electrical transport measurements [22]. Although the superconducting transition temperature is low (< 1 K), the resistive behavior of the nanotube bundle can serve as a prototype for the resistive transition in quasi-1D superconducting wires with a finite number of transverse conduction channels.

Figure 1 shows the two-probe resistance data for a SWNT bundle that consists of about 350 tubes [22]. One can see that the resistance starts to drop below about 0.5 K, decreases more rapidly below $T_{c0} \simeq 0.44$ K and saturates to a value of 74 Ω. From the saturated value of $R_0 = 74$ Ω and the relation: $R_0 = R_Q/tN_{ch} + R_c$, one can easily find that more than 174 transverse channels are connected to the electrodes and participate in electrical transport. This implies that more than 87 metallic-chirality superconducting SWNTs take part in electrical transport. Considering the fact that one third of tubes should have metallic chiralities and become superconducting, we find the total number (N_m) of the superconducting tubes to be 117, implying that $t \geq 0.74$.

The value of N_m can be also deduced from the measured current I_c^* at which the last resistance jump occurs. The I_c^* corresponds to the critical current for a superconducting wire without disorder and with the same number of transverse channels [22]. According to the BCS theory, the mean-field critical current in the clean limit is given by [17] $I_c^*(0) = en_s A\Delta(0)/\hbar k_F$. The superfluid density n_s is equal to the normal-state carrier density n which is given by $n = 2N_m N(0)E_F = 2N_m N(0)\hbar v_F k_F = 4N_m k_F/A\pi$. Here we have

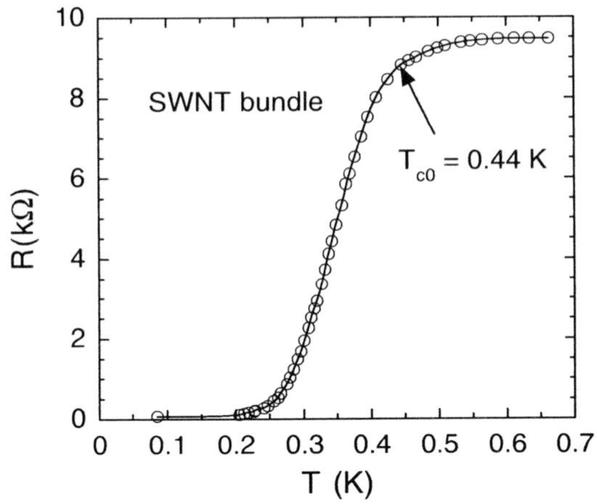

Figure 1. The temperature dependence of the two-probe resistance for a SWNT bundle that consists of about 350 tubes. The data are extracted from Ref. [22].

used the relations for single metallic tube: $N(0)A = 4/3\pi a_{C-C}\gamma_\circ$, $\hbar v_F = 1.5a_{C-C}\gamma_\circ$, and $E = \hbar v_F|k|$. Using $\Delta(0)/k_BT_{c0} = 1.76$, we finally get $I_c^* = 7.04k_BT_{c0}N_m/eR_Q$. With $I_c^* = 2.4$ μA (Ref. [22]) and $T_{c0} = 0.44$ K, we have $N_m = 116$, in remarkably good agreement with the value (117) deduced above.

In Fig. 2, we fit the resistance data below $0.88T_{c0}$ by

$$R = R_\circ + \alpha \exp[-\frac{3\beta T_c^*}{T}(1 - \frac{T}{T_c^*})^{3/2}]. \tag{10}$$

Here the first term $R_\circ = 74$ Ω is the sum of the tunneling and contact resistances and the second term is the on-tube resistance, which has a similar exponential dependence on T as Eq. 6 but with a temperature independent prefactor. We can see that the fit is excellent with the fitting parameter $\beta = 2.99\pm0.05$ and $T_c^* = 0.394\pm0.002$ K. Reducing or increasing the temperature region for the fit tends to worsen the fit quality. Therefore, the on-tube resistance goes to zero exponentially below $T_c^* = 0.89T_{c0}$. The microscopic origin of this simple exponential dependence up to $0.89T_{c0}$ is not clear, so we consider Eq. 10 only as an empirical formula.

We also try to fit the resistance below $0.88T_{c0}$ by

$$R = R_\circ + \frac{m}{T^{1.5}}(1 - \frac{T}{T_{c0}})^{9/4} \exp[-\frac{3cT_{c0}}{T}(1 - \frac{T}{T_{c0}})^{3/2}]. \tag{11}$$

Here the first term $R_\circ = 74$ Ω and the second term is the on-tube resistance which is the same as Eq. 6 predicted by the LAMH theory. We find that the fit is not good and the fitting parameters have no quantitative agreement with the LAMH theory. This is because the LAMH theory is only applied to the temperature region where the barrier height is far larger than k_BT so that current carrying states involved are truly metastable [23]. The estimated region of validity for the LAMH theory is below 0.07 R_N for dirty wires where the mean free path $l << \xi_{BCS}$ (Ref. [23]). For the SWNT bundle, the condition of $l << \xi_{BCS}$ is well satisfied, as seen below. In Fig. 3, we fit the resistance data below 0.06 R_N by

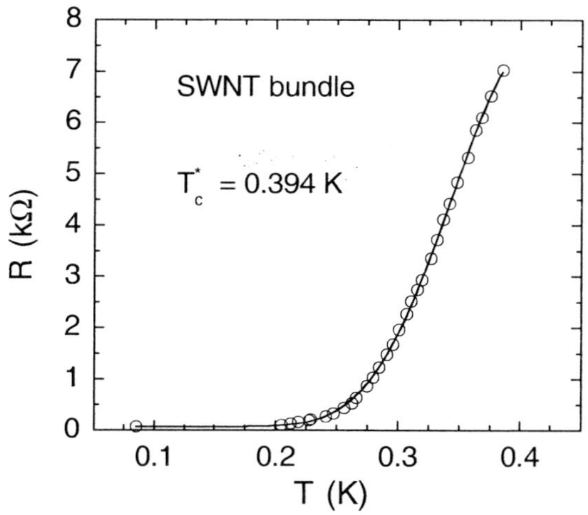

Figure 2. The temperature dependence of the two-probe resistance for a SWNT bundle below $0.88T_{c0}$. The solid line is the curve best fitted by Eq. 10 with β = 2.99±0.05 and T_c^* = 0.394±0.002 K. Reducing or increasing the temperature region for the fit tends to worsen the fit quality. It is striking that the on-tube resistance below 0.88 T_{c0} decreases exponentially to zero.

Eq. 11. Here we have ignored the normal conduction because R_{TA} is a factor of about 20 smaller than R_N. One can see that the fitting is very good with the fitting parameters: m = 26.6±4.7 k$\Omega K^{1.5}$ and c = 3.08±0.13. It is remarkable that the value of c is nearly the same as the value of β (2.99±0.05) deduced above from a simple exponential fit (Eq. 10). We will see below that the fitting parameters c and m are in quantitative agreement with the LAMH theory.

From the values of m, c, and T_{c0}, we can evaluate the zero-temperature coherence length $\xi(0)$ using Eq. 7. Substituting m = 26.6 kΩK$^{1.5}$, c = 3.08, T_{c0} = 0.44 K, and L = 10000 Å into Eq. 7, we obtain $\xi(0)$ = 850 Å. From the measured R_N(0.25K)/L = 12 kΩ/μm (Ref. [22]) and the relation $R_N/L = R_Q/2N_m l$ (Ref. [24]), we can calculate the mean free path l at 0.25 K. With N_m = 117, we get l = 46 Å. Substituting l = 46 Å and $\xi(0)$ = 850 Å into the dirty-limit formula [23]: $\xi(0) = 0.85\sqrt{\xi_{BCS}l}$, we have ξ_{BCS} = 21739 Å. With ξ_{BCS} = 21739 Å, $\xi(0)$ = 850 Å, and $N_{ch} = 2N_m$ = 334, we calculate c = 3.11 from Eq. 9. It is remarkable that the calculated value of c from Eq. 9 is in quantitative agreement with the value (3.08) deduced from the fitting.

We can also estimate the Fermi velocity v_F from the deduced value of ξ_{BCS} and the formula $\xi_{BCS} = 0.18\hbar v_F/k_B T_{c0}$. With ξ_{BCS} = 21739 Å and T_{c0} = 0.44 K, we get $\hbar v_F$ = 4.6 eVÅ. Then we estimate $\gamma_\circ$ = 2.16 eV from $\hbar v_F = 1.5a_{C-C}\gamma_\circ$. This value is very close to the value (2.4 eV) estimated from the first-principle calculation [25].

The deduced value of $\xi(0)$ = 850 Å is also in excellent agreement with the measured critical current I_c for this SWNT bundle. For a diffusive superconducting wire, the critical current I_c is given by [26]

$$I_c = \frac{\Delta(0)}{eR_N}\frac{L}{\xi(0)}. \tag{12}$$

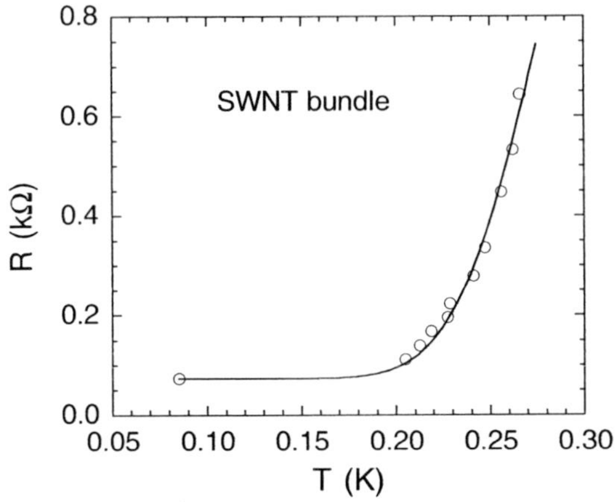

Figure 3. The temperature dependence of the two-probe resistance for a SWNT bundle below $0.06R_N$. The solid line is the curve fitted to the data below $0.06\,R_N$ by Eq. 11 predicted by the LAMH theory. The estimated region of validity for the LAMH theory is below $0.07R_N$ for dirty wires where $l << \xi_{BCS}$ (Ref. [23]). The condition of $l << \xi_{BCS}$ is well satisfied in the SWNT bundle (see text).

Using $\Delta(0) = 1.76k_BT_{c0}$ and substituting R_N(0.1K)/L = 12.5 kΩ/μm (Ref. [22]) and $\xi(0)$ = 850 Å into Eq. 12, we find I_c = 62.4 nA, which is very close to the measured value (62 nA) at 0.1 K (Ref. [22]). Thus, the resistance data of the SWNT bundle agrees with the LAMH theory in a quantitative way.

2.3. The Resistive Transition in an Individual MWNT with T_{c0} = 262 K

In 1996, Ebbesen *et al.* made four-probe resistance measurements on individual multi-walled carbon nanotubes [27]. Four 80-nm-wide tungsten leads were patterned by ion-induced deposition of tungsten from $W(CO)_6$ carrier gas. This technique makes it possible for electrodes to connect multi-shells of the tubes [28]. It is interesting that the electrical properties vary significantly from samples to samples. Some tubes show abrupt jumps in resistivity when the temperature increases. In some other tubes, the resistance at room temperature is very small (e.g., 200 Ω) but a metal-insulator transition occurs below about 200 K. We will show that the resistive behavior in the former case is in quantitative agreement with that expected for quasi-1D superconductivity. The resistive behavior in the latter case may be explained by a superconductor-to-insulator transition in dirty quasi-1D systems.

Figure 4 shows the four-probe resistance data for a single MWNT with a diameter of 12±2 nm. The inner contact distance is L = 5000 Å. One can see that the resistance drops more rapidly below about 262 K and saturates to a value of about 8.80 k Ω below 160 K. The resistive behavior of this single tube is similar to the resistive transition for a quasi-1D superconductor with $T_{c0} \simeq 262$ K. The finite resistance far below the superconducting transition temperature is due to a finite number of transverse channels, which is estimated to be about 54 for this tube (see below). Using R_0 = 8.80 kΩ, N_{ch} = 54, and the relation: $R_0 = R_Q/tN_{ch}$, we estimate the average t for each channel to be about 0.013. The small value of t suggests that the electrical contacts to the tube are rather poor. Assuming a negligible

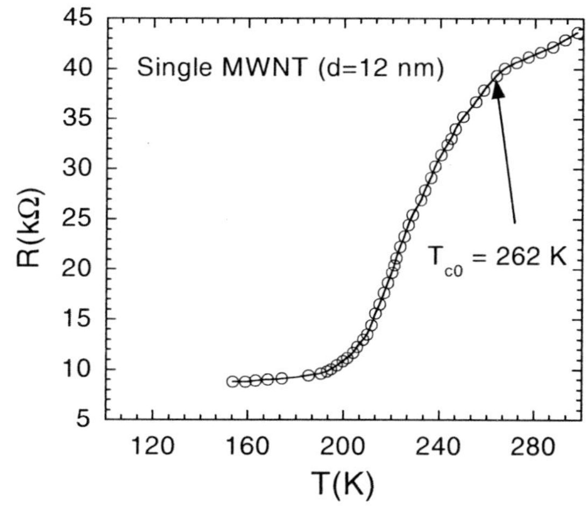

Figure 4. The temperature dependence of the resistance for a single MWNT with $d \simeq 12$ nm. The data are extracted from Ref. [27].

on-tube resistance below 160 K, we estimate that the on-tube resistance in the normal state (at 300 K) is about 34 kΩ, leading to $R_N/L = 68$ kΩ/μm.

In Fig. 5 we fit the resistance data below $T = 233$ K $= 0.89T_{c0}$ by Eq. 10 with a fixed $R_0 = 8.80$ kΩ. One can see that the fitting is excellent with the fitting parameters: $\beta = 11.71\pm0.12$, α=19.0±0.1 kΩ, and $T_c^* = 234$ K. It is interesting that $T_c^* = 0.89T_{c0}$, which is the same as that for the SWNT bundle with $T_{c0} = 0.44$ K.

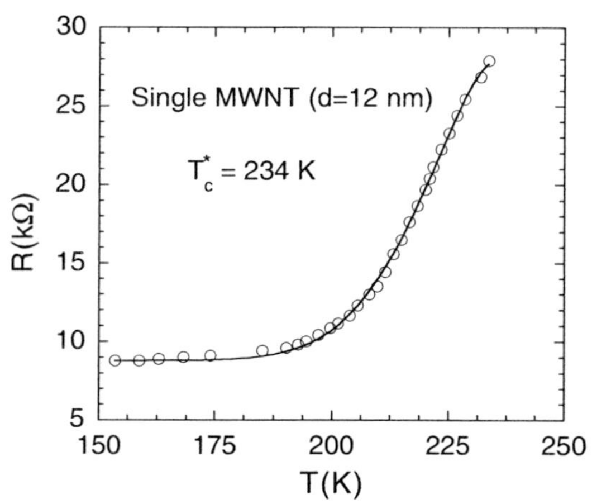

Figure 5. The temperature dependence of the resistance for the 12-nm MWNT below 233 K $= 0.89T_{c0}$. The solid line is the curve best fitted by Eq. 10 with the fitting parameters: $\beta = 11.71\pm0.12$, $\alpha = 19.0\pm0.1$ kΩ, and $T_c^* = 234$ K, and with a fixed $R_0 = 8.80$ kΩ. Reducing or increasing the temperature region for the fit tends to worsen the fit quality.

In Fig. 6 we fit the resistance data below about 0.15 R_N by

$$R = R_0 + \frac{m}{T^{1.5}}(1 - \frac{T}{T_{c0}})^{9/4} \exp[-\frac{3cT_{c0}}{T}(1 - \frac{T}{T_{c0}})^{3/2}]. \quad (13)$$

We fit the resistance data in this region because $l \simeq \xi_{BCS}$, as seen below. We can see that the fitting is excellent between 190 K and 210 K. There is a small deviation between 160 and 190 K, which may arise from quantum phase slips. The fitting parameters are $m = (1.64\pm0.14)\times10^7$ kΩK$^{1.5}$ and $c = 10.27\pm0.23$. It is striking that the values of β and c are also very close, similar to the case of the SWNT bundle with $T_{c0} = 0.44$ K.

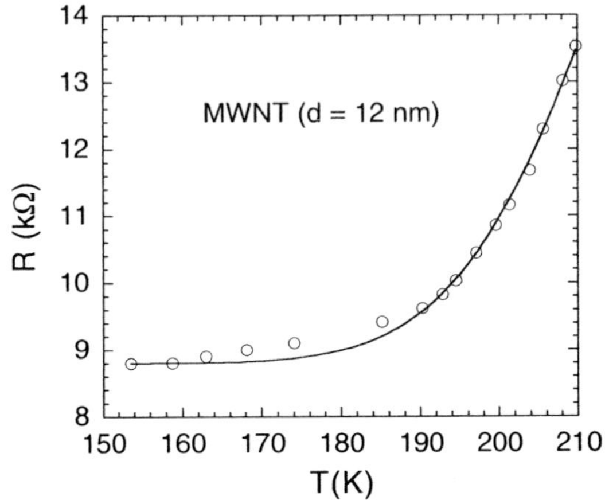

Figure 6. The temperature dependence of the resistance for the 12-nm MWNT below about $0.15R_N$. The solid line is the curve best fitted by Eq. 13 with the fitting parameters: $m = (1.64\pm0.14)\times10^7$ kΩK$^{1.5}$ and $c = 10.27\pm0.23$.

From the values of m, c, and T_{c0}, we can evaluate the zero-temperature coherence length $\xi(0)$ using Eq. 7. Substituting $m = 1.64\times10^7$ kΩK$^{1.5}$, $c = 10.27$, $T_{c0} = 262$ K, and $L = 5000$ Å into Eq. 7, we obtain $\xi(0) = 18.3$ Å. From the measured $R_N/L = 68$ kΩ/μm and the relation $R_N/L = R_Q/N_{ch}l$ (Ref. [24]), we calculate $N_{ch}l = 1897$ Å. Substituting c =10.27 into Eq. 9, we get $N_{ch} = 30.2\xi_{BCS}/\xi(0)$. By solving the three equations: $N_{ch}l =$ 1897 Å, $N_{ch} = 30.2\xi_{BCS}/\xi(0)$, and $\xi(0) = 0.74\xi_{BCS}\sqrt{\chi(0.882\xi_{BCS}/l)}$, we finally obtain $l = 35$ Å, $N_{ch} = 54$, and $\xi_{BCS} = 33.0$ Å. Here the Gorkov function $\chi(x)$ is defined as [23]

$$\chi(x) = \sum_{n=0}^{\infty} \frac{0.95}{(1+2n)^2(1+2n+x)}. \tag{14}$$

The fact that $l \simeq \xi_{BCS}$ indicates that the LAMH theory (Eq. 13) is valid only below 0.17 R_N (Ref. [23]). This justifies the region of the data we select to fit.

We can also evaluate the Fermi velocity v_F from the deduced value of ξ_{BCS} and the formula $\xi_{BCS} = 0.18\hbar v_F/k_BT_{c0}$. With $\xi_{BCS} = 33$ Å and $T_{c0} = 262$ K, we get $\hbar v_F =$ 4.14 eVÅ, which is about 10% smaller than that deduced above for the SWNT bundle. Using the value of $\gamma_\circ = 2.4$ eV estimated from the first principle calculation [25] and the relation $\hbar v_F = 1.5a_{C-C}\gamma_\circ$ (Ref. [20]) for the first subband of metallic chirality tubes, we obtain $\hbar v_F = 5.1$ eVÅ. The value of $\hbar v_F$ deduced from the LAMH theory is about 20% lower than the value expected for the first subband of metallic chirality tubes. Such a small discrepancy may be explained by a fact that the Fermi level of some outer shells is crossing the second or higher subbands where [29] the Fermi velocity is smaller than that for the first subband. This quantitative agreement provides compelling evidence for high-temperature superconductivity at 262 K in this 12-nm MWNT.

Now we would like to show that the deduced N_{ch} = 54 is also reasonable. For the MWNT with d = 12 nm, the total number of shells can be estimated to be about 17 using the fact that the intershell distance is 0.34 nm. This implies that the average number of conducting channels per shell is about 3.2. This is possible when the Fermi level of some outer shells crosses their second subband (see above), which have 6 channels for a metallic chirality shell and 4 channels for a semiconducting chirality shell [20].

We can estimate the lower limit of the average number of conducting channels per shell for an 18-nm MWNT from the measured room-temperature four-probe resistance, which is 200 Ω (Ref. [27]). Using the relation $R = R_Q/tN_{ch} + R_{tube} \geq R_Q/N_{ch}$ and $R = 200$ Ω, we get $N_{ch} \geq 64$. The total number of shells for the 18-nm MWNT should be about 26. This implies that the average number of conducting channels per shell for the 18-nm MWNT is larger than 2.5, in agreement with that (3.2) deduced independently for the 12-nm MWNT.

The other puzzling feature in the resistive behavior of this 18-nm MWNT is the metal-insulator transition below about 200 K (Ref. [27]). We can attribute this to a superconductor-insulator transition in dirty quasi-1D systems. It is shown that when the thermal length is larger than the localization length below a temperature T_{loc} in quasi-1D systems, the Anderson localization sets in and the ground state becomes insulating [30].

2.4. The Resistive Transition in a SWNT Mat with T_{c0} = 710 K

Single-walled carbon nanotubes, prepared by metal-catalysed laser ablation of graphite, form closely-packed crystalline bundles. The bulk samples, or mats consist of entangled bundles that are contacted with each other and oriented randomly [31]. If close-packed crystalline bundles are superconductors, the bulk samples should behave like granular superconductors. Depending on the Josephson coupling strength between the superconducting "grains", the ground state could be metallic, insulating, or superconducting [32]. It is interesting that the contact barrier resistance of a granular superconductor follows a rather unusual exponential temperature dependence in a certain temperature range [32], that is, $R_b(T) = R_b(0)\exp(BT)$, where B could be positive, negative, or zero. This temperature dependence of the barrier resistance was also suggested [33] for the intertube barrier resistance in MWNTs. The barrier resistance extrapolated to zero temperature is finite even if the T-dependence of the resistance behaves like an insulator. This unique resistive behavior makes a clear distinction from that for conventional semiconductors where the resistance at zero temperature goes to infinity.

Figure 7 shows the temperature dependence of the resistivity for a SWNT mat. The data are extracted from Ref. [31]. Below 200 K the resistivity is nearly temperature independent while above 200 K the resistivity increases suddenly and starts to flatten out above 550 K. This behavior is similar to that for a granular superconductor. Below we will show that this is indeed the case.

We fit the resistivity data by

$$\rho = \rho_0(T) + \alpha \exp[-\frac{3\beta T_c^*}{T}(1 - \frac{T}{T_c^*})^{3/2}]. \quad (15)$$

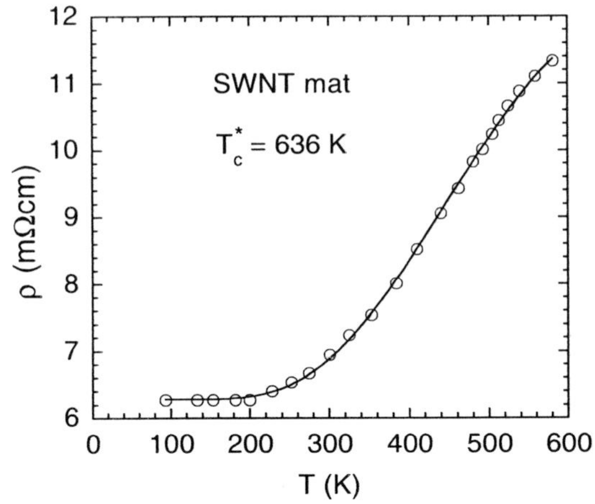

Figure 7. The temperature dependence of the resistivity for a SWNT mat. The data are extracted from Ref. [31]. The solid line is the curve fitted by Eq. 15 with the fitting parameters: $\rho_0 = 6.277\pm0.018$ mΩcm, $\beta = 0.90\pm0.04$, and $T_c^* = 636.3$ K.

Here the first term is the intertube contact barrier resistivity, which could be temperature dependent, and the second term is the on-tube resistivity that follows a simple exponential form same as Eq. 10. In the present case, the intertube barrier resistivity appears to be independent of temperature in the temperature region we are interested in, i.e., $B \simeq 0$. It is striking that the data can be well fitted by Eq. 15 with the fitting parameters: ρ_0= 6.277±0.018 mΩcm, $\beta = 0.90\pm0.04$, and $T_c^* = 636.3$ K. Using the relation $T_c^* = 0.895T_{c0}$ deduced empirically above, we obtain T_{c0} = 710 K. It is remarkable that the T_{c0} value obtained from the resistivity data is very close to that (665 K) inferred from the Raman data for a similarly prepared SWNT mat [13, 14] (see section 4.).

By extrapolation of the data shown in Fig. 7 to $T = 710$ K, we estimate the normal-state on-tube resistivity at 710 K to be $\rho_N^{exp} \simeq 1.1\rho_N^{exp}(T_{c0})$ = 6270 $\mu\Omega$cm. Since about one-third of tubes have metallic chiralities and the mat consists of crystalline bundles that are oriented randomly, the intrinsic normal-state on-tube resistivity ρ_N^i of the superconducting tubes should be much smaller than ρ_N^{exp}, that is, $\rho_N^i = \rho_N^{exp}/f$, where f is the reduction factor that should be close to 3(1/0.33) = 9. The intrinsic mean free path l is related to ρ_N^i by

$$l = \frac{R_Q}{2}\frac{A_\circ}{\rho_i}, \tag{16}$$

were $A_\circ$ is the area of single tube, which is equal to 1.54×10^{-18} m^2 for $d = 1.4$ nm. If we take $f = 9$, we get $l = 14$ Å at $T = 710$ K from Eq. 16. If the resistivity jump above 200 K were due to inelastic scattering, the inelastic mean free path would be about 14 Å at about 700 K. This is inconsistent with any electrical transport mechanism for SWNTs.

Using $\hbar v_F = 4.5$ eVÅ and $T_{c0} = 710$ K, we find that the BCS coherence length along the tube-axis direction $\xi_{BCS} = 13.2$ Å. The zero-temperature coherence length along the tube-axis direction $\xi(0)$ should be smaller than the clean-limit value: $0.74\xi_{BCS} = 10$ Å. Then the coherence length perpendicular to the tube axis should be order of 1 Å. Such a short coherence length implies that only those superconducting tubes that are adjacent to each other can have enough Josephson coupling to form a superconducting bundle. A simulation [34] indicates that the average number of the metallic-chirality tubes that are adjacent to each other is about 2. This implies that the average number of the metallic-chirality tubes comprising a superconducting bundle is also about 2, and that there are a number of independent superconducting bundles within a physical bundle.

If we assume that the normal-state resistivity is linearly proportional to T above 200 K, the average mean-free path between 200 K and 580 K should be about 26 Å, significantly larger than ξ_{BCS}. Then we estimate $\xi(0) \simeq 8$ Å. Using Eq. 9 and $c \simeq \beta = 0.9$, we obtain $N_{ch} \simeq 4.5$. This implies that, on average, about two metallic chirality tubes are adjacent to each other and form a superconducting bundle, in quantitative agreement with the simulation [34].

2.5. The Resistive Transition in a MWNT Mat with T_{c0} = 752 K

A multi-walled carbon nanotube is packed in such a way that each shell is concentric with each other. If each shell has phase-incoherent superconductivity, MWNTs are almost optimally packed to maximize the Josephson coupling and phase coherence. Individual MWNTs can be closely packed into bundles. The bulk samples, or mats are made of entangled bundles that are contacted with each other and oriented randomly. The bulk samples should also behave like granular superconductors.

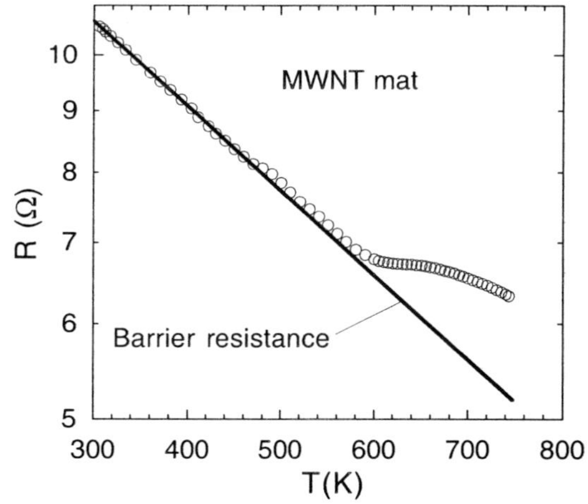

Figure 8. The temperature dependence of the resistance for a MWNT mat. The data are the same as those in Ref. [8] but less data points are plotted for charity. The resistance between 300 K and 450 K can be excellently described by $17.3\exp(-T/618.3)$ Ω, which represents the intertube barrier resistance [33]. The average diameter of the tubes is about 10 nm.

Figure 8 shows the temperature dependence of the resistance for a MWNT mat. It is interesting that the resistance decreases monotonically with increasing temperature below about 570 K. Above 570 K, the resistance tends to turn up. The resistance between 300 K and 450 K can be excellently described by $17.3\exp(-T/618.3)$ Ω. This temperature dependence is expected for an intertube barrier resistance [33]. This implies that the on-tube resistance between 300 K and 450 K is negligible, in agreement with several independent experiments which consistently show a negligible on-tube resistance at room temperature in many individual MWNTs [35, 36, 37, 38]. The observed finite and very small on-tube resistances in the individual MWNTs [37] are consistent with quasi-1D room-temperature superconductivity with finite quantum phase slips.

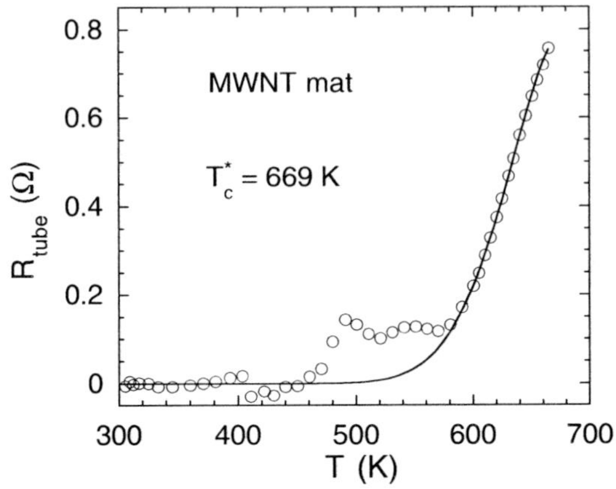

Figure 9. The on-tube resistance by subtracting the barrier resistance from the total resistance of the MWNT mat. The solid line is the fitted curve by Eq. 10 by excluding the data points between 450 K and 600 K. The fitting parameters are $\beta = 11.34$ and $T_c^* = 669$ K.

If we assume that this temperature dependence for the intertube barrier resistance remains valid up to 750 K, we then obtain the on-tube resistance by subtracting the barrier resistance from the total resistance. The resultant on-tube resistance below 665 K is shown in Fig. 9. The solid line is the fitted curve by Eq. 10 by excluding the data points between 450 K and 600 K. The shoulder feature between 450 K and 600 K may be caused by quantum phase slips. The fitting parameters are $\beta = 11.34$ and $T_c^* = 669$ K. Using the empirical relation $T_c^* = 0.89T_{c0}$, we obtain $T_{c0} = 752$ K.

It is interesting that the value of $\beta = 11.34$ for this MWNT mat is slightly smaller than that (11.71) for the 12-nm MWNT. This implies that, on average, the total number of transverse channels for each superconducting bundle (which may contain one or more individual MWNTs near T_{c0}) is comparable with that for the single 12-nm MWNT. As discussed in section 7., the Josephson coupling among individual MWNTs has not been established near T_{c0}, so most of the superconducting bundles actually consist of only single MWNT. Only for those small-diameter MWNTs, the Josephson coupling among them is established near T_{c0} such that several individual tubes are formed into a superconducting bundle. Therefore, the average number of conduction channels per shell should be about 3 for the mat sample, similar to that for the single 12-nm MWNT.

In Fig. 10, we plot the normalized on-tube resistance versus T/T_{c0} for the 12-nm MWNT ($T_{c0} = 262$ K) and for the MWNT mat ($T_{c0} = 752$ K). It is remarkable that the on-tube resistive transitions for the two systems are nearly identical although they have different T_{c0}'s and the electrical measurements were done by independent groups. This agreement also suggests that both sets of data are reliable and that the procedure to extract the on-tube resistance for the MWNT mat is justified.

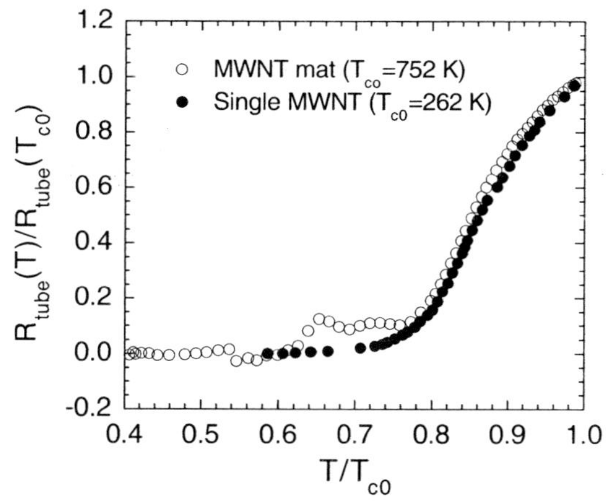

Figure 10. The normalized on-tube resistance versus T/T_{c0} for the 12-nm MWNT (T_{c0} = 262 K) and for the MWNT mat (T_{c0} = 752 K). The on-tube resistive transitions for the two systems are nearly identical although they have different T_{c0}'s.

3. Magnetic Properties of MWNTs

From the quantitative analyses of the electrical transport data in several carbon nanotubes, we can clearly see that the superconducting transition temperatures can vary from 0.4 K to 750 K for different samples. We believe that the T_{c0} variation may be associated with the differences in the doping level, the chirality and diameter of tubes, the screening of the long-range Coulomb interaction, and in disorders (see section 7. below). In order to unambiguously show that high-temperature superconductivity in carbon nanotubes is real, one needs to provide magnetic evidence such as the existence of the Meissner effect. However, the Meissner effect may be less visible because the diameters of the tubes may be much smaller than the magnetic penetration depth. Further, the orbital diamagnetic susceptibility in the magnetic field perpendicular to the graphite sheet is large, making it difficult to separate the Meissner effect from the large orbital diamagnetic susceptibility. Fortunately, the orbital diamagnetic susceptibility of carbon nanotubes in the magnetic field parallel to the tube axis is predicted to be very small at room temperature [40]. This makes it possible to extract the Meissner effect from the measured susceptibility in the parallel field.

Figure 11 shows the temperature dependence of the susceptibility for physically separated and aligned MWNTs in a magnetic field parallel to the tube axis. The diameters of the tubes are 10$\pm$5 nm, and the lengths are on the order of 1 μm. It is apparent that the diamagnetic susceptibility is significant up to 265 K. Because the orbital diamagnetic susceptibility in the parallel field is negligible at room temperature [40], the observed diamagnetic susceptibility at 265 K should mainly contribute from the Meissner effect due to superconductivity. Thus, the Meissner effect at 265 K is about -0.8×10^{-5} emu/g, which is significant. This result clearly indicates that the superconducting transition temperature should be higher than 300 K.

For superconducting tubes of radius r in the magnetic field parallel to the tube axis, the

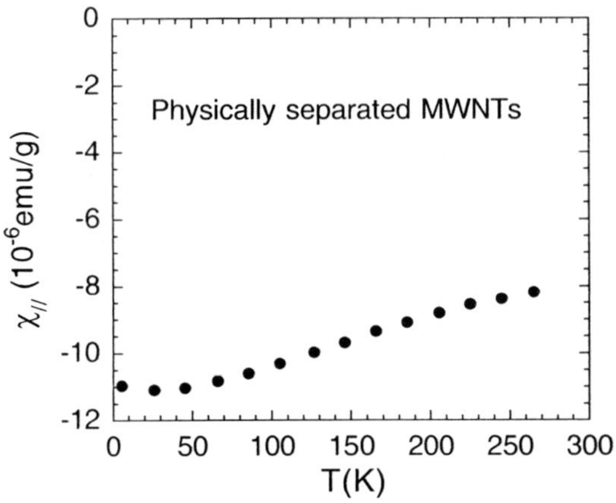

Figure 11. The temperature dependence of the susceptibility for physically separated and aligned MWNTs in a magnetic field parallel to the tube axis. The data are extracted from Ref. [39].

diamagnetic susceptibility due to the Meissner effect is given by

$$\chi_{\parallel}^{S}(T) = -\frac{\bar{r^2}}{32\pi\lambda_{\theta}^{2}(T)}. \tag{17}$$

Here $\bar{r^2}$ is the average value of r^2, and $\lambda_\theta(T)$ is the penetration depth when carriers move along the circumferential direction. The above equation is valid only if $\lambda_\theta(0)$ is larger than the maximum radius of tubes, which should be the case for carbon nanotubes. Eq. 17 indicates that the Meissner effect is inversely proportional to $1/\lambda_\theta^2(T)$. Assuming an isotropic gap and taking $T_{c0} = 752$ K, we find that $1/\lambda_\theta^2(T)$ and thus $\chi_{\parallel}^{S}(T)$ are nearly independent of temperature below 265 K. Then we have $\chi_{\parallel}^{S}(0) = \chi_{\parallel}$ (265 K) $\simeq -0.8 \times 10^{-5}$ emu/g. If we assume that the radii of tubes are equally distributed from 0 to 100 Å, we find $\bar{r} = 50$ Å and $\bar{r}^2 = 3333$ Å^2. With the weight density of 2.17 g/cm^3 (Ref. [41]) and $\chi_{\parallel}^{S}(0) = -0.8 \times 10^{-5}$ emu/g, we calculate $\lambda_\theta(0) \simeq 1380$ Å. This value of the penetration depth corresponds to $n/m_\theta^* = 1.48 \times 10^{21}/\text{cm}^3 m_e$, where n is the carrier density, m_θ^* is the effective mass of carriers along the circumferential direction. If we take $m_\theta^* = 0.012\ m_e$, typical for graphites [42], we estimate $n = 1.78\times10^{19}/\text{cm}^3$, in good agreement with the Hall effect measurement [33] which gives $n = 1.6\times10^{19}/\text{cm}^3$. It is worthy of noting that the Hall coefficient in the physically separated MWNTs does not go to zero below T_{c0}. This is because the on-tube resistance does not exactly go to zero due to quantum phase slips and because the magnetic field is almost penetrated into the whole volume of the tubes.

From Eq. 17, we can see that $\chi_{\parallel}^{S}(T)$ will increase linearly with increasing $\bar{r^2}$ or cross-sectional area. For Josephson coupled MWNT bundles in unprocessed mats, the effective $\bar{r^2}$ is larger than that for physically separated tubes. As the temperature decreases, the Josephson coupling strength increases so that the effective $\bar{r^2}$ and thus $\chi_{\parallel}^{S}(T)$ also increase. This can naturally explain why the diamagnetic susceptibility for physically coupled MWNTs is larger than that for physically separated MWNTs and why the enhancement in the diamagnetic susceptibility increases significantly with decreasing temperature (see Fig. 11).

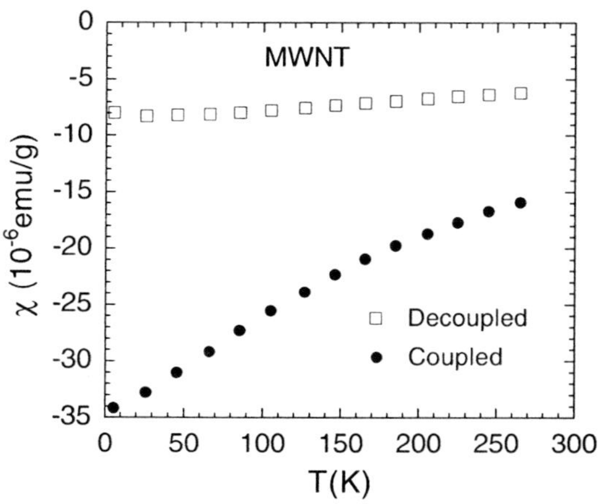

Figure 12. The temperature dependence of the angle-averaged susceptibility for physically separated MWNTs (open squares) and for physically coupled MWNT mat (solid circles). The physically coupled mat sample is not processed so that the tubes are closely packed into bundles. The data are extracted from Ref. [39].

At the lowest temperature, the enhancement factor is about 4.3. Without superconductivity in these MWNTs, it is very difficult to account for such a large enhancement in the diamagnetic susceptibility upon bundling of the tubes.

4. Raman Spectroscopic Evidence for Superconductivity at 665 K

It is known that Raman scattering has provided essential information about the electron-phonon coupling and the electronic pair excitation energy in the high-T_c cuprate superconductors [43, 44, 45]. The anomalous temperature-dependent broadening of the Raman active B_{1g}-like mode of 90 K superconductors $RBa_2Cu_3O_{7-y}$ (R is a rare-earth element) allows one to precisely determine the superconducting gap [44]. The pronounced softening observed only for the B_{1g} mode is due to the fact that the phonon energy of the B_{1g} mode is very close to $2\Delta(0)$ and the mode is strongly coupled to electrons [44, 46, 47].

The temperature dependence of the frequency for the Raman-active B_{1g} mode of an optimally doped $YBa_2Cu_3O_{7-y}$ is shown in Fig. 13a. It is apparent that the frequency decreases linearly with increasing temperature above T_c and that the mode starts to soften below about $0.95T_c$. The temperature dependence of the frequency above T_c is caused by thermal expansion. The temperature dependence of the frequency will become more pronounced at higher temperatures because the magnitude of the slope $-d\ln\omega/dT$ is essentially proportional to the lattice heat capacity that increases monotonically with temperature. The significant softening of the mode below T_c occurs only if the energy of the Raman mode is very close to $2\Delta(0)$ and the electron-phonon coupling is substantial [46, 47], as it is the case in the optimally doped $YBa_2Cu_3O_{7-y}$ [43, 44, 45]. In order to see more clearly the softening of the mode, we show in Fig. 13b the difference of the measured frequency and the linearly fitted curve above T_c. It is clear that the softening starts at about 89 K $\simeq$

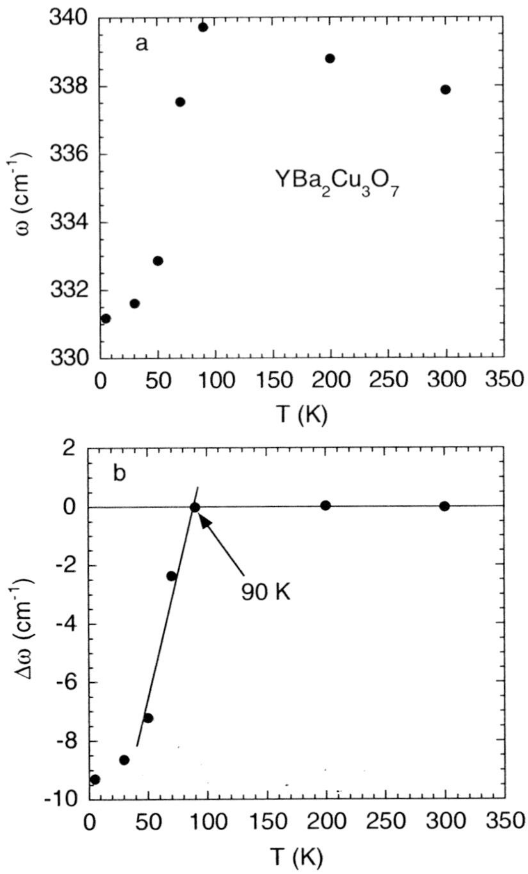

Figure 13. a) The temperature dependence of the frequency for the Raman-active B_{1g} mode of an optimally doped $YBa_2Cu_3O_{7-y}$. The data are extracted from Ref. [43]. b) The difference between the measured frequency and the linearly fitted curve above T_c.

$0.95T_c$.

Figure 14a shows the temperature dependence of the frequency for the Raman active graphitic mode of a SWNT mat. It is striking that the frequency data show a clear tendency of softening below about 630 K. Above 630 K, the frequency decreases linearly with increasing temperature with a slope much larger than that in $YBa_2Cu_3O_{7-y}$. This is due to a much larger thermal expansion at such high temperatures in SWNTs. The slope will decrease with decreasing temperature and eventually level off at low temperatures. The leveling off between 450-630 K must be due to the competition between the mode hardening due to the normal thermal effect and the mode softening due to a transition to the charge/spin density wave state or to the superconducting state. The transition to the charge/spin density wave state will lead to a metal-insulator transition upon cooling, in contrast to the transport data in Fig. 7 which is only consistent with the superconducting transition.

In order to see more clearly the softening of the mode, we show in Fig. 14b the difference between the measured frequency and the linearly fitted curve above the kink temperature (~630 K). One can see that the result shown in Fig. 14b is similar to that shown in

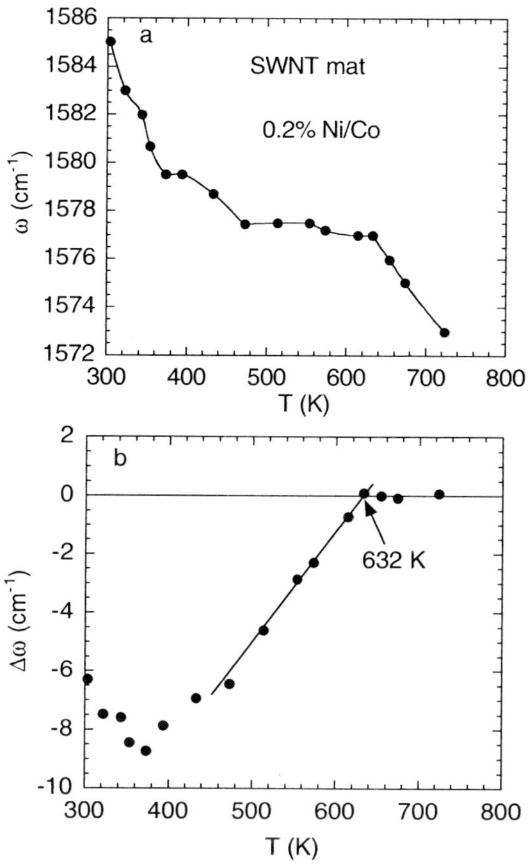

Figure 14. a) The temperature dependence of the frequency for the Raman active graphitic mode of a SWNT mat with 0.2% Ni/Co catalyst. The data are from Ref. [48]. b) The difference of the measured frequency and the linearly fitted curve above the kink temperatures (see text).

Fig. 13b. This suggests that the softening of the Raman active graphitic mode in the SWNTs may have the same microscopic origin as the softening of the Raman active B_{1g} mode in $YBa_2Cu_3O_{7-y}$. This explanation is plausible only if the phonon energy of the graphitic mode is very close to $2\Delta(0)$. We are fortunate that the phonon energy of the graphitic mode is 197 meV, very close to $2\Delta(0) \simeq 200$ meV expected for a T_{c0} in the range of 600-700 K. If T_{c0} is away from this range, the phonon self-energy effect will disappear. This can explain why this effect is not visible for another SWNT mat with 1.3 % Ni/Co catalyst, as shown in Fig. 15. It is likely that the T_{c0} of this sample is suppressed to a much lower value by the magnetic impurities.

From Fig. 14, we can clearly see that the softening starts at about 632 K. Using the fact that the softening starts at $0.95T_{c0}$ (Ref. [46]), we can assign the mean-field transition temperature $T_{c0} = 665$ K. It is interesting to note that there is a clear minimum at $T^* = 370$ K $= 0.57T_{c0}$ in Fig. 14b. Such a minimum is also seen at about $0.6T_{c0}$ in a calculated curve for a superconductor (see Fig. 8 of Ref. [46]). This shallow minimum in the frequency shift is related to a sharp minimum in the real part of the polarization $\Pi(\omega, T)$, which occurs at

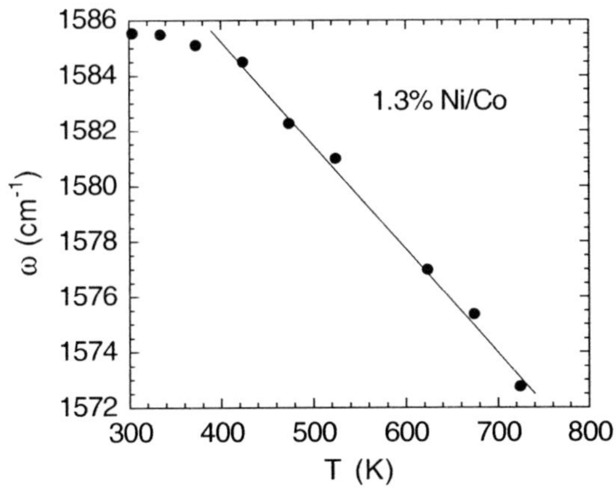

Figure 15. The temperature dependence of the frequency for the Raman active graphitic mode of a SWNT mat with 1.3% Ni/Co catalyst. The data are from Ref. [48]. The phonon self-energy effect is not visible in this sample because the T_{c0} of the sample might be far away from the range of 600-700 K, which is optimal for the observation of the phonon self-energy effect.

$\hbar\omega/2\Delta(T^*) = 1$ for weak coupling [46, 47]. From the temperature dependence of the BCS gap [17], we find that $\Delta(T^*) = \Delta(0.57T_{c0}) = 0.93\Delta(0)$. With $\hbar\omega/2\Delta(T^*) = 1$ and $\hbar\omega = 197$ meV, we finally obtain $\Delta(0) = 106$ meV. Then we calculate $2\Delta(0)/k_BT_{c0} = 3.7$, which is in excellent agreement with the prediction based on the plasmon-mediated pairing mechanism [4]. It is worth noting that $\Delta(0)$ would be much larger if the magnetic impurities would not suppress superconductivity.

We can evaluate the electron-phonon coupling constant for the Raman-active graphitic phonon mode λ^g using a relation [44]: $\Delta\omega/\omega = 0.5\lambda^g \text{Re}\Pi/N(0)$. In the weak-coupling limit, the value of $\text{Re}\Pi/N(0)$ at zero temperature is calculated to be -7.0 using Eq. 16 of Ref. [47] and $\hbar\omega/2\Delta(0) = 0.93$. Strong coupling and impurity scattering tend to reduce the magnitude of $\text{Re}\Pi/N(0)$ [46, 47]. Because $\Delta\omega/\omega$ is nearly independent of temperature below $0.45T_{c0}$ [46], we take the value of $\Delta\omega/\omega$ at zero temperature to be the same as that (-0.378%) at room temperature. Then we obtain $\lambda^g = 0.0011$ using the relation: $\Delta\omega/\omega = 0.5\lambda^g \text{Re}\Pi/N(0)$. The value of $\lambda^g = 0.0011$ may be slightly underestimated since the coupling to bosonic modes is not in the weak limit ($2\,\Delta(0)/k_BT_{c0} = 3.7$).

Recently, Lazzeri *et al.* [49] have used accurate density functional theory (DFT) to compute with high accuracy the electron-phonon coupling for optical phonons in (6, 6) and (11, 11) armchair SWNTs. From their results, we can readily evaluate the electron-phonon coupling constant λ^g for the Raman-active graphitic mode in metallic-chirality SWNTs with different diameters. For $d = 1.4$ nm, we find that $\lambda^g = 0.0014$, in quantitative agreement with that (0.0011) deduced from the superconductivity induced phonon self-energy effect. Such quantitative agreement justifies the above data analyses that suggest a superconducting transition at about 665 K in the SWNT mat sample.

5. Tunneling Spectra

One of the important signatures for superconductivity is the existence of the single-particle excitation gap. From a single-particle tunneling spectrum obtained through two high-resistance contacts (see Fig. 1b of Ref. [50]), we can clearly see a pseudo-gap feature which appears at an energy of about 220 meV. The pseudo-gap feature could be related to the superconducting gap. Considering the broadening of the gap feature due to large quantum phase slips and the double tunneling junctions in series, we estimate the superconducting gap $\Delta(0)$ to be about 100 meV. The scanning tunneling microscopy [51] on individual

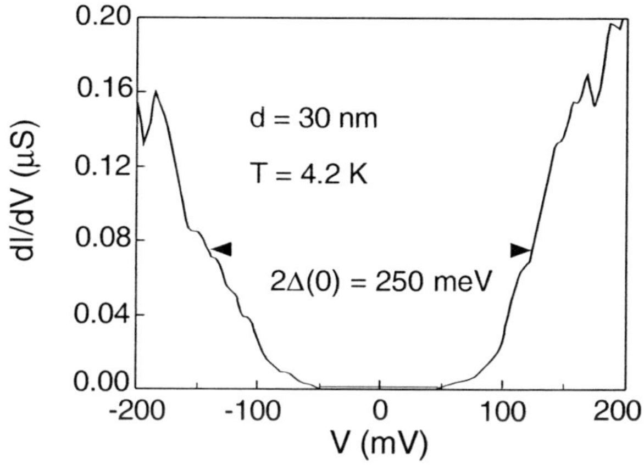

Figure 16. The tunneling spectrum for an "insulating" MWNT with $d = 30$ nm. The figure is reproduced from Ref. [52]. The energy scale has been reduced by a factor of two compared with the original spectrum for double junctions [52].

single-walled nanotubes also show pseudo-gap features with $\Delta(0) > 100$ meV in doped metallic SWNTs (the Fermi level E_F is about 0.2 eV below the charge neutrality point). From the Raman data for a SWNT mat, we have quantitatively deduced $\Delta(0) = 106$ meV from the phonon self-energy effect. These data consistently suggest that the magnitude of $\Delta(0)$ can be larger than 100 meV in SWNTs.

For MWNTs, the electrical transport data indicate superconductivity at about 750 K. This implies that the magnitude of $\Delta(0)$ should be about 120 meV according to $2\Delta(0)/k_B T_{c0} = 3.7$. The expected single-particle excitation gap has been seen in the tunneling spectrum for an "insulating" MWNT with $d = 30$ nm, which is shown in Fig. 16. In the figure, the energy scale has been reduced by a factor of two compared with the original spectrum for double junctions. This spectrum would imply a semiconducting gap of about 200 meV if the outermost shell had a semiconducting chirality. However, according to a formula $E_g = 2a_{C-C}\gamma_\circ/d$ (Ref. [20]), the predicted semiconducting gap E_g for a semiconducting chirality tube with $d = 30$ nm should be about 22 meV, which is about one order of magnitude smaller than the observed value for the "insulating" tube. In fact, the tunneling spectrum for another "conducting" tube with the same diameter [52] is consistent with $E_g \simeq 20$ meV (see Fig. 17), suggesting that the "conducting" tube has a semiconducting chirality.

In order to consistently explain the tunneling spectra, we must assume that the "insu-

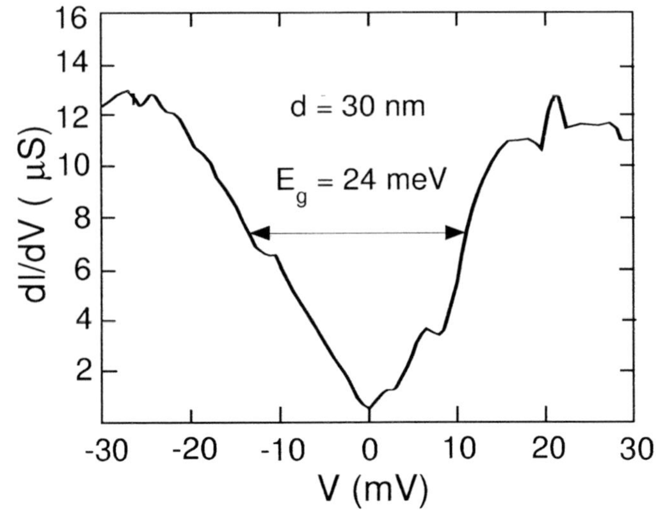

Figure 17. The tunneling spectrum for a "conducting" MWNT with $d = 30$ nm. The figure is reproduced from Ref. [52]. The magnitude of the gap is consistent with a semiconducting chirality tube.

lating" tube has a metallic chirality and that the gap is related to either the superconducting state or the insulating CDW state. If the gap is related to the insulating ground state, one should be able to observe a rectification effect in a heterojunction between this "insulating" MWNT and another "conducting" MWNT. Indeed, the rectification effect is observed [52] below about 192 K. However, the rectification effect disappears [52] above 192 K, which cannot be explained if the metallic-chirality tube remains insulating. Therefore there must be a transition from the superconducting to the insulating state below $T_{loc} \simeq 192$ K. A similar transition at about 200 K has been seen in an 18-nm MWNT [27] (see section 2.). Such a transition can occur in dirty quasi-1D systems [30] where the Cooper pairs get localized below T_{loc}. It is clear that the single-particle gap remains essentially unchanged across T_{loc}. Therefore, the superconducting gap would be about 125 meV if this 30-nm MWNT were clean enough to avoid the Anderson localization.

The single-particle tunneling spectrum for another MWNT with $d = 17$ nm is displayed in Fig. 18. It is apparent that both the intensity and the peak position in this spectrum is inconsistent with the normal-state density of states for any types of metallic-chirality tubes [53, 54]. On the other hand, this spectrum is very similar to the one for a superconductor with a superconducting gap of about 54.4 meV.

We can quantitatively explain this spectrum by invoking superconductivity. The peak positions denoted by 1, 2, 3, 4, 5 are located at < -117 mV, -54.4 mV, 54.4 mV, 76.5 mV, and $\sim$120 mV, respectively. Below T_{c0}, the Van Hove singularities are located at $\epsilon_{VH} = \pm\sqrt{[\Delta(0)]^2 + (E_{VH} - E_F)^2}$. Here the sign in the prefactor of the square root is the same as the sign of $E_{VH} - E_F$. In order to get a consistent explanation to the spectrum, we assume that the tube is hole doped (i.e., $E_F < 0$) and that the positive voltage corresponds to the negative electron energy.

It is interesting that peak 2 is broader than peak 3. This indicates that peak 2 corresponds

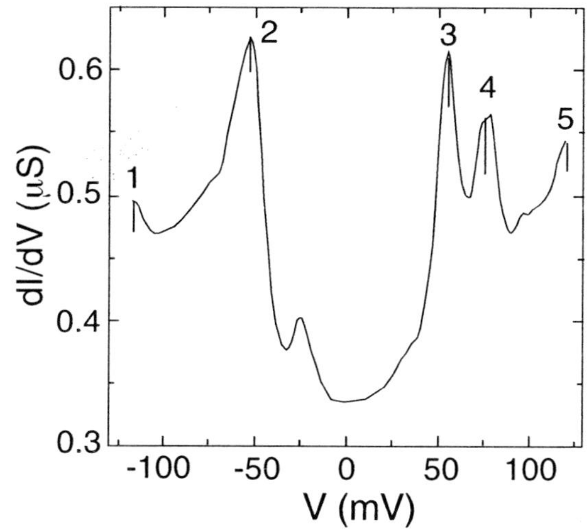

Figure 18. Single-particle tunneling spectrum for a single MWNT with $d \simeq 17$ nm, which is reproduced from Ref. [21].

to two overlapped peaks: one is a superconducting quasi-particle peak and another is a Van Hove singularity located slightly above the Fermi level. Peak 3 corresponds to another superconducting quasiparticle peak. From the positions of peak 2, 3, and 4, we deduce that $\Delta(0) = 54.4$ meV, $E_{VH1} = \pm 54.4$ meV, $E_{VH2} = \pm 108.5$ meV $\simeq 2E_{VH1}$, and $E_F = -55.0$ meV. The positions of Van Hove singularities agree with the theoretical prediction for an armchair tube [20]. With the values of E_{VH1}, E_{VH2} and $\Delta(0)$, we can predict the other peak positions. For example, we predict $V_1 = -121.6$ mV and $V_5 = +121.6$ mV, in excellent agreement with the measured results. Using $E_{VH1} = 54.4$ meV, $d = 17$ nm, and the relation $E_{VH1} = 3a_{C-C}\gamma_\circ/d$ (Ref. [20]), we find that $\gamma_\circ = 2.2$ eV, in quantitative agreement with another independent measurement [55]. It is remarkable that the value of $\gamma_\circ = 2.2$ eV also agrees quantitatively with those deduced from the resistive transitions for both SWNT bundle and individual MWNT (see section 2.).

Since the Fermi level is right on the first Van Hove singularity in this MWNT, the large enhancement in the density of states due to the Van Hove singularity leads to a large increase in the condensation energy, and thus a large suppression of quantum phase slips [10]. The observed sharp quasiparticle peak at the gap edge should be related to substantially reduced quantum phase slips. The much lower gap in the outermost layer of this MWNT is due to the fact that electron-plasmon coupling is significantly reduced when the second subband is crossed [4].

6. Negligible on-Tube Resistances in MWNTs

In order to confirm superconductivity at room temperature in a 3D bulk sample, one needs to show zero on-tube resistance at room temperature. However, the on-tube resistance for a quasi-1D superconducting tube with a finite number of transverse channels will never go to zero at finite temperatures due to quantum phase slips [16]. The resistance at room temperature should be negligibly small if N_{ch} is not too small. Indeed, negligible on-tube resistances have been observed in a multi-walled nanotube bundle consisting of two MWNTs with a diameter d = 16 nm (Ref. [35]), in an individual MWNT with d = 40 nm (Ref. [36]), in over 50 individual MWNTs [37], as well as in a single-walled nanotube bundle consisting of two SWNTs with d = 1.4 nm (Ref. [56]). These electrical transport data are either consistent with ballistic transport or superconductivity at room temperature.

In order to distinguish between ballistic transport and superconductivity at room temperature, it is essential to reliably estimate the lower limit of the room-temperature resistivity due to inelastic scattering. If we consider inelastic scattering only by the twist mode, the calculated on-tube resistance per unit length per shell is the lower limit. If this lower limit is far larger than the measured values, the ballistic transport mechanism is ruled out.

The on-tube resistance per unit length due to the scattering by the twist mode has been deduced for armchair tubes (n, n) [57]. For a Luttinger parameter $K_c \leq 1$, the on-tube resistance per unit length is given by [57]

$$R_{tube}/L = \frac{R_Q}{2}\frac{\gamma(K_c)\Lambda^2}{a\hbar v_F C_t}(\pi a k_B T/\hbar v_F)^{(1+K_c)/2}, \quad (18)$$

where $\Lambda = n\times 2.92$ eVÅ, $C_t = n^3 \times 18$ eVÅ, $a = 2.46$ Å, and

$$\gamma(K_c) = 4\pi^{-2}(K_c)^{1+K_c/2}\sin[\pi(1+K_c)/4]\int_0^\infty dz z/[\sinh(z)]^{(3+K_c)/2}. \quad (19)$$

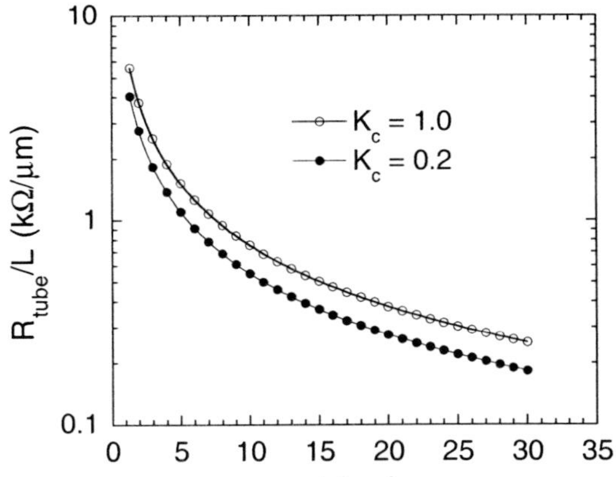

Figure 19. The calculated lower limit of the room-temperature on-tube resistance per unit length R_{tube}/L as a function of the tube diameter d for the Luttinger parameter K_c = 1.0 and 0.2. The lower limit is set by the resistivity contributed only from the scattering of a twist mode.

One can easily show that $\gamma(1) = 2/\pi$ and $\gamma(0.2) = 0.125$. Using the relation $d = (\sqrt{3}/\pi)an$ and $\hbar v_F = 5.3$ eVÅ, and from Eq. 18, we calculate R_{tube}/L at room temperature as a function of d for K_c = 1 and 0.2. The result is shown in Fig. 19. We can see that for d = 20 nm, the on-tube resistance per unit length at room temperature would be at least 275 Ω/μm if the tube were not superconducting. Other inelastic scattering will further increase the resistivity. For example, the phonon energy of the breathing mode for the 20-nm tube is only 1.4 meV and this mode is strongly coupled to electrons [58]. Such a soft mode will produce a large inelastic scattering at room temperature.

Poncharal *et al.* [37] have made extensive transport studies on over 50 individual MWNTs with diameters ranging from 5 to 25 nm. These MWNTs, which protrude from unprocessed arc produced nanotube fibers and are contacted with liquid metals, show very small per unit length resistances at room temperature: $R_{tube}/L = 31 \pm 61$ Ω/μm. For example, $R_{tube}/L = 14$ Ω/μm for a MWNT with $d \simeq 20$ nm. Since the tunneling resistance is close to R_Q and the electrical contacts are nearly ideal, the current would only flow the outermost shell if the MWNT were not a superconductor. This implies that the value of R_{tube}/L for the outermost shell ($d \simeq 20$ nm) would be about 14 Ω/μm, which is over one order of magnitude smaller than the lower limit (275 Ω/μm) expected from the ballistic transport mechanism. Therefore, the ballistic transport mechanism cannot explain such a small on-tube resistance at room temperature.

Using the average $R_{tube}/L = 31$ Ω/μm and the average $d = 15$ nm, and assuming that the current flows through all the shells, we estimate the room-temperature resistivity to be 5.5×10^{-7} Ωcm. The resistivity is about 3 orders of magnitude smaller than the normal-state resistivity in the SWNT mat and the single MWNT (see section 2.). The very small but non-zero resistivity is consistent with the existence of quantum phase slips due to a finite number of transverse channels. The quantum-phase slip theory [16, 10] can also account for semiconductor-like temperature dependence of the resistance below room temperature in some MWNTs that are lithographically contacted [21]. The lithographically contacted MWNTs may contain high density of defects or imperfections which are introduced through purification and other processing steps [37]. Defects and disorders enhance quantum-phase slips and localization of Cooper pairs, which would lead to a large on-tube resistance and a semiconductor-like temperature dependence of the resistance well below T_{c0} (Ref. [16, 10]).

The other important issue to be addressed is why the the tunneling resistance is close to R_Q rather than $R_Q/2$ for the ideal electrical contacts to the outermost shell [35]. A possible explanation is that the metal electrode drives the contact region of the outermost shell to the normal state through the proximity effect, which can destroy the Josephson coupling between the outermost shell and the inner shells in the contact region. Because the strong coupling between the liquid metal and the outermost shell, the band with π character is pushed down below the Fermi level, and thus only one channel with π^* character contributes to electrical transport [59].

7. Microscopic Pairing Mechanism

It has been shown that the low energy physics of an individual metallic single-walled carbon nanotube is equivalent to a ladder system [60] and can be described by Luttinger liquid with an effective interaction parameter K_c. For unscreened Coulomb interactions, K_c is calculated to be about 0.2 in the absence of electron-phonon interactions [61]. On the other hand, if electron-phonon interactions are strong and the Coulomb interactions are effectively screened off, K_c will become larger than 1. Recent theoretical calculation [58] indicates that the strong electron-phonon interaction with the long-wave breathing mode produces a large attractive interaction, which leads to the following formula

$$K_c = \frac{K_c^\circ}{\sqrt{1-(K_c^\circ)^2 r_B/r}}, \tag{20}$$

where K_c° is the unrenormalized parameter in the absence of the electron-phonon interaction with the breathing mode and r_B = 2.4±0.9 Å. If the Coulomb interaction is completely screened off by external gate electrodes, we have $K_c^\circ \simeq 1$. If we take r = 7 Å, K_c° =1, and r_B = 2.4 Å, we find K_c =1.23.

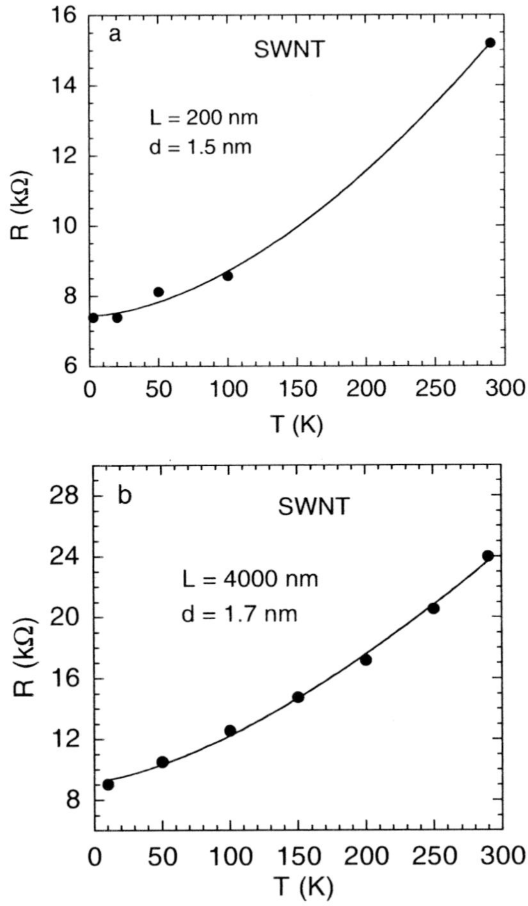

Figure 20. The temperature dependences of the resistance at zero gate voltage for two SWNTs with L = 200 and 4000 nm. The data are extracted from Refs. [62, 63].

The value of K_c can be determined from the measured temperature dependence of the on-tube resistance $R_{tube}(T)$ on the basis of the theoretical prediction of $R_{tube}(T)$ for a ladder system [30]. The theory [30] predicts that $R_{tube}(T) \propto T^{2K_c-2}$. So the two-probe resistance is given by

$$R(T) = R_\circ + CT^{2K_c-2}. \tag{21}$$

In Fig. 20, we show the temperature dependences of the resistance at zero gate voltage for two metallic-chirality SWNTs with d = 1.5 nm and 1.7 nm, respectively. The contacts to the nanotubes are nearly ideal with the transmission probability close to 1 (Ref. [62, 63]). In this case, the two-probe resistance approaches 6.45 kΩ if the on-tube resistance approaches zero. It is remarkable that the temperature dependence of the resistance can be well fit by Eq. 21 with K_c = 1.86±0.12 for d = 1.5 nm and K_c = 1.74±0.06 for d = 1.7 nm. The large K_c values for these individual SWNTs suggest that the long-range Coulomb interaction is effectively screened off by the external gate electrodes. Moreover, since $K_c >$ 1.5, the on-tube resistance is predicted to be zero at zeo temperature [30], in good agreement with experiment [62, 63]. Using Eq. 20 and the measured K_c values, we can deduce r_B = 5.5±0.2 Å if we take K_c° =1, which is significantly larger than that (2.4 Å) estimated theoretically [58].

In order to resolve this discrepancy, we should also consider the electron-phonon interactions with other phonon modes (e.g., short-wave acoustic and optic modes) and the coupling with electronic excitations such as acoustic plasmons [4, 64]. These additional interactions will renormalize K_c° such that the quantity K_c° in Eq. 20 may be replaced by the renormalized $\bar{K}_c^\circ$, that is,

$$K_c = \frac{\bar{K}_c^\circ}{\sqrt{1-(\bar{K}_c^\circ)^2 r_B/r}}. \tag{22}$$

If we substitute r_B = 2.4 Å, and the measured K_c values into Eq. 22, we obtain $\bar{K}_c^\circ$ = 1.28 in both cases. Using this parameter, we can estimate K_c to be 1.40 and 1.33 for a single-walled tube with r = 25 Å and 50 Å, respectively.

When individual single-walled nanotubes are closely packed into crystalline bundles, or formed into a multi-walled nanotube, the long-range Coulomb interaction could be reduced by two orders of magnitude [65] depending on the number of tubes comprising the bundles [66]. This implies that the long-range Coulomb interaction in the bulk mat samples can be effectively screened off. Because the short-range on-site U is reduced by a factor of 10 for an armchair tube with d = 1.4 nm, and the reduction factor is proportional to d (Ref. [60]), the effective on-site U becomes very small for a tube with $d > 1$ nm. Therefore, K_c° in SWNT mat samples should be close to 1. Moreover, since the intertube coupling will increase K_c further [58], we expect that K_c should be about 2 in SWNT mats. For MWNTs, intershell coupling will significantly increase the pairing interaction [5], so the average K_c in a MWNT with an average shell radius of 25 Å should be much larger than 1.4. For the outermost shell with $r >$ 50 Å, we may expect $1.3 < K_c < 1.5$. If $K_c < 1.5$ and disorder is significant, the outermost shell will become insulating at low temperatures [30].

Although the strong electron-phonon coupling produces a large attractive interaction, this coupling alone cannot lead to superconductivity above room temperature because of low density of states and low bosonic energies. We believe that electron-plasmon coupling along with strong electron-phonon coupling should be responsible for high-temperature superconductivity [4, 64]. A SWNT bundle or an individual MWNT is equivalent to a multimode wire. Because the Fermi velocities of the acoustic plasmons in neighboring tubes/shells are comparable, the attractive interaction mediated by undamped acoustic plasmons in the adjacent tubes/shells should be much stronger than that mediated by acoustic phonons [64]. A similar idea has been proposed to explain high-temperature superconductivity in layered cuprates [5]. Within this scenario, high-temperature superconductivity can occur in a multi-layer electronic system due to an attraction of charge carriers in the same conducting layer via exchange of virtual plasmons in neighboring layers [5]. Multi-walled carbon nanotubes and SWNT crystalline bundles are ideal candidates for realization of plasmon-mediated high-temperature superconductivity.

The most favorable conditions for superconductivity are that all the individual tubes comprising MWNTs or SWNT bundles must have metallic chiralities and the chiralities of the adjacent tubes are different. In this situation, the single-particle hopping between neighboring tubes is strongly suppressed [65]. The suppression of the single-particle hopping stems from the obstruction to having precise momentum conservation in the hopping between the misaligned lattices of the nanotubes. This ensures that the single-particle electronic state in individual tubes remains one-dimensional nature such that acoustic plasmon modes remain undamped. This also suppresses any long-range spin or charge ordering because the spin or charge correlation remains one dimensional. On the other hand, the tunneling of Cooper pairs, which is proportional to the square of transverse hopping integral t_T, turns out to be significant. Because Cooper pairs are formed at zero total momentum, they do not find an obstacle in the tunneling processes from the misalignment of the nanotube lattices. For this reason, one has to take the t_T value to be the same as that for the case of the perfect alignment [65], that is, $t_T \simeq 10$ meV for a crystalline bundle composed of 1.4-nm SWNTs [67]. For a bundle consisting of 10 nm SWNTs, t_T is reduced by a factor of $\sqrt{10/1.4} = 2.67$ (Ref. [67]), that is, $t_T \simeq 3.78$ meV. For MWNTs, t_T should be similar to that for graphites, i.e., $t_T \simeq 100$ meV (Ref. [67]).

Now a question arises: Is the Josephson coupling among the individual tubes in a bundle or a MWNT strong enough to sustain phase-coherent superconductivity above room temperature? In order to answer this question, one needs to consider two characteristic temperatures, one is T_J below which the phase coherence among the tubes takes place, and another is T_{c0} below which single-particle density of states starts to suppress due to the superconducting correlation in individual tubes. If $T_J > T_{c0}$ the superconducting transition occurs below T_{c0}. Otherwise, if $T_J < T_{c0}$ the superconducting transition occurs below T_J. We will show that the T_J values calculated from realistic parameters are significantly higher than 700 K in both SWNT bundles and MWNTs.

The analytical expression for T_J has been derived within the random-phase- approximation (RPA) [68]. Since the RPA underestimates [68] T_J by about 20%, we need to multiply the RPA formula by a factor of 1.2 to get a more accurate estimate of T_J. The modified T_J formula is then given by [68]

$$T_J = \frac{1.2\Delta(0)}{2\pi}[\frac{2zJ}{\Delta(0)}\sin\pi\delta\frac{\Gamma^2(\delta/2)\Gamma^2(1-\delta)}{\Gamma^2(1-\delta/2)}]^{1/(2-2\delta)}, \tag{23}$$

where z is the number of nearest neighbor tubes, $\delta = 1/(2K_c)$, $\Delta(0)$ is the single-particle excitation gap (pairing gap), and J is the intertube Josephson coupling energy which is given by [68]

$$J \simeq (\frac{\Delta(0)}{W})^{1/K_c-1}\frac{t_T^2}{\Delta(0)}, \tag{24}$$

where W is the order of the original bandwidth. For the SWNT mat with $T_{c0} = 710$ K, we have shown that, on average, only two superconducting tubes are formed into a bundle, that is, only one of the six nearest neighbor tubes contribute to the Josephson coupling J. Thus,

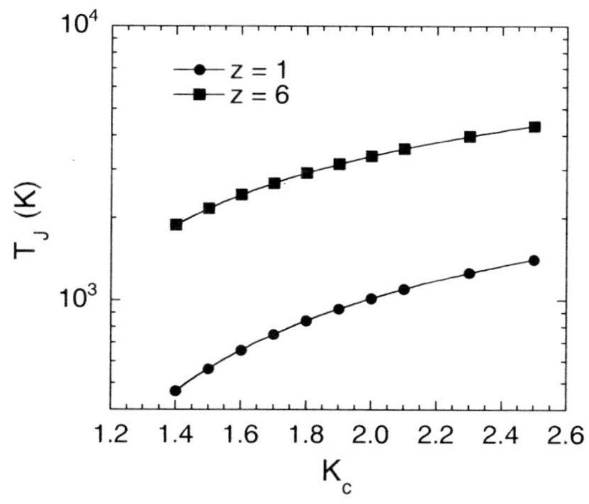

Figure 21. The calculated T_J as a function of K_c for SWNT bundles with $z = 1$ and 6. In the calculation, the following parameters are used: $\Delta(0)/k_B = 1200$ K, $t_T = 10$ mV, and $W = 10$ eV. The value of $t_T = 10$ meV is suitable for a bundle consisting of SWNTs with $d = 1.4$ nm [65, 67]. $z = 1$ should correspond to the SWNT mat with $T_{c0} = 710$ K (see section 2.).

we can take $z = 1$ for this special case. From the Raman data, we have found that $\Delta(0)/k_B$ = 1200 K.

Fig. 21 shows the calculated T_J as a function of K_c for SWNT bundles with $z = 1$ and 6. In the calculation we have taken the following parameters: $\Delta(0)/k_B = 1200$ K, $t_T = 10$ mV, and $W = 10$ eV. It is apparent that in order to have $T_J > T_{c0} = 710$ K for $z = 1$, one needs $K_c > 1.66$. As we discussed above, K_c can be larger than 2 for SWNT mats if the long-range Coulomb interaction can be effectively screened off. This implies that the Josephson coupling is strong enough to sustain superconductivity below 710 K. If all the tubes have metallic chiralities and over a hundred tubes are formed into a bundle, quasi-3D superconductivity can be realized well above room temperature. In order to have truly macroscopic superconductivity above room-temperature, one needs over one thousand tubes to comprise a bundle.

For individual MWNTs, $t_T \simeq 100$ meV if all the shells are superconducting. Even if one third of shells are superconducting, t_T should be larger than 10 meV. In Fig. 22, we show the calculated T_J as a function of K_c with $t_T = 10$ meV and 100 meV, respectively. In the calculation we have taken the following parameters: $\Delta(0)/k_B = 1400$ K, $z = 2$, and $W = 10$ eV. For MWNTs, $z = 2$ and $\Delta(0)/k_B \simeq 1400$ K (see section 5.). We can see that, in the case of $t_T = 10$ meV, $T_J > T_{c0} = 750$ K for $K_c > 1.43$. Since the average K_c in MWNTs with $d = 10$ nm is at least 1.4, the Josephson coupling is strong enough to sustain superconductivity below 750 K. If all the shells have metallic chiralities and over 50 shells are formed into a MWNT, quasi-3D superconductivity can be realized well above room temperature.

For a bundle composed of individual MWNTs with outer diameter of 10 nm, the Josephson coupling temperature T_J can be calculated with $t_T = 3.78$ meV (see the above discussion). In Fig. 23, we show the calculated T_J as a function of K_c with $z = 2$ and 6. Because

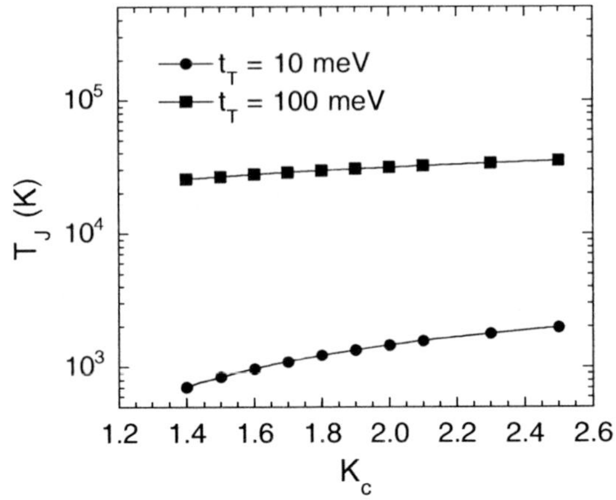

Figure 22. The calculated T_J as a function of K_c for individual MWNTs with $t_T = 10$ meV and 100 meV. In the calculation, the following parameters are used: $\Delta(0)/k_B = 1400$ K, z = 2, and W = 10 eV.

K_c for the outermost shell should be below 1.5, as discussed above, T_J is below 500 K even for $z = 6$. This implies that the Josephson coupling among the individual MWNTs has not been established near T_{c0} = 750 K. In order to have truly macroscopic superconductivity above room temperature, one needs to bundle the individual MWNTs. But the outer diameters of these tubes should not exceed a critical value, which is about 10 nm according to the result in Fig. 23.

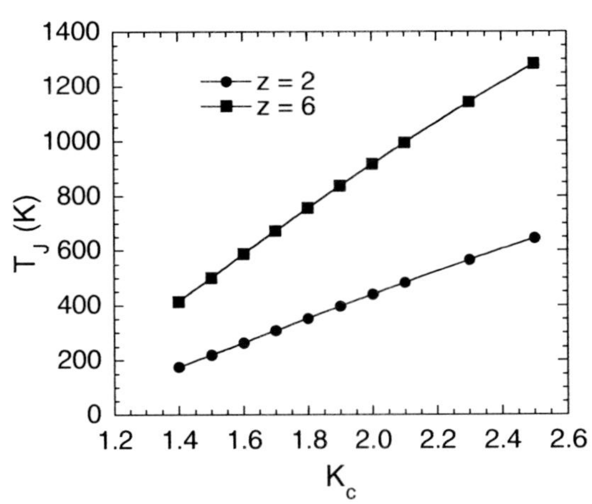

Figure 23. The calculated T_J as a function of K_c for a bundle with z = 2 and 6. In the calculation, the following parameters are used: $\Delta(0)/k_B = 1200$ K, $t_T = 3.78$ meV, and W = 10 eV. The value of t_T = 3.78 meV is suitable for a bundle composed of tubes with an outer diameter of 10 nm.

Now let's explain why the T_{c0}'s in some single-walled nanotube bundles are below 1 K (Ref. [22]). If we check the measured temperature dependence of the normal-state resistance, we find that these bundles behave like semiconductors in the normal state. Assuming that the normal-state resistances of these samples follow Eq. 21, we can estimate the K_c values from the measured resistances at 1 K and 300 K. We obtain $K_c \simeq 0.93$ for a SWNT bundle (Pt2) with $T_{c0} = 0.44$ K, and $K_c \simeq 0.98$ for another SWNT bundle (Pt1) with $T_{c0} = 0.15$ K. The values of K_c indicate that the effective interactions in both bundles are repulsive, leading to d-wave superconductivity [30]. Because the K_c values are very close to 1, the pairing interaction is weak, leading to low T_{c0} superconductivity. Since the K_c value for sample Pt1 is closer to 1 than for sample Pt2, one expects that the repusive interaction in the former sample is weaker than the latter sample. This can qualitatively explain why the T_{c0} for sample Pt1 is significantly lower than for sample Pt2. Further, disorder will effectively suppress T_{c0} for d-wave superconductivity [30]. Because the mean free path of sample Pt2 is one order of magnitude smaller than that of sample Pt1, the T_{c0} suppression in the former sample is much stronger than in the latter one. Thus, the T_{c0} difference between samples Pt1 and Pt2 would even be much greater if the two samples could have similar disorders.

The other important question to be addressed is why these SWNT bundles have K_c values close to 1. One possibility is that the long-range Coulomb interaction is not effectively screened off for these free-standing samples. The other possibility is that the Fermi levels of these bundle samples are close to the charge neutrality point where the repulsive umklapp interaction cannot be fully ignored and the attractive interaction produced by exchanging short-wave acoustic phonon modes is essentially zero [65, 69]. Indeed, the transport data [62] show that as the Fermi level shifts away from the neutrality point, the K_c for individual SWNTs can continuously change from $<$1 to $>$1. Therefore, to obtain a large K_c and high-temperature superconductivity, sufficient doping and effective screening of the long-range Coulomb interaction are essential. Without screening the long-range Coulomb interaction, high-temperature d-wave superconductivity would also occur if the materials were extremely clean. In this case, the electron-phonon and electron-plasmon interactions behave like pair breakers. Moreover, impurities and disorders can easily break pairs and cause Anderson localization of the Cooper pairs [30].

8. Pairing Symmetry

The remaining issue to be clarified is the pairing symmetry. In highly oriented pyrolitic graphite (HOPG) samples, there appear to be an interplay between superconductivity and ferromagnetism [70]. Such an interplay suggests that the pairing symmetry in HOPG samples should be of p-wave, similar to the case of Sr_2RuO_4. Considering the similarity between carbon nanotubes and HOPG, we may expect that the gap symmetry in carbon nanotubes should also be p-wave. An essential attribute for p-wave superconductivity is that the spin susceptibility is the same in the superconducting and normal states. The spin susceptibility for aligned and physically separated MWNTs has been derived from electron-spin-resonance (ESR) signals. Fig. 24 shows the temperature dependence of the spin susceptibility for the MWNTS. The data are extracted from Ref. [39]. It is interesting that the spin susceptibility is independent of temperature between 100 K and 200 K. Below 100 K, the spin susceptibility suddenly turns up and follows a Curie law ($\propto 1/T$) below 40 K. Above 200 K, there appears to be a tendency of increase in the spin susceptibility. From the magnitude of the spin susceptibility between 100 K and 200 K (6.0×10^{-9} emu/g), one finds that the density of states at the Fermi level $N(0) = 2.2\times 10^{-3}$ states/eV atom. The

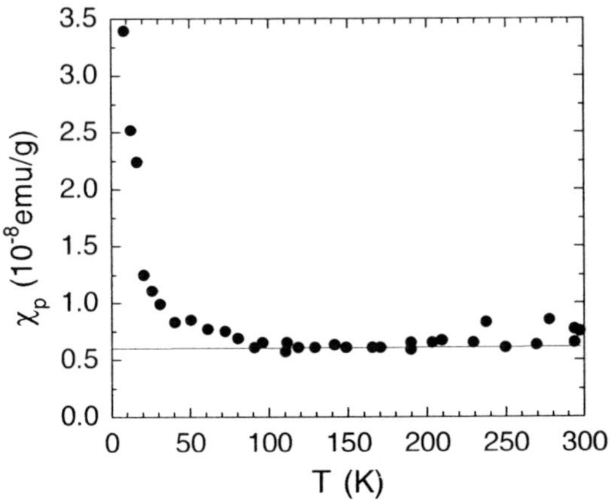

Figure 24. Electron spin susceptibility for a physically separated MWNT film, derived from electron-spin-resonance (ESR) signals. The data are extracted from Ref. [39].

observed finite density of states below 300 K may argue against the interpretation of superconductivity above room temperature in the case of s-wave gap symmetry. Nevertheless, the finite density of states is also consistent with s-wave gap symmetry if some of the tubes (e.g., semiconducting chirality tubes or shells) are nonsuperconducting.

In order for the s-wave interpretation to be plausible, the density of states in the normal state must be significantly larger than that in the superconducting state. Since the spin susceptibility for the film sample was measured only up to room temperature, the normal-state density of states was undetermined. However, if the doping level of the film sample is similar to that for the MWNT mat sample with T_{c0} = 752 K, the normal-state density of states in the film sample should be similar to that for the mat sample. Using the deduced β value from the resistive transition, we have found that the average number of transverse conduction channels per shell is about 3 for the superconducting MWNTs with an average outer diameter of 10 nm (see section 2.). Then $N(0)$ in the normal state is $1.5(2\sqrt{3}/\pi^2)(1/\gamma_\circ)(a_{C-C}/\bar{d})$ (Ref. [20]), where a_{C-C} = 0.142 nm and $\gamma_\circ$ = 2.4 eV. The average diameter for all the shells $\bar{d}$ should be a half of the average outer diameter, that is, $\bar{d}$ = 5 nm. Substituting these numbers into the above formula yields $N(0) = 6.2 \times 10^{-3}$ states/eV atom. It is clear that the $N(0)$ in the superconducting state is about one third of that in the normal state, in agreement with s-wave gap symmetry.

The finite spin susceptibility in the superconducting state could be partly due to some insufficiently doped semiconducting chirality shells which do not undergo the superconducting transition. Especially for those semiconducting chirality shells in which the Fermi level is in the vicinity of the first Van Hove singularity (near the semiconducting gap edge), the contribution to the density of states is huge. If one of such shells in a MWNT is present, the density of states contributed from this shell should be comparable with the measured value below 300 K. Moreover, since the Fermi level is also very close to the mobility edge of the semiconducting chirality shell, the charge localization at low temperatures is inevitable. This can naturally explain the upturn of the spin susceptibility below 100 K.

Because the ESR linewidth, which is proportional to the transport scattering rate, also shows upturn below 100 K, there should be one type of carriers responsible for the ESR signals [39]. Since the localized carrier density [39] is only about $1 \times 10^{18}/\text{cm}^3$ while the mobile carrier density [33] above 100 K is over $2 \times 10^{19}/\text{cm}^3$, the majority of the mobile carriers must be ESR-silent, that is, the majority of the mobile carriers are bound into spin singlet pairs.

On the other hand, if there were only one-type of carriers (no superconductivity interpretation), the ESR and Hall effect would measure the same type of carriers. If this were the case, the Hall coefficient would jump to a very large number below 40 K where all the carriers are localized [39]. This is in contrast to the Hall coefficient of the same sample, where no such a jump is seen below 40 K [33]. Moreover, the Fermi energy E_F would be equal to $n/N(0)$ = 4.5 meV if there were only one-type of carriers. With such a small Fermi energy, one would predict the magnitude of the orbital diamagnetic susceptibility in the perpendicular field [40] to be at least 1.4×10^{-5} emu/g, which is significantly larger than the measured one ($< 0.7 \times 10^{-5}$ emu/g) [39]. Therefore, there must be two types of carriers; one type of carriers is ESR silent below 300 K due to formation of spin singlet pairs and the other is from nonsuperconducting tubes. A small increase in the spin susceptibility above 200 K may be caused by the thermally excited quasiparticles. If this interpretation is relevant, the minimum superconducting gap is approximately equal to 200 K/0.2 = 1000 K, in good agreement with the tunneling spectrum (see Fig. 16).

It is worth noting that the observation of the finite spin susceptibility in the superconducting state does not necessarily imply an unconventional pairing symmetry. Even for conventional superconductors such as mercury and tin, the spin susceptibility in the superconducting state is about two thirds of the value in the normal state [71, 72]. For vanadium and aluminum, the spin susceptibility is nearly the same in the superconducting and normal states [72, 73].

9. Conclusion

It is well known that copper-based perovskite oxides rightly enjoy consensus as high-temperature superconductors on the basis of two signatures: the resistive transition and the Meissner effect. Here we have provided evidence for room-temperature superconductivity in carbon nanotubes. Although the superconducting transition temperatures can vary from 0.44 K to 750 K, the resistive transitions in these superconducting carbon nanotubes are in quantitative agreement with the Langer-Ambegaokar-McCumber-Halperin theory. The quantitative agreement between the resistive transitions and well established quantum theory leaves little room for alternative explanations. Because of a finite number of transverse conduction channels of these carbon nanotubes, the four-probe resistance will never go to zero in the superconducting state. This would lead to a false impression that these tubes are not superconducting. Nonetheless, the on-tube resistance at room temperature has been found to be indistinguishable from zero for many individual multi-walled nanotubes. The very small but finite room-temperature on-tube resistance is consistent with quantum phase slips due to the finite number of transverse channels. We have also observed the Meissner effect in aligned and physically separated multi-walled nanotubes up to room temperature. The Meissner effect agrees quantitatively with the predicted magnetic penetration depth from the measured carrier density. Furthermore, the bundling of individual MWNTs into closely packed bundles leads to a large enhancement in the diamagnetic susceptibility, which is the hallmark of the Josephson coupling among the tubes in bundles. Further, Raman data and tunneling spectra consistently show single particle excitation gaps of larger than 100 meV. Therefore, these results consistently indicate quasi-1D room-temperature superconductivity in carbon nanotubes.

For practical applications, zero on-tube resistance at room temperature is required. Although the room-temperature resistivity for individual MWNTs, prepared by arc discharge, is significantly smaller than that for simple elemental metals, we can further suppress the resistivity down to nearly zero by bundling these individual tubes into a big closely packed bundle. But the outer diameters of the tubes should not exceed a critical value above which the magnitude of t_T is not large enough to sustain the Josephson coupling among the tubes at room temperature. Sufficient doping, effective screening of the long-range Coulomb interaction, and proper bundling are essential to achieve high-temperature superconductivity. Impurities and disorders tend to destroy superconductivity especially in the case of $K_c <$ 1.5. It is very likely that the K_c of the outermost shell of a MWNT is less than 1.5 if d $>$ 15 nm. These tubes, if not clean enough, will undergo the superconductor-to-insulator transition below a localization temperature T_{loc}. For $K_c > 1.5$, such a transition should not occur and the ground state is always superconducting.

References

[1] Alexandrov, A. S. and Mott, N. F. *Polarons and Bipolarons*, World Scientific, Singapore, 1995.

[2] Ginzburg, V. L. in: Ginzburg, V. L. and Kirzhnits, D. A. (Eds) *High-Temperature Superconductivity*, Consultants Bureau, New York, 1982.

[3] Little, W. A. *Phys. Rev.* **164**, A1416 (1964).

[4] Lee, Y. C. and Mendoza, B. S. *Phys. Rev. B* **39**, 4776 (1989).

[5] Cui, S. M. and Tsai, C. H. *Phys. Rev. B* **44**, 12500 (1991).

[6] Saito, R., Fujita, M., Dresselhaus, G., and Dresselhaus, M. S. *Appl. Phys. Lett.* **60**, 2204 (1992).

[7] Ajiki, H. and Ando, T. *J. Phys. Soc. Jpn.* **62**, 1255 (1992).

[8] Zhao, G. M. and Wang, Y. S. *cond-mat/0111268.*

[9] Zhao, G. M. *cond-mat/0208197.*

[10] Zhao, G. M. *cond-mat/0208198.*

[11] Zhao, G. M. *cond-mat/0208200.*

[12] Zhao, G. M. *cond-mat/0208201.*

[13] Zhao, G. M. *cond-mat/0307770.*

[14] Zhao, G. M. *Molecular Nanowires and Other Quantum Objects*, Alexandrov, A. S., Demsar J., and Yanson, I. K. (Eds), Nato Science Series, Kluwer Academic Publishers, Netherlands, 2004, page 95-106.

[15] Langer, J. S. and Ambegaokar, V. Phys. Rev. **164**, 498 (1967); McCumber, D. E. and Halperin, B. I. *Phys. Rev. B* **1**, 1054 (1970).

[16] Zaikin, A. D., Golubev, D. S., Van Otterlo, A., and Zimanyi, G. T. *Phys. Rev. Lett.* **78**, 1552 (1997).

[17] Tinkham, M. *Introduction to Superconductivity*, McGraw-Hill, 1996.

[18] Lau, C. N., Markovic, N., Bockrath, M., Bezryadin, A., and Tinkham, M. *Phys. Rev. Lett.* **87**, 217003 (2001).

[19] Roche, S. and Saito, R. *Phys. Rev. B* **59**, 5242 (1999).

[20] Mintmire, J. W. and White, C. T. *Phys. Rev. Lett.* **81**, 2506 (1998).

[21] Schönenberger, C., Bachtold, A., Strunk, C., Salvetat, J.-P., and Forro, L. *Appl. Phys. A* **69**, 283 (1999).

[22] Kociak, M., Kasumov, A. Yu., Gueron, S., Reulet, B., Khodos, I. I., Gorbatov, Yu. B., Volkov, V. T., Vaccarini, L., and Bouchiat, H. *Phys. Rev. Lett.* **86**, 2416 (2001).

[23] Trucker, J. R. and Halperin, B. I. *Phys. Rev. B* **3**, 3768 (1971).

[24] Bendedict, L. X., Crespi, V. H., Louie, S. G., and Cohen, M. L. *Phys. Rev. B* **52**, 14935 (1995).

[25] Mintmire, J. W., Dunlap, B. I., and White, C. T. *Phys. Rev. Lett.* **68**, 631 (1992).

[26] Ferrier, M., De Martino, A., Kasumov, A., Gueron, S., Kociak, M., Egger, R., and Bouchiat, H. *cond-mat/0405449.*

[27] Ebbesen, T. W., Lezec, H. J., Hiura, H., Bennett, J. W., Ghaemi, H. F., and Thio, T. *Nature (London)* **382**, 54 (1996).

[28] *Private communications* with Prof. T. W. Ebbesen.

[29] Saito, R., Fujita, M., Dresselhaus, G., and Dresselhaus, M. S. Phys. Rev. B **46**, 1804 (1992).

[30] Orignac, E. and Giamarchi, T. *Phys. Rev. B* **56**, 7167 (1997).

[31] Lee, R. S., Kim, H. J., Fischer, J. E., Thess, A., and Smalley, R. E. *Nature* (London) **388**, 255 (1997).

[32] Merchant, L., Ostrick, J., Barber, Jr. R. P., and Dynes, R. C. *Phys. Rev. B* **63**, 134508 (2001).

[33] Baumgartner, G., Carrard, M., Zuppiroli, L., Bacsa, W., De Heer, W. A., and Forro, L. *Phys. Rev. B* **55**, 6704 (1997).

[34] Stahl, H., Appenzeller, J., Martel, R., Avouris, Ph., and Lengeler, B. *Phys. Rev. Lett.* **85**, 5186 (2000).

[35] Frank, S., Poncharal, P., Wang, Z. L., De Heer, W. A. *Science* **280**, 1744 (1998).

[36] De Pablo, P. J., Graugnard, E., Walsh, B., Andres, R. P., Datta, S., and Reifenbergera, R. *Appl. Phys. Lett.* **74**, 323 (1999).

[37] Poncharal, P., Berger, C., Yi, Y., Wang, Z. L., de Heer, W. A. *J. Phys. Chem. B* **106**, 12104 (2002).

[38] Urbina, A., Echeverra, I., Perez-Garrido, A., Daz-Sanchez, A., and Abellan, J. *Phys. Rev. Lett.* **90**, 106603 (2003).

[39] Chauvet, O., Forro, L., Bacsa, W., Ugarte, D., Doudin, B., and De Heer, W. A. *Phys. Rev. B* **52**, R6963 (1995). The aligned nanotube films were produced by a process in which the tubes are ultrasonically separated.

[40] Lu, J. P. *Phys. Rev. Lett.* **74**, 1153 (1995).

[41] Qian, D., Dickeya, E. C., Andrews, R., and Rantell, T. *Appl. Phys. Lett.* **76**, 2868 (2000).

[42] Bayot, V., Piraux, L., Michenaud, J.-P., and Issi, J.-P. *Phys. Rev. B* **40**, 3514 (1989).

[43] Krantz, M., Rosen, H. J., Macfarlane, R. M., and Lee, V. Y. *Phys. Rev. B* **38**, 4992 (1988).

[44] Friel, B., Thomsen, C., and Cardona, M. *Phys. Rev. Lett.* **65**, 915 (1990).

[45] Ham, K. M., Kim, J. T., Sooryakumar, R., and Lemberger, T. R. *Phys. Rev. B* **47**, 11 439 (1993).

[46] Zeyher, R. and Zwicknagl, G. *Z. Phys. B* **78**, 175 (1990).

[47] Zeyher, R. and Zwicknagl, G. *Solid State Commun.* **66**, 617 (1988).

[48] Walter, R. *et al.*, *Bull Am. Phys. Soc.* **47**, 361 (2002); The Raman data were sent by Dr. R. Walter from the University of North Carolina.

[49] Lazzeri, M., Piscanec, S., Mauri, F, Ferrari, A. C., and Robertson, J. *Phys. Rev. Lett.* **95**, 236802 (2005).

[50] Yao, Z., Kane, C. L., and Dekker, C. *Phys. Rev. Lett.* **84**, 2941 (2000).

[51] Wildoer, J. W. G., Venema, L. C., Rinzler, A. G., Smalley, R. E., and Dekker, C. *Nature* (London) **391**, 59 (1998).

[52] Kim, J., Lee, J.-O., Oh, H., Yoo, K.-H., and Kim, J.-J. *Phys. Rev. B* **64**, 161404R (2001).

[53] Kwon, Y. K. and Tomanek, D. *Phys. Rev. B* **58**, R16001 (1998).

[54] Sanvito, S., Kwon, Y.-K., Tomanek, D., and Lambert, C. J. *Phys. Rev. Lett.* **84**, 1974 (2000).

[55] Collins, P. G., Arnold, M. S., and Avouris, P. *Science* **292**, 706 (2001).

[56] Bachtold, A., Fuhrer, M. S., Plyasunov, S., Forero, M., Anderson, E. H., Zettl, A., and McEuen, P. L. *Phys. Rev. Lett.* **84**, 6082 (2000).

[57] Komnik, A. and Egger, R. *cond-mat/9906150 (1999).*

[58] De Martino, A. and Egger, R. *Phys. Rev. B* **67**, 235418 (2003).

[59] Choi, H. J. Ihm, J., Yoon, Y.-G., and Louie, S. G. *Phys. Rev. B* **60**, R14009 (1999).

[60] Balents, L. and Fisher, M. P. A. *Phys. Rev. B* **55**, 11973 (1997).

[61] Kane, C., Balents, L., and Fisher, M. P. A. *Phys. Rev. Lett.* **79**, 5086 (1997).

[62] Kong, J., Yenilmez, E., Tombler, T. W., Kim, W., Dai, H. J., Laughlin, R. B., Liu, L., Jayanthi, C. S., and Wu, S. Y. *Phys. Rev. Lett.* **87**, 106801 (2001).

[63] Mann, D. Javey, A. Kong, J. Wang, Q., and Dai, H. *Nano Lett.* **3**, 1541 (2003).

[64] Loss, D. and Martin, T. *Phys. Rev. B* **50**, 12 160 (1994).

[65] Gonzalez, *J., Phys. Rev. B* **67**, 014528 (2003).

[66] Alvarez, J. V., and Gonzalez, J. *Phys. Rev. Lett.* **91**, 076401 (2003).

[67] Maarouf, A. A., Kane, C. L., and Mele, E. J. *Phys. Rev. B* **61**, 11 156 (2000).

[68] Carr, S. T. and Tsvelik, A. M. *Phys. Rev. B* **65**, 195121 (2002).

[69] Sedeki, A., Caron, L. G., and Bourbonnais, C. *Phys. Rev. B* **65**, 140515R (2002).

[70] Kopelevich, Y., Esquinazi, P., Torres, J. H. S., and Moehlecke, S. J. *Low Temp. Phys.* **119**, 691 (2000).

[71] Reif, F., *Phys. Rev.* **106**, 208 (1957).

[72] Androes, G. M. and Knight, W. D. *Phys. Rev. Lett.* **2**, 386 (1959).

[73] Noer R. J. and Knight, W. D. *Rev. Mod. Phys.* **36**, 177 (1964).

In: New Research on Superconductivity
Editor: Barry P. Martins, pp. 59-103

ISBN 978-1-59454-197-1

Chapter 2

PHASE DIAGRAM FOR $NA_x(H_3O)_zCoO_2 \cdot yH_2O$

Yoshihiko Ihara[1,*] *Kenji Ishida*[1,2],
Hiroya Sakurai[3] *and Eiji Takayama-Muromachi*[3]
[1]Department of Physics,
Graduate school of Science, Kyoto University, Kyoto 606-8502, Japan
[2]International Innovation Center,
Kyoto University, Kyoto 606-8502, Japan
[3]Advanced Nano Materials Laboratory,
National Institute for Materials Science, Ibaraki 305-0044, Japan

Abstract

Since the discovery of superconductivity in the hydrous cobaltate $Na_xCoO_2 \cdot yH_2O$, intensive studies, particularly for the synthesis of superconductors with high transition temperature, have been performed. As a result, various phase diagrams have been reported, in which the Na content and the period of time after intercalation of water have been employed as tuning parameters. Recently, the valence of Co was investigated and the presence of oxonium H_3O^+ was recognized. It has been accepted that the chemical formula for the superconducting cobaltate can be expressed as $Na_x(H_3O)_zCoO_2 \cdot yH_2O$, which includes at least three parameters x, y, and z. Magnetic ordering was discovered in a sample with a large z content, but with the same crystal structure as the superconducting sample. Much attention has been paid to the relationship between superconductivity and magnetism. In order to investigate the relationship, the development of the phase diagram is desirable. It is necessary to find the crucial physical parameters that are responsible for the superconductivity, and give consideration to the superconducting mechanisms.

We succeeded in controlling the chemical composition by a modified soft-chemical method, and established a phase diagram using samples with various chemical compositions. It was found that two compositional parameters, Na content (x) and Co valence ($s = 4 - x - z$), are necessary to determine the physical properties of the compound. In the case of $s = +3.48$, the superconducting region has a dome-shape in the range of $0.33 < x < 0.38$, while in the case of $s = +3.40$, the magnetically ordered phase appears in the range of $0.33 < x < 0.35$, sandwiched by two superconducting regions at $x < 0.33$ and $x > 0.35$. The previously reported phase diagrams

*E-mail address: ihara@scphys.kyoto-u.ac.jp

may be certain projections of the $x-s-T$ phase diagram onto the $x-T$ or $s-T$ plane.

Microscopic experiments are powerful tools used for the study of superconductivity. We found that the nuclear quadrupole resonance frequency of Co, which is strongly sample-dependent, is a good parameter to use for representation of the phase diagram. With the increasing frequency of Co, the superconducting transition temperature increases, and the magnetically ordered phase appears. Our analysis strongly suggests that the change in the frequency corresponds to pressing of the CoO_2 layer along the c axis, and therefore the microscopic phase diagram indicates that the superconductivity and magnetism of this system are governed by the thickness of the CoO_2 layer. This scenario is consistent with the chemical phase diagram, because the ion exchange between Na and H_3O ions would change the thickness of the CoO_2 layer. In addition, the spin-lattice relaxation rate indicates that the superconductivity is closely related to quantum critical magnetic fluctuations, which is a key factor for understanding the superconductivity.

In this article, we summarize the sample dependence of various physical parameters and present the phase diagrams, which are based on chemical analyses and nuclear quadrupole measurements. The superconducting mechanisms are also discussed on the basis of our experimental results.

1. Introduction

Superconductivity in a cobalt oxide compound was discovered in 2003 by Takada *et al.* [1]. The hydrous cobaltate $Na_x(H_3O)_zCoO_2 \cdot yH_2O$ demonstrates superconductivity when water molecules are sufficiently intercalated into the compound by a soft chemical procedure. In contrast, anhydrous Na_xCoO_2 does not demonstrate superconductivity down to 40 mK [2]. The water molecules and related oxonium ions, which are included in the compound, are considered to induce superconductivity. This compound is the first superconductor which shows superconductivity only in the hydrous phase.

The parent compound Na_xCoO_2 is synthesized by the solid state reaction of Na_2CO_3 and Co_3O_4. After this reaction, Na is subtracted and water is intercalated by a soft chemical procedure. As a result, the c-axis length of $Na_x(H_3O)_zCoO_2 \cdot yH_2O$ becomes twice as long as that of the parent compound. However, the conductive CoO_2 layer maintains the crystal structure of the parent compound, which is a two dimensional triangular lattice. This enormous elongation of the c axis while preserving the crystal structure of the CoO_2 layer is one of the remarkable features of the soft chemical procedure. Due to the elongation, each CoO_2 layer is separated by approximately 10 Å. Thus, the CoO_2 layer, in which superconductivity is realized, is considered to have a highly two-dimensional nature. The cuprate and ruthenate superconductors, which have been intensively investigated [3,4], also have two-dimensional superconducting planes. In the case of cuprate and ruthenate superconductors, the superconducting planes are square-lattice CuO_2 and RuO_2 layers, respectively. By contrast, in the cobaltate superconductor, the CoO_2 layer forms a triangular lattice. The superconductivity on the triangular lattice is interesting from both a theoretical and an experimental perspective.

Although a highly two-dimensional triangular lattice CoO_2 layer is obtained by the soft chemical procedure, the sample quality is easily degraded under ambient conditions. This is a disadvantage of the soft chemical procedure. In addition, the complex chemical formula, in which three variable parameters are included, x, y and z, results in strong sample dependence of physical quantities, such as spin susceptibility and specific heat. The sample dependence of the superconducting transition temperature has been studied by several groups with respect to the Na content [5, 6], but the results cannot be consistently explained, because three stoichiometric parameters are intimately connected. Therefore, in order to develop a consistent phase diagram for $Na_x(H_3O)_zCoO_2 \cdot yH_2O$, we need to find a parameter which directly reflects the sample properties [7, 8]. The c-axis length is one candidate for such a parameter, because the elongation along the c axis is intimately related to the water-molecule and oxonium-ion content. In this article, we introduce the resonant frequency of the nuclear quadrupole resonance (NQR) as another promising parameter used to determine ground states.

The discovery of magnetic order in the hydrous compound with the same crystal structure as the superconducting bilayered hydrate (BLH) compound also motivated efforts to establish a phase diagram [9]. The main differences between magnetic ordering samples and the superconducting samples are the oxonium content z and the c-axis lattice parameters, which can be controlled by the preparative conditions described in § 3.. Investigations were concentrated on the differences among these BLH samples, and also on the difference between the superconducting BLH sample and non-superconducting mono-layered hydrate (MLH) sample, A phase diagram was then constructed, with regard to chemical and microscopic points of view. The relationship between the superconductivity and magnetism in the BLH compound is revealed on the basis of these phase diagrams, and is presented in § 3. and § 4..

In this article, the fundamental physical properties of $Na_x(H_3O)_zCoO_2 \cdot yH_2O$ in both normal and superconducting states are first summarized in § 2.. The chemical approach used to control the compositions is explained in § 3., and on the basis of the modified soft chemical procedure, phase diagrams including the magnetic phase are developed. In § 4., the phase diagram constructed using the results of nuclear quadrupole resonance measurements are presented, and the relationship between magnetism and superconductivity is discussed. In the last part of this article, the mechanisms for superconductivity in the hydrate cobaltate are discussed from an investigation of the phase diagrams.

ŒÚ □´ °ø ª ¨ø¥–Æ ªÆª ± "ø$_x$ ÿ 3—˜$_z$› ±—2 · yÿ 2—

ŒŒÚ › Æ ¨ø¥Õ˙Æ ¨´ Æ

The cobaltate compound has three types of crystal structure, depending on the water content. The parent compound Na_xCoO_2, which has $y = 0$ and $z = 0$, contains a randomly occupied Na layer between the CoO_2 layers. This composition does not show superconductivity down to 40 mK, even in a single crystal [2]. When Na ions are deintercalated, and water molecules are intercalated between the CoO_2 layers, the crystal structure changes to a BLH structure, in which the Na layer is sandwiched with double water layers to form a H_2O-Na-H_2O block layer. Due to the formulation of this thick block layer, the CoO_2 layers are separated by approximately 10 Å, and the structure is considered to have a highly two-dimensional nature. Superconductivity is observed below 5 K in this composition. The crystal structure of the superconducting BLH compound changes to a mono-layered hy-

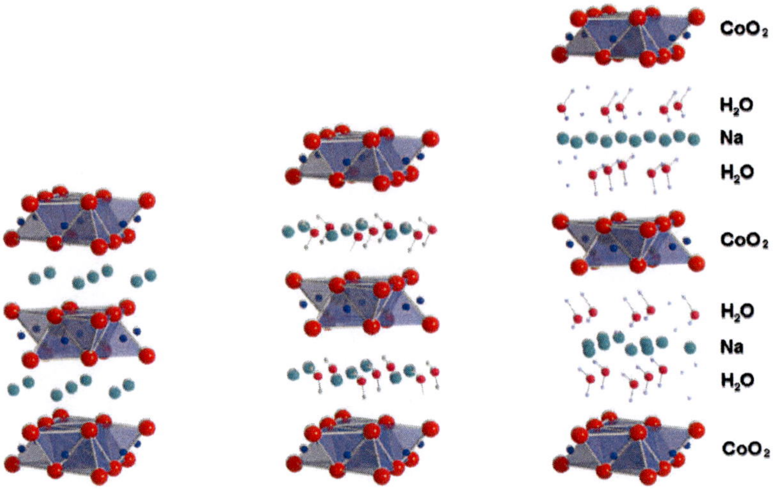

Figure 1. Crystal structure of the non-hydrate, mono-layer hydrate (MLH) and bilayer hydrate (BLH) compounds.

drate structure under ambient conditions. The water content of the MLH compound, in which Na-H_2O mixed layers are formed between the CoO_2 layers, is less than that of the BLH compound. The water molecules between the CoO_2 layers are considered to have an important role in the appearance of superconductivity. However, H_2O in the crystal is easily evaporated into the air under ambient conditions, and the physical properties of the BLH compound are seriously affected by the evaporation. The superconducting transition temperature of the samples preserved at room temperature gradually decreases, depending on the duration [10, 11]. Water evaporation should be prevented by storing samples in liquid nitrogen or in a freezer, thereby maintaining the sample properties. Although the water content is a promising candidate as a tuning parameter for the sample properties, it is difficult to directly measure the H_2O content, because of the diffusive nature of the H_2O-Na-H_2O block layer. The c-axis length, which sensitive to the water content, is considered to be useful for classifying the sample dependence of physical properties. The a-axis is almost constant during water intercalation, in contrast to the anomalous change in the c-axis length, and due to the elongation of the c-axis, the conducting CoO_2 layer is compressed along the c direction. With the intercalation of water, negative O^{2-} ions are attracted by positive Co ions more strongly than by Na ions, because the thick block layer separates the positive Na ions from the O^{2-} ions. The crystal field due to the trigonal distortion of the oxygen octahedtra split the triplet t_{2g} orbital of Co-3d electrons to singlet a_{1g} and doublet e'_g orbitals.

In addition to water intercalation, oxonium ions are intercalated between the CoO_2 layers at the same time. The Co valence determined from the redox titration measurement was in the range of 3.4 to 3.5, which corresponds to a Na content of $x = 0.6 \sim 0.5$, if the oxonium content is not taken into consideration. However, inductively coupled plasma-atomic emission spectroscopy (ICP-AES) measurements have indicated that the Na content is less than 0.4. In order to compensate the lack of Na ions, the oxonium ion content is determined from the relation $z = 4 - v - x$, where z and v are the oxonium ion content and the Co valence, respectively. Although the existence of oxonium ions has been reported from redox titration measurements and Raman spectroscopy, the exact arrangement of the oxonium ions, which has been investigated with Neutron diffraction measurement, has not been settled yet. The crystallographic position of water molecules and oxonium ions seems to have no effect on the physical properties in the CoO_2 layer of the BLH compound, because the water molecules act as spacers that elongate the c axis and compress the CoO_2 layer. The difficulty in determining the chemical compositions is also derived from the difficulty in controlling them. There are numerous variable parameters that exist for the preparation of $Na_x(H_3O)_zCoO_2 \cdot yH_2O$ samples by chemical procedures, such as temperature and humidity. A useful process for control of the oxonium content is presented in § 3., where a phase diagram was constructed using systematically controlled samples.

2.2. Superconductivity

A peculiarity of the superconductivity in the $Na_x(H_3O)_zCoO_2 \cdot yH_2O$ compound is the necessity for a sufficient amount of water intercalation between the CoO_2 layers. The superconducting transition temperatures vary from 2 K to 4.9 K, depending on the water content. The BLH compounds with sufficient water molecules have high superconducting transition temperatures, and the superconducting transition temperatures decrease with decreasing water content. A BLH compound left in a vacuum space for three days becomes a MLH compound, and does not demonstrate superconductivity. In contrast, the crystal structure of a MLH compound stored in high-humidity atmosphere reverts back to the BLH structure, and superconductivity is recovered, although the transition temperature of a BLH compound after such a dehydration-hydration cycle is lower than that of a freshly prepared BLH compound. Superconducting-state and normal-state properties in various kinds of samples have been investigated using several experimental methods.

In the normal state, spin susceptibility is almost temperature independent above 100 K, which is a unique behavior irrespective of samples. Sample dependence appears below 100 K, that is, the spin susceptibility increases with decreasing temperature for some samples, but does not for other samples. From the temperature dependence of spin susceptibility below 100 K, temperature independent spin susceptibility χ_0, effective moment $\mu_{\rm eff}$ and Weiss temperature $\Theta_{\rm W}$ can be estimated. These values are reported to be $\chi_0 = 3.02 \times 10^{-4}$ emu/mol, $\mu_{\rm eff} \sim 0.3\mu_{\rm B}$ and $\Theta_{\rm W} = -37$ K in Ref. [12]. However, different values are reported in Ref. [13], where the increase of susceptibility toward low temperature is hardly observed, and correspondingly, $\mu_{\rm eff}$ is rather small. Although the low temperature behavior of susceptibility is strongly sample dependent, we believe that the slight increase below 100 K is an intrinsic behavior, because it is observed in most high-quality powder samples, in addition to being observed in the Knight shift measured by nuclear magnetic resonance [14] and muon spin rotation measurements [15].

In the superconducting state, specific heat has been intensively measured by several groups [16–22]. The specific heat jump at the superconducting transition temperature $\Delta C/\gamma T_c$ is estimated to be approximately 0.7, which is half of the BCS (Bardeen, Cooper, and Schrieffer) value 1.43. The small jump suggests that the quality of the sample is insufficient, and/or that the superconductivity is an unconventional type with nodes. Below the superconducting transition temperature, C/T does not follow exponential temperature dependence but follows power-law behavior, which is universally observed in unconventional superconductors. The power-law behavior observed from the nuclear-spin-lattice relaxation rate $1/T_1$ measurement also supports unconventional superconductivity [23,24]. The results of $1/T_1$ are presented in § 4.. The values for the Sommerfeld constant, which are 12 $\sim$ 16 mJ/mol K^2 depending on the sample, are comparable to those of the anhydrous compound $Na_{0.3}CoO_2$ and less than those of the parent compound $Na_{0.7}CoO_2$. It is curious that the BLH compound, which has smaller density of state compared to the parent compound, demonstrates superconductivity, while a single crystal of $Na_{0.7}CoO_2$ with larger density of state is not a superconductor.

The field dependence of the superconducting transition temperature is measured using magnetization, specific heat and resistivity measurements, and an $H-T$ phase diagram has been constructed [12, 17, 18, 25, 26]. These measurements were performed on both single and poly crystalline samples. The highest T_c obtained for a fine powder sample reaches to 4.8 K [7]. The resistivity measurement was performed on a single crystal with a size of $1 \times 1 \times 0.02$ mm^3, and zero resistivity was achieved at 4 K with the width of the transition temperature $\Delta T_c \sim 0.5$ K. Although zero resistivity is observed in a single crystal, the transition temperature is lower than that for powder samples. This is because H_2O can penetrate into the compound more uniformly and easily in a small size sample [26]. For this reason, high-quality powder samples were used for measurements in the present study.

2.3. Magnetism

A magnetic order is observed from the Co-NQR measurement of a BLH compound without superconducting transition [9]. A magnetic anomaly has been already observed in the parent compound Na_xCoO_2 with $x > 0.75$ at 20 K [27]. This magnetism is considered to possess A-type magnetic structure, which has ferromagnetic couplings in the ab plane and antiferromagnetic couplings along the c axis. In the BLH compound, however, the magnetic structure is not specified because of the difficulty in preparing a large single crystal for neutron scattering measurements. From the Co-NQR measurements on powder samples, it was found that the value of $1/T_1T$ diverges at $T_M \sim 6$ K, and that a distributing internal field appears at Co sites below T_M. These two results are evident of static magnetic ordering. Details of the NQR measurements on the magnetically ordered sample are given in § 4.. After the report of the Co-NQR measurement, the magnetism in the BLH compound was also identified with magnetization and μSR (muon spin rotation) measurements [7, 28].

In addition to magnetism, we found a sample that displayed both superconducting and magnetic anomalies at 3 K and 6 K, respectively. The sample emphasizes the intimate relationship between superconductivity and magnetism in the hydrate cobaltate compound. To date, many kinds of superconductivity, which are realized near magnetism, have been investigated, and most of these have been classified as unconventional type, such as that for cuprate, ruthenate and heavy fermion superconductors. Superconductivity in $Na_x(H_3O)_zCoO_2 \cdot yH_2O$ is also considered to be unconventional type, and magnetic fluctuations near spin ordering are intimately related to the realization of superconductivity.

3. Chemical Phase Diagram

3.1. Synthesis

In order to establish the chemical phase diagram, it is necessary to understand the synthesis of the compound. In this article, we show the synthesis method for the BLH compound, with an explanation regarding the reaction mechanism, which indicates the existence of oxonium ions in the compound. In addition, our synthesis method provides a basis for control of the chemical-composition, as will be shown later.

The synthesis of the BLH can be divided into three steps. The first is to prepare the anhydrate Na_xCoO_2. For this compound, three types of structure are known, depending on the Na content and the synthesis temperature: α-, β- and γ-Na_xCoO_2, which have different stacking sequences of the CoO_2 layers. The γ-type compound is the most widely studied, because it can be easily synthesized by conventional solid state reaction from very stable materials, Na_2CO_3 and Co_3O_4. The water-intercalated compound with the BLH structure can be obtained from the α-, β- and γ-type compounds [29–33]. In this review, we focused on the BLH compound derived from the γ-type compound, because it also has been extensively investigated. The synthesis temperature for γ-Na_xCoO_2 is usually between 800 and 950 °C. It was reported that placing a mixture of the raw materials into a furnace preheated at 750 °C was effective in reducing the vaporization of Na atoms [34]. This method is called a rapid-heat-up technique, but the effectiveness of the technique seems dubious, because the melting point of Na_2CO_3, at 851 °C, is substantially higher than 750 °C. Even assuming that CO_2 gas is produced from Na_2CO_3 at around 500 °C, Na_2O resulting from this decomposition has high reactivity with Co_3O_4 at this temperature. Therefore, a slow heat-up would be more effective. It is known that the Na content does not change significantly when the temperature increases is very slow [35]. Moreover, it should be noted that an excess amount of Na atoms, against the solid-solution limit of Na_xCoO_2, still remains in the sample after heating at 950 °C [35]. Usually, a gradual increase of temperature suppresses the vaporization of an element more effectively, as long as the reaction occurs below the melting point of the raw material including the element.

The second step of the synthesis is to deintercalate Na ions, that is, to oxidize the anhydrate Na_xCoO_2 ($x \sim 0.7$). Chemical oxidation and electrochemical oxidation have been reported to achieve oxidation of the anhydrate [16, 36–38]. Electrochemical deintercalation had been already established [39], because it was expected to be industrially useful as a material for rechargeable batteries, such as $LiCoO_2$. However, the BLH was not given attention, or was perhaps undiscovered because the deintercalation was performed in certain organic liquids, such as acetonitrile; it was widely known that water was unsuitable as a solvent, because of the high reduction-oxidation (redox) potential of the deintercalated compound. Chou *et al.* first applied this method to prepare BLH compound [16]. On the other hand, in chemical oxidation, some oxidizers, such as $Na_2S_2O_6$ and $NaMnO_4$, were used to produce the BLH compound [36–38]. The bromine in acetonitrile is the first, and most frequently used among these oxidizers [1]. Thus, oxidation by bromine is discussed here. The reaction equation may be represented as $Na_xCoO_2 + (x - x')/2\ Br_2 \rightarrow Na_{x'}CoO_2 + (x - x')NaBr$. Even if an excess amount of bromine ($\gg (x - x')/2$) is used, the final Na content does not reach $x' = 0$, because the oxidizer does not work for a highly oxidized compound. The oxidation will not proceed when the redox potential of Co^{4+}/Co^{3+} in Na_xCoO_2 becomes equal to that of Br_2/Br^-. Thus, there is a lowest limit of Na content that is dependent on the oxidizer, which was estimated to be $x = 0.4$ for bromine [40]. Of course, the limit decreases when a stronger oxidizer is used [18].

The last step is the reaction in water, which is very complicated, but governs the Na content and the Co valence of the final product. At first, it was thought that the reaction was simply the intercalation of water molecules. However, it was finally proven by Raman spectroscopy [40] and ^{1}H NMR [41] that a significant amount of oxonium ions are included in the BLH compound. Table 1 shows the Na content and the Co valence, which were chemically determined by ICP-AES and redox titration for two samples just after each synthesis step [40]. It should be noted that both the Na content and the Co valence decrease at the last step, which clearly indicates that ion-exchange between Na^+ and H_3O^+ ions and reduction by water molecules occur simultaneously in water. The reaction formulas may be represented as

$$\mathrm{Na}_x\mathrm{CoO}_2 + a\mathrm{H}_2\mathrm{O} \rightarrow \mathrm{Na}_{x-a/2}(\mathrm{H}_3\mathrm{O})_{a/2}\mathrm{CoO}_2 + a/2\mathrm{NaOH} \quad \text{(A)}$$

$$\mathrm{Na}_x\mathrm{CoO}_2 + b\mathrm{H}_2\mathrm{O} \rightarrow \mathrm{Na}_x(\mathrm{H}_3\mathrm{O})_{2b/3}\mathrm{CoO}_2 + b/6\mathrm{O}_2 \uparrow . \quad \text{(B)}$$

The water in which the samples were immersed becomes alkaline with pH = 11, and some bubbles, probably of oxygen gas, emerge from the sample in the water. Since the equilibrium state of the former reaction depends on the activity of NaOH, the volume of the water is one of the crucial parameters for successful synthesis. Too small or too large amount of water degrades the sample with impurity phases of anhydrous Na_xCoO_2 or CoOOH, respectively. In some reports [42, 43], the samples were hydrated after being left in air with high humidity. Foo *et al.* reported that the BLH and MLH compounds coexisted in the humidity range between 52 and 75% at 298 K [42], when they were prepared from $Na_{0.3}CoO_2$. This clearly indicates that other reactions are simultaneously occurring with water-intercalation. Since the humidity was kept constant (the number of water molecules in the system was free), either BLH or MLH would be obtained as a more stable compound for a certain humidity condition if the reaction with water were a simple intercalation. In other words, coexistence of both compounds would be against the Gibbs' phase rule ($f = c - p + 2$), if there were no component except water and the remaining part of the oxide (Na_xCoO_2) ($c = 2$). In this case, the number of degrees of freedom becomes unity ($f = 1$) for three-phase coexistence ($p = 3$: BLH, MLH and gas phase), which means that the BLH-MLH coexistence range has no width to the humidity for a fixed temperature. Thus, the side reactions shown above are considered to take place even when the BLH compound is prepared by storage in air. In that case, there is probably NaOH aqueous solution present on the surface of the sample particles.

In some cases, the second and the last steps have been combined; that is, electrochemical or chemical oxidation has been performed in an aqueous solution [11, 36–38]. Even in this case, ion exchange and reduction by water take place. However, the situation becomes more complex, because the equilibrium state of the ion exchange must be influenced by the activity of sodium ions, which is in turn influenced by the number of Na ions deintercalated by the oxidizer, as well as that by the ion exchange itself. Thus, the use of separate synthesis steps seems to be more suitable for precise control of the composition.

Table 1. Results of chemical analyses [40]. The Na content was determined by ICP-AES, and the Co valence by redox titration. At each synthesis step, two samples were analyzed.

	Na content, x	Co valence, s
After 1st step	0.695	+3.30
	0.701	+3.29
After 2nd step	0.406	+3.56
	0.402	+3.57
After 3rd step	0.337	+3.43
	0.349	+3.42

Single crystals should be prepared by the same method. However, as far as we know, no crystal with a single domain has been reported thus far. There are two major difficulties in making a good crystal. Firstly, it is not a simple task to obtain good single crystal of the parent compound, γ-Na_xCoO_2, because the Na atoms tend to volatilize at high temperature. In addition, the compound supposedly tends to include stacking disorders, determined by the fact that the stacking sequence of the oxygen layers in Na_xCoO_2 are dependent on and sensitive to the Na content and the synthesis temperature. Secondly, a large particle is basically unsuitable for the intercalation/deintercalation reaction, because the driving force for the reaction is the chemical or electrochemical potential gradient. The larger the crystal is, the smaller the gradient becomes, because the difference in the potentials between the inside and outside of the particle is constant. Thus, in case of a larger particle, the reaction requires a longer time to complete, and sometimes stops halfway, due to extrinsic potentials such as crystal defects and so on.

3.2. Control of the Composition

Based on the analogous high T_c cuprate, it is quite natural to study the carrier-content dependence of T_c. Before oxonium ions were discovered in the compound, much effort was made to control the Na content by adjusting the concentration or the amount of bromine [5, 6,44], because the Na content was considered to be a unique parameter for the determination of the Co valence. The amount of bromine is not a suitable parameter to control the Na content so long as an excess amount of bromine is used, because the redox potential of the bromine does not depend on the amount, but on the concentration. Controlling the Na content by the bromine concentration could be possible, although the reaction with the water should be taken into account for the control of the Na content and the Co valence.

Thus, it seems easier to control the composition at the final step than at the second step. $Na_x(H_3O)_zCoO_2 \cdot yH_2O$ has three adjustable parameters: x, y and z. The y value is controlled by the humidity of the air while the sample is in storage. The x value can be varied by reaction A, while the z value can be varied by both reactions A and B. Reaction A is expected to be reversible, because it is a simple ion exchange reaction, and the equilibrium state would be adjustable using the pH, according to Le Chatelier's law; the

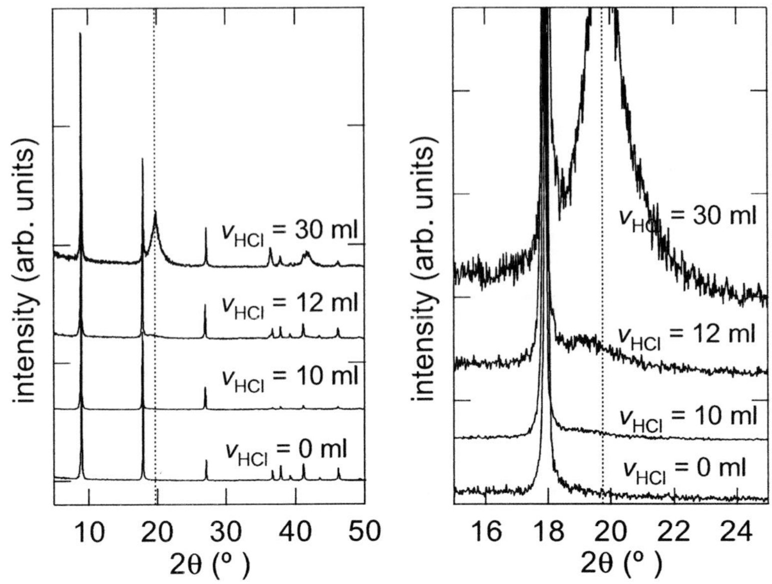

Figure 2. XRD patterns of the HCl samples. The broken line indicates the 002 peak position of CoOOH. Details of part of the pattern at left are shown in the right-side figure.

reaction is promoted in an acidic environment. On the other hand, reaction B is irreversible because oxygen gas is generated, and the reaction is affected by the redox potential of water, which is dependent on pH. Since the half reaction formula is represented as $4OH^- \rightarrow 2H_2O + O_2 + 4e^-$, the reaction is considered to be promoted in an alkaline environment. Although the equilibrium state of reaction B is, of course, determined also by the redox potential of the sample, which could depend on the composition, it is naively expected that the ion substitution proceeds without any change in the valence by the addition of an acid after the immersion of samples in water. Based on this idea, two sets of samples, (1) HCl series and (2) NaOH series, were prepared [7, 8] as shown below. Systematic control of the composition has never been reported except for the following.

1. HCl series [7]

   At first, 1 g of $Na_{0.70}CoO_2$ prepared and annealed in an oxygen atmosphere [35], was immersed in a 5 vol.% Br_2 acetonitrile solution for 5 days at room temperature to deintercalate a portion of the Na^+ ions, and the filtrated sample was then immersed in 400 ml of distilled water for 3 days. Then, v_{HCl} ml of 0.1 M HCl aqueous solution was added to the water. Lastly, the sample was filtrated and dried in air, then stored for three weeks in air with 70% humidity. X-ray diffraction (XRD) measurements, chemical analyses and magnetization measurements of all the samples were performed almost simultaneously, to minimize the storage duration effect [10, 11, 45] shown later.

   Figure 2 shows the XRD patterns of the samples. For $v_{HCl} \leq 10$ ml, a single phase sample was obtained, while for $v_{HCl} \geq 12$ ml, a secondary phase of CoOOH or its related compound was recognized. The x value and the valence estimated from the

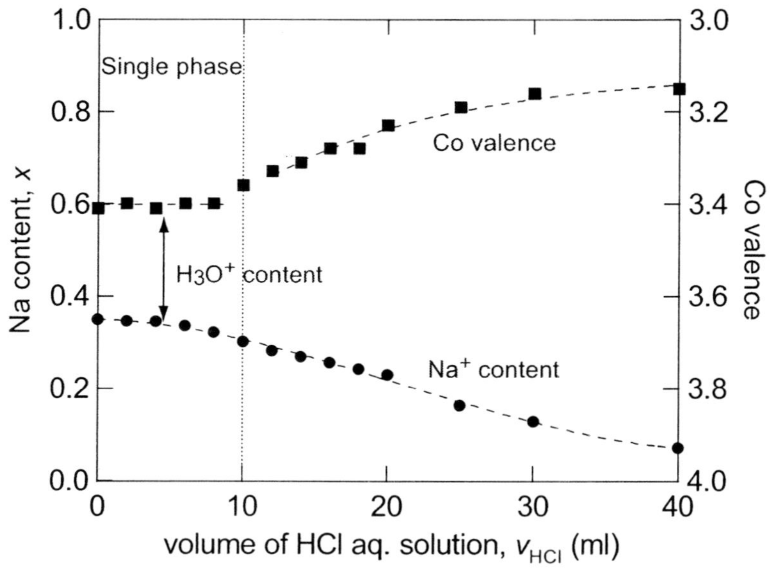

Figure 3. Na content and Co valence of the HCl samples, as determined by the chemical analyses.

chemical analyses are shown in Fig. 3, which clearly indicates that the Na content decreases with increasing $v_{\rm HCl}$, while the valence is kept constant at +3.40 across the single phase region. The oxonium ions compensate the decrease in the Na content as explained above. The c-axis length increases with increasing $v_{\rm HCl}$, as shown in Fig. 4, which is consistent with ion substitution, because the oxonium ion is much larger than the Na^+ ion.

2. NaOH series [8]

   The sample preparation procedure was basically the same as that for the HCl series, but $v_{\rm NaOH}$ ml of 0.1 M NaOH aqueous solution was used instead of HCl (10 and 50 ml of 1M solution were actually used for the $v_{\rm NaOH} = 100$ and 500 ml samples). Only in the $v_{\rm NaOH} = 500$ ml sample, a tiny amount of secondary phase is observed, as shown in Fig. 5. The x value increases with increasing $v_{\rm NaOH}$, while the valence is kept almost constant at +3.48 as shown in Fig. 6. This result cannot be expected by the simple considerations given above; there is a possibility that reaction B proceeded. However, the addition of NaOH solution changes the Na content with the valence almost unchanged. On the other hand, it is unclear at the present, why the valence of the $v_{\rm NaOH} = 0$ ml sample is different from that of the $v_{\rm HCl} = 0$ ml sample. Some possible reasons for this will be given in the next subsection. Although this may indicate that the reaction in water has not been completely understood, these two sets of samples are very useful for understanding the superconductivity phase diagram as shown later.

It may be thought that the valence is controllable if aqueous solutions with various pH are used instead of using distilled water, because the redox potential of the water depends

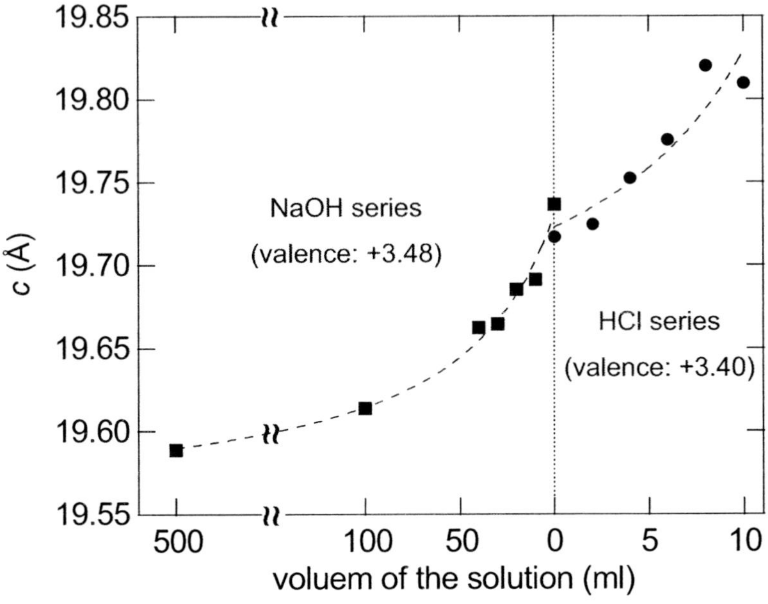

Figure 4. Lattice parameters c for the HCl/NaOH samples.

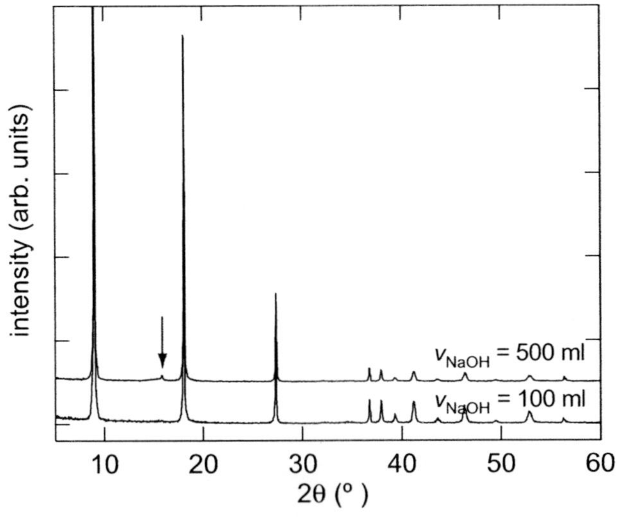

Figure 5. XRD patterns of the NaOH samples. The arrow indicates the peak due to a secondary phase.

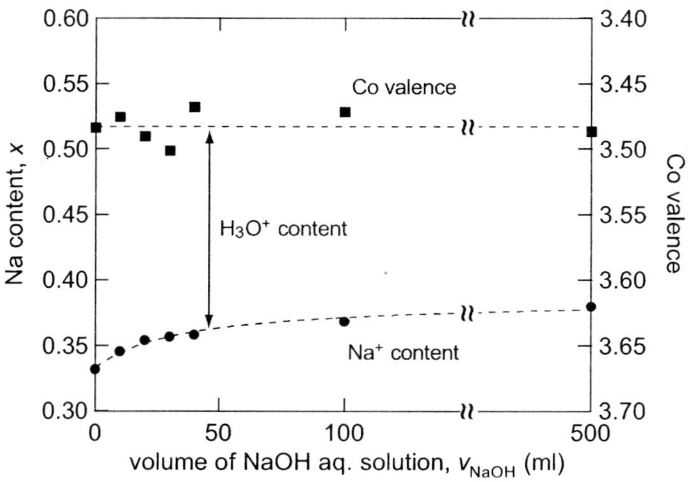

Figure 6. Na content and Co valence of the NaOH samples, as determined by the chemical analyses.

on the pH. However, such an attempt did not provide clear success. Sakurai *et al.* used HCl and NaOH solutions with various pH between $pH_i = 1$ and 14 [46]. For $pH_i = 3 \sim 11$, the single phase BLH was obtained, but there was no clear systematic change in the Na content nor the valence, and for other pH_i, each sample included a secondary phase of Na_xCoO_2 ($x \sim 0.4$ or 0.5). In the former case, the pH of the water after immersion, pH_f, became approximately 11, as shown in Fig. 7, and the valence seemed to correspond to pH_f. Thus, the basic idea is probably correct, but more precise control of pH_f is required in order to obtain systematic samples. These results are consistent with the very early report by Park *et al.*, which stated that the single phase BLH is obtained for $pH_f = 10.5$.

3.3. Caution from a Chemical Perspective

The BLH is prepared by a soft-chemical method, which is not so common in the superconductivity research. As mentioned above, the understanding of the reaction mechanism and control of the composition are not perfect. In this section, some possible reasons for the reduction in the reproducibility of synthesis or measurements are discussed from a chemical perspective.

Reaction with water in the air is the most inconvenient problem. As clearly seen from the synthesis method, the BLH will react with water in the air while stored, although humidity is necessary for preservation, low humidity changes the BLH into MLH [42,47,48]. The surface of the sample particles can be dressed with a certain kind of solution, such as NaOH or $NaHCO_3$ aqueous solution, which would provide a means for the change in the Na content and the valence of the BLH compound. In other words, it is necessary to consider the reaction between the BLH compound, a liquid on the surface, and the air, even after the synthesis has apparently finished.

The precursor Na_xCoO_2 ($x \sim 0.7$) can also react with water in the air to form $Na_{x'}(H_3O)_zCoO_2$ and NaOH [49]. Indeed, the color of the sample becomes darker even

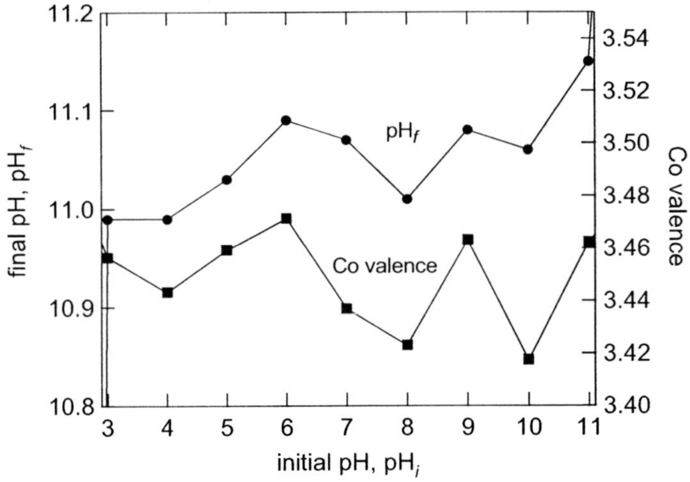

Figure 7. Relation between the final pH and Co valence.

while the sample is stored in a desiccator with silica-gel, because the reaction of water with the sample is faster than absorption of the water by the silica-gel. The aging effect of surface degradation, observed by Raman spectroscopy measurement [50], may be due to this reaction. If the Na content in the precursor is smaller from the start, the Na content in the BLH compound would be much smaller than expected. This can certainly be prevented by storing the precursor in an inert gas, but this does not seem to be so widely known; sometimes single crystals are washed with water to remove the flux [51, 52].

Reaction with carbon dioxide is another problem. Carbon dioxide dissolves easily in water to form H_2CO_3 acid. Thus, if this occurs during the sample immersion in water, then the same effect is expected as for the addition of HCl solution. The first magnetically ordered sample was discovered among samples that were prepared by an extremely long immersion in water. Water containing CO_2 may have been unknowingly used as distilled water.

As far as we know, there has been no report where the temperature during synthesis has been precisely conditioned. According to the Nernst equation, the redox potential is a function of temperature, and also, generally speaking, the equilibrium constant, even for an ion exchange, depends on temperature. In the case of many compounds made by conventional solid state reaction, small thermal fluctuations do not usually have a significant influence on the final product, but for the soft chemical method, this can sometimes cause a crucial difference in the sample quality.

Oxygen nonstoichiometry would be a key factor if it was significantly present, because the Co valence changes due to oxygen nonstoichiometry. However, unfortunately, there is no crucial evidence for oxygen nonstoichiometry in the BLH compound. Chemical analyses (ICP and redox titration) and neutron diffraction measurements were performed to estimate oxygen nonstoichiometry [11, 43], but the analysis of the data is insufficient, because the presence of oxonium ions was not taken into account. In other words, the chemical analyses did not distinguish between oxygen nonstoichiometry and the oxonium ion content, and the oxygen site occupancy estimated by the neutron diffraction experiment would be the wrong due to the wrong chemical composition assumed. Similar experiments have also been carried out on anhydrous $Na_xCoO_{2-\delta}$, with inconsistent results having been reported. Molenda ª¨ø¥Úand Chou ª¨ø¥Úusing thermogravimetric (TG) experiments, reported that a $\delta \sim 0.08$ lack of oxygen content was present in $Na_{0.7}CoO_2$ and $Na_{0.75}CoO_2$ synthesized in the air [53, 54]. However, nonstoichiometry was not detected, within an accuracy of $\delta = 0 \pm 0.02$, from the ICP and redox titration experiments on samples with $x = 0.70 \sim 0.78$, prepared in air [43]. In addition, on the basis of precise TG measurements [55], it was found that nonstoichiometry appears above 600 °C and approximately 450 °C under oxygen partial pressures of 1.0 atm and 0.20 atm (corresponding to the air), respectively, but it reaches only $\delta \sim 0.01$ and 0.02 at 800 °C, respectively. For samples with $x = 0.74$ and 0.38, prepared under flowing oxygen gas, neutron diffraction experiment also indicated no nonstoichiomentry within an accuracy of $1 \sim 2$ % [56]. The results in Table 1 also indicate $\delta \sim 0.00$ and 0.02 for $x = 0.70$ and 0.40, respectively, if they are estimated from the Co valence. Thus, it is likely that oxygen nonstoichiometry is negligible for anhydrous Na_xCoO_2, which indicated significant rigidity of the CoO_2 layer. This is expected, because the oxygen atoms in the layer are arranged in the closest packed manner and are so tightly bound with the Co atoms that the bond length is very short probably due to high covalency between them. Thus, oxygen nonstoichiometry in the BLH compound is also reasonably expected to be negligibly small. It is very difficult to expect that more oxygen defects appear in the BLH compound where the Co valence is significantly lower than that in $Na_{0.4}CoO_2$, because generally, transition-metal oxides in higher valence states tend to have more oxygen defects.

The last topic in this subsection is the duration effect; the changes in the sample quality while it is stored. The sample gradually becomes more fluffy the longer it is stored, and the superconductivity also changes as reported in Ref. [10, 45]. However, there is little information regarding the speed of the change or the origin of the change. Viciu ª¨ø¥Ú attributed the change in the superconductivity to increasing oxygen nonstoichiometry from neutron diffraction data [56]. However, their analyses did not take into account the oxonium ions, which could lead to the possibly erroneous results. In addition, the time dependence of the nonstoichiometry was rather scattered and had a large margin of error. In consideration of the fact that oxygen nonstoichiometry is unlikely, the duration effect seems to be due to the reaction with water or carbon dioxide in the air and/or the rearrangement of H_2O molecules, Na^+ and H_3O^+ ions, which would take a long time because of weak driving forces. Also, the phase of the sample sometimes changes; when a BLH sample including anhydrous Na_xCoO_2 ($x \sim 0.4$) as an impurity is preserved in a humid atmosphere, the amount of impurity decreases, and sometimes disappears completely [45]. In contrast, a single phase sample is decomposed into the BLH and CoOOH after a while. These results clearly indicate that some reactions occurred while the samples were stored. They also seem to suggest that the Na ions in the BLH compound tend to become condensed just after the synthesis, which apparently corresponds to the duration dependency of the NQR frequency and T_c [10]; the resonant frequency decreased and then T_c followed a certain relation with the frequency. The concentration of Na ions is possibly accompanied with a change in the Co valence.

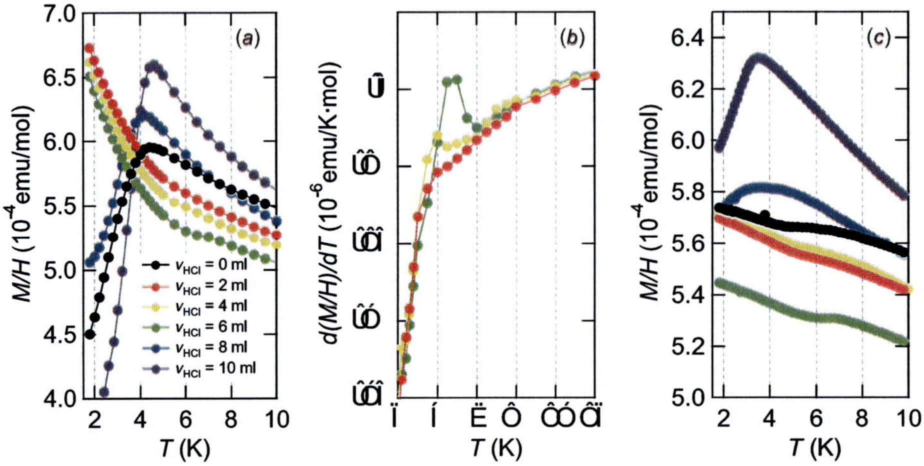

Figure 8. (a) Magnetic susceptibilities of the HCl samples measured under 1 T, (b) their derivatives, and (c) magnetic susceptibilities under 7 T.

ÌŨÚ Õ ªÆ± º´ ¨ – ø ª ‹ ø Æ ©¨ › ± ± ¨± ø¥–øÆ ª·ªÆ

In order to establish the phase diagram, the samples used should be carefully characterized based on the previous cautions. Unfortunately, no sample reported thus far is suitable for the determination of the phase diagram, except the HCl and NaOH series of samples.

The magnetic susceptibilities of the HCl samples are shown in Fig. 8. As mentioned above, the Na content of these samples decreases with increasing v_{HCl} under a constant Co valence at $s = +3.40$ across the single phase region. Despite the systematic change of the Na content and the c-axis length in these samples, superconductivity was observed in the $v_{\mathrm{HCl}} = 0$, 8, and 10 ml samples. The other samples show an anomaly in the magnetic susceptibility and its derivative at around 6 K. This is considered to be a sign of magnetic ordering, which was confirmed by NQR measurement. Thus, the superconducting phase diagram for the $s = +3.40$ section can be depicted as shown in Fig. 9. The superconducting phases, SC1 and SC2, are divided by the magnetically ordered phase, M. In other words, the superconductivity appears at both edges of the M phase, which strongly suggests that it is induced by magnetic fluctuations. On the other hand, all the samples made using NaOH solution exhibit superconductivity as shown in Fig. 10, and so the dome-shaped superconducting region is drawn as in Fig. 11. From Na content x dependency of superconducting transition temperatures in HCl and NaOH series of samples, shown in Fig. 9 and Fig. 11, a two-dimensional phase diagram is depicted in Fig. 12. It is notable that no M phase is detected for the $s = +3.48$ section in the range of $0.33 \leq x \leq 0.35$, where the M phase appears for $s = +3.40$. By the way, the M phase seems to be located at a lower x region on the $s = +3.48$ section, because the transformation from a superconducting to the magnetically ordered state was observed for $v_{\mathrm{NaOH}} = 0$ ml samples [57], as shown later.

The physical properties of the BLH clearly depend on both the Na content and the valence. As in the case of the high T_c cuprate, the valency dependence is expected, because the change in the valence affects many physical parameters related to superconductivity, such

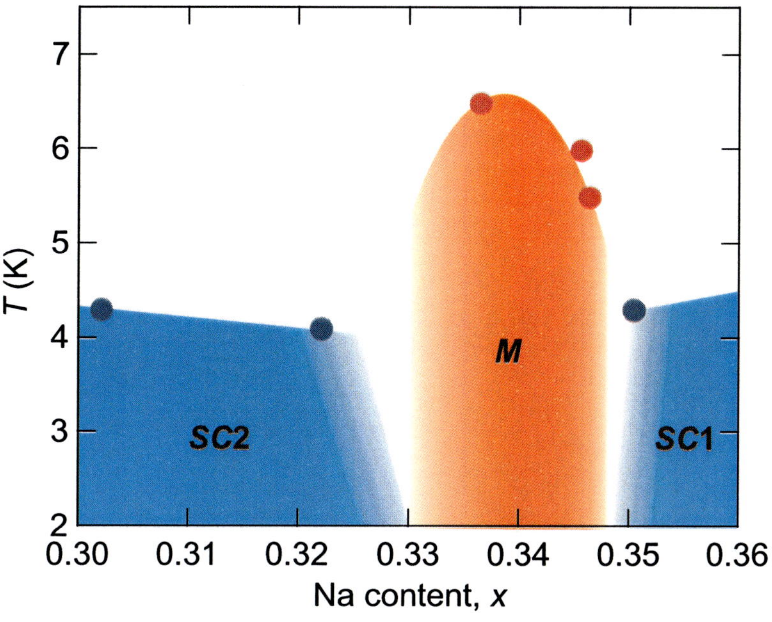

Figure 9. Na content vs. temperature phase diagram determined using the HCl samples. The SC and M phases represent the superconducting and magnetic ordering phases, respectively.

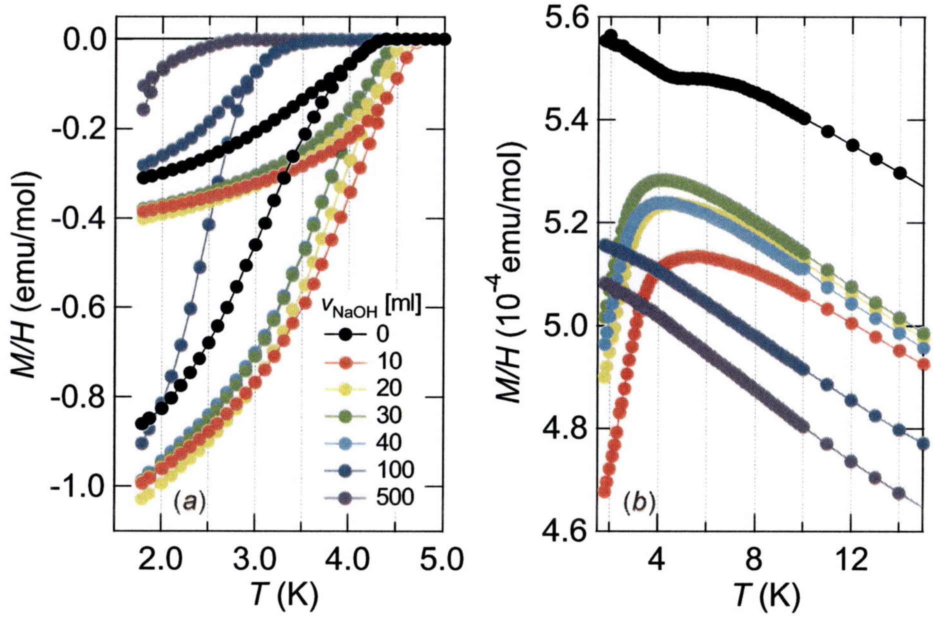

Figure 10. Magnetic susceptibility of the NaOH samples measured under (a) 0.001 T and (b) 7 T.

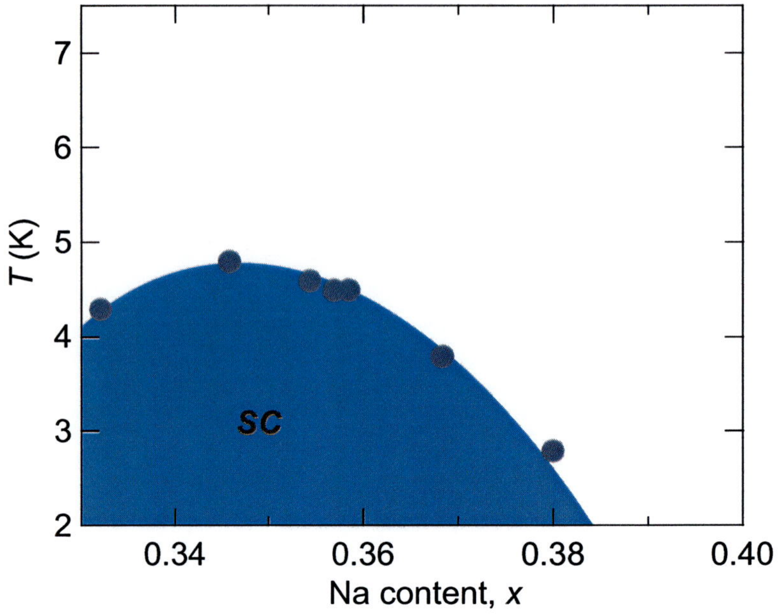

Figure 11. Na content vs. temperature phase diagram, determined using the NaOH samples.

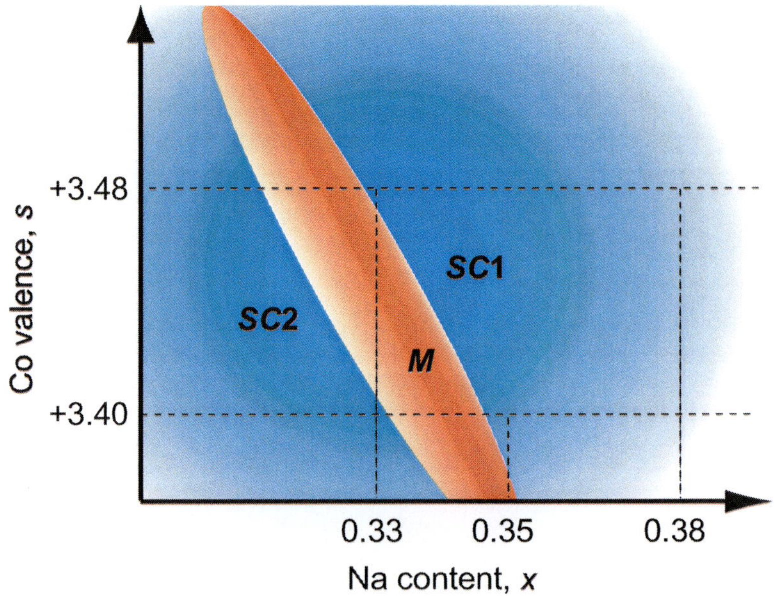

Figure 12. Expected Na content vs. Co valence phase diagram at 0 K.

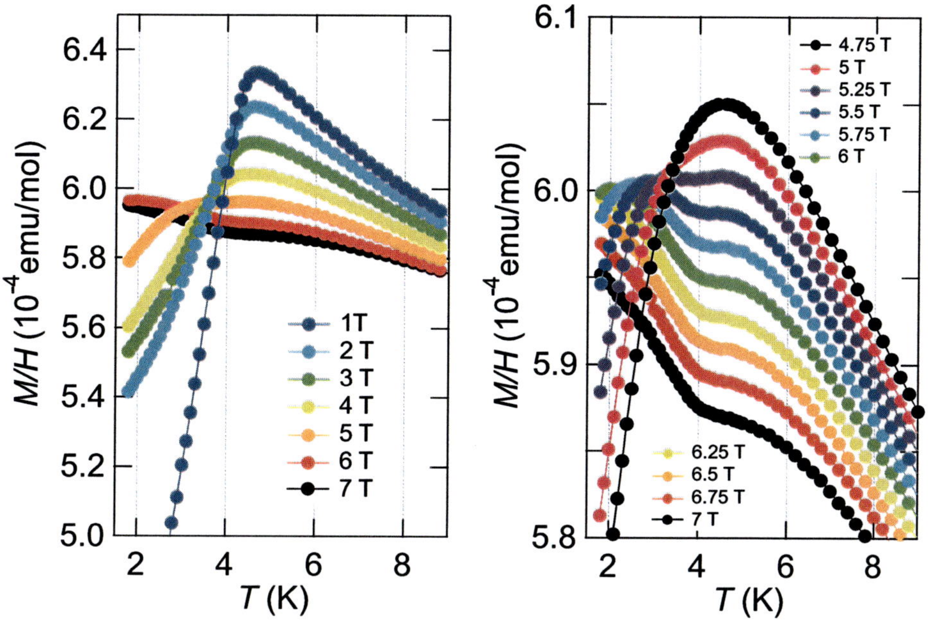

Figure 13. Magnetic susceptibility of the v_{HCl} = 0 ml sample measured under various magnetic fields.

as the density of states or magnetic fluctuations. On the other hand, dependence on the Na content may be expected from consideration of the analogous $Sr_{2-x}Ca_xRuO_4$ [58], because the BLH compound can be regarded as a solid-solution system between $Na_{x+z}CoO_2 \cdot yH_2O$ and $(H_3O)_{x+z}CoO_2 \cdot yH_2O$. However, in case of the BLH compound, the structure of the CoO_2 layer seems to be very rigid, and hardly influenced by ion substitution, as deduced by the misfit-layer compounds of $[Bi_{0.87}SrO_2]_2[CoO_2]_{1.82}$ [59], $Ca_3Co_4O_9$ [60, 61], $[Sr_2O_{2-\delta}]_qCoO_2$ [62, 63], and $[Ca_xOH]_qCoO_2$ [64–66]. That is why the effect of ion exchange on the $Na_x(H_3O)_zCoO_2 \cdot yH_2O$ system is rather surprising. The electronic band structure must be significantly affected by the ion exchange. A band calculation has indicated that two kinds of Fermi surface are formed; a large cylindrical surface centered at Γ point, and the hole pockets around K point [67] or the electron pocket inside the larger pocket [68]. Since the appearance of the latter Fermi surface is believed to be very sensitive to the change in the Co valence or to the CoO_2 layer thickness [67], the $x-s$ phase diagram strongly suggests that the pocket-like Fermi surface is the key to the superconductivity and magnetic fluctuations. Indeed, the hole pockets become more stabilized when the CoO_2 layer becomes thinner [69, 70], and ion substitution with H_3O^+ to Na^+ ions would make the CoO_2 layer thinner due to the enlarged c axis, which weakens the Coulomb interaction between Na^+/H_3O^+ and oxygen ions in the layer.

Next, the $H-T$ phase diagram (H: magnetic field) is discussed. It is interesting to note that the $v_{\mathrm{HCl}} = 0$ ml sample does not show a superconducting transition under a magnetic field of 7 T. Instead, as seen in Fig. 8, it has a magnetic anomaly, which is due to the magnetic transition seen in the $v_{\mathrm{HCl}} = 2$, 4 and 6 ml samples. Thus, the sample shows a transformation from the superconducting state to a magnetically ordered state. In order

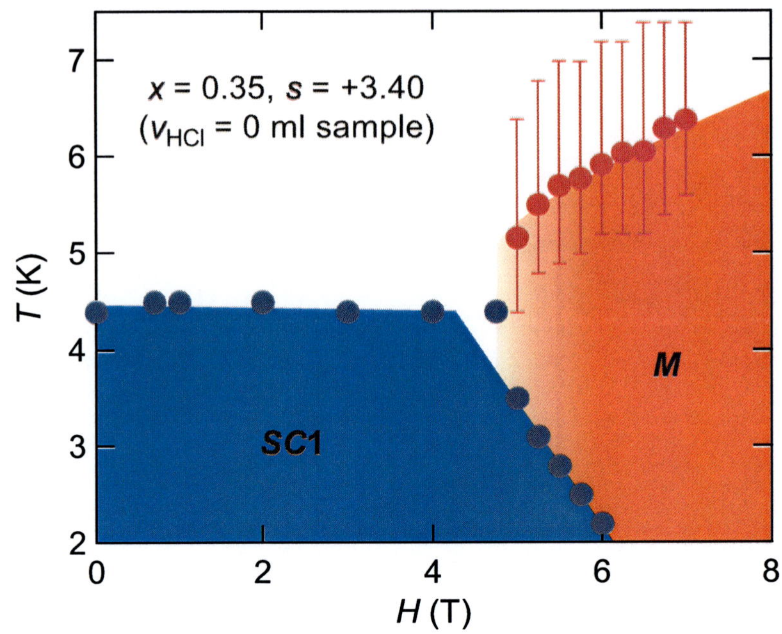

Figure 14. Magnetic field vs. temperature phase diagram, determined using the $v_{\rm HCl} = 0$ ml sample.

to investigate the boundary between these states, the magnetic susceptibility was measured under various fields, as shown in Fig. 13. Below 4.75 T, only the superconducting transition is seen, but above 5 T, the susceptibility has two peaks due to the superconducting and the magnetic transitions. As an example, two peaks are recognized at 3 K and 5.7 K for the 5.5 T curve. The former peak moves to a lower temperature, while the latter seems to shift slightly to a higher temperature as the magnetic field increases. The $H - T$ phase diagram for this sample is given in Fig.14. The superconducting transition steeply decreases with the increasing magnetic field above 5 T, suppressed by the M phase, which clearly indicates that the transformation is intrinsic to the BLH compound, not of extrinsic origin, such as sample inhomogeneity. The anomaly was also observed in the magnetization curve, shown in Fig. 15 [71], although the transformation field of 5.4 T at 1.8 K is slightly lower than that determined from the susceptibility. On the other hand, the $v_{\rm NaOH} = 0$ ml sample shows a transformation at 1.8 K under a lower field of 4.5 T, as shown in Fig. 16, while the $v_{\rm NaOH} =$ 100 and 50 ml samples do not show any anomaly in the magnetization curves [57], or in the susceptibility curves (see Fig. 10). Thus, the $x - s$ phase diagram in Fig. 12 was obtained. Since the susceptibility peak of the $v_{\rm HCl} = 8$ ml sample becomes less pronounced under a field of 7 T (see Fig. 8) compared with that of the $v_{\rm HCl} = 0$ ml sample under a field of 5 T, the M phase may expand, not only to a higher x region, but also to a lower x region under magnetic field.

The $v_{\rm HCl} = 0$, 8, and 10 ml samples and the $v_{\rm NaOH} = 0$, 10, 20, 30 and 40 ml samples have almost the same T_c under 1 mT, but only the $v_{\rm HCl} = 0$ ml (and possibly 8 ml) sample(s) and the $v_{\rm NaOH} = 0$ ml sample show the transformations while the other samples remain

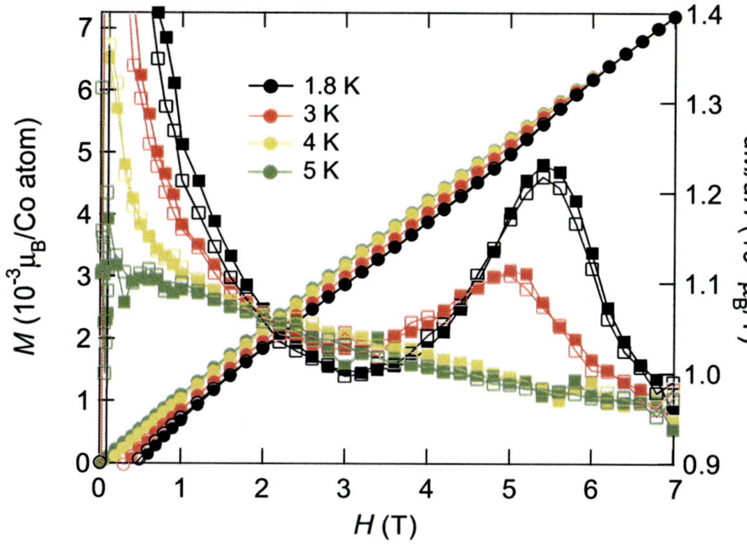

Figure 15. Magnetization curves (circles) and their derivatives (squares) for the $v_{\mathrm{HCl}} = 0$ ml sample measured at various temperatures. The full and open markers represent the data collected with increasing and decreasing fields, respectively.

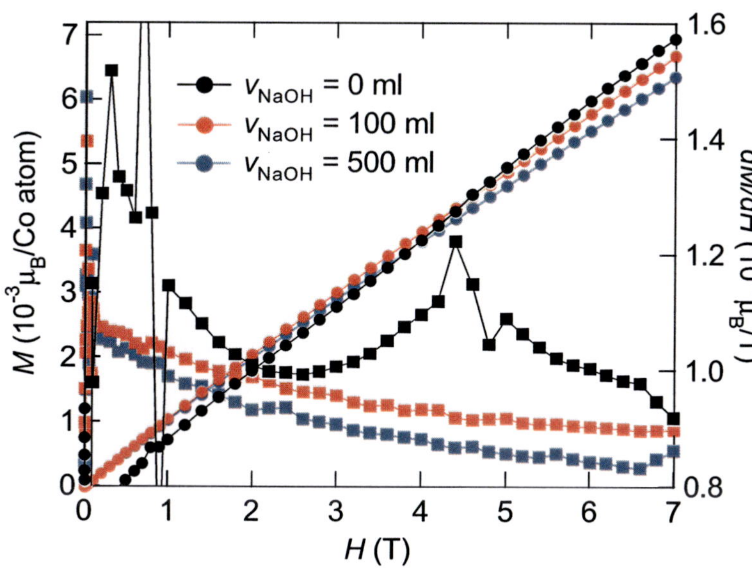

Figure 16. Magnetization curves (circles) and their derivatives (squares) for the NaOH samples at 1.8 K.

superconducting under 7 T with $T_c = 3 \sim 5$ K. Thus, it is quite reasonable that for the BLH compound the upper-critical field, H_{c2}, is independent of T_c itself. Chou a¨ ø¥Úand Sasaki a¨ ø¥Ú reported that the $H_{c2}(0)$ perpendicular to the c axis is approximately 8 T [16, 25] for those samples with $T_c = 4.2$ K and 4.5 K, which corresponds to a Pauli limit of $H_P = 1.84T_c$ [72, 73]. However, much larger $H_{c2}(0)$ have been reported by many researchers, such as Sakurai a¨ ø¥Ú(61 T for $T_c = 4.6$ K) [12], Maśka a¨ ø¥Ú(28 T for $T_c = 4.3$ K) [74], and Jin a¨ ø¥Ú(17.1 T for $T_c = 4.7$ K) [18]. Since the $H_{c2}(0)$ can be extrapolated to approximately 8 T for the $v_{\rm HCl} = 0$ ml sample (see Fig. 14), the former estimations might have been performed using samples in the vicinity of the M phase. On the other hand, the latter estimations were based on the WHH model [75], which may not be applicable, but it is clear that the $H_{c2}(0)$ values of some samples are significantly larger than H_P. The T_c of $\sim$ 4 K for the $v_{\rm NaOH} = 10$ ml sample under 7 T is significantly higher than the superconducting transition temperatures reported by Chou a¨ ø¥Úand Sasaki a¨ ø¥Ú($\sim$ 2.6 K and $\sim$ 1.7 K, respectively). The superconducting transition was actually observed under a magnetic field of more than 8 T, using specific heat [76] and NMR measurements, as shown later. Thus, the premise for the unconventional superconductivity of the BLH compound becomes more promising.

In this section, we have presented chemical techniques for the control of the compositions, and the phase diagram made using these techniques. However, it was difficult to obtain crucial information regarding the exact origin of the superconducting transition. In the next section, we would like to present the phase diagram determined from microscopic experiments, in addition to universal scaling plots, which will provide more direct information regarding the mechanism for the superconductivity.

ÏÚ “´ ¥øÆœ´ø° Æ ±¥ Œ ± ø a ” aø ´Æ a ¨

ÏÚ –Æ ¥ ± ¨ a “œŒ” aø ´Æ a ¨

In this section, the fundamental principles of nuclear quadrupole resonance (NQR) are briefly explained. Resonance phenomena are observed between split nuclear states using an radio frequency field with energy comparable to the splitting width. To observe the resonance, degenerated nuclear spin states have to be split by a magnetic field and/or an electric-field gradient (EFG). For NQR measurements, a magnetic field is not required because the nuclear levels are split by the electric-field gradient. Under zero magnetic field, nuclear levels are determined by the electric quadrupole Hamiltonian $\mathcal{H}_Q$, which describes the interaction between the electric quadrupole moment of the nuclei Q and the EFG at the nuclear site. In general, $\mathcal{H}_Q$ is expressed as

$$\mathcal{H}_Q = \frac{\nu_{\rm zz}}{6}\left[(3I_z^2 - I^2) + \frac{1}{2}\eta(I_+^2 + I_-^2)\right], \quad \left(\nu_{\rm zz} = \frac{6e^2qQ}{4I(2I-1)}\right) \tag{1}$$

where $eq(= V_{zz})$ and $\eta = (V_{yy} - V_{xx})/V_{zz}$ are the EFG along the principal axis (z axis) and the asymmetry parameter, respectively. The resonant frequency is calculated by solving the Hamiltonian. The nuclear spin of Co nuclei is 7/2. Within the second order perturbation against η, the resonant frequency is derived as

$$\begin{aligned}
\nu_1 &= \nu_{zz}\left(1+\frac{109}{30}\eta^2\right) \quad (\text{for} \pm 1/2 \leftrightarrow \pm 3/2), \\
\nu_2 &= 2\nu_{zz}\left(1-\frac{17}{30}\eta^2\right) \quad (\text{for} \pm 3/2 \leftrightarrow \pm 5/2), \text{ and} \\
\nu_3 &= 3\nu_{zz}\left(1-\frac{1}{10}\eta^2\right) \quad (\text{for} \pm 5/2 \leftrightarrow \pm 7/2).
\end{aligned} \tag{2}$$

From the measurement of two resonant frequencies, ν_{zz} and η are estimated separately. These two quantities provide information concerning the Co- $3d$ electronic state and subtle crystal distortions around the Co site, because the EFG at the Co site is determined by on-site $3d$ electrons and ionic charges surrounding the Co site. The ionic charge contribution is estimated from a calculation, in which the ions are assumed to be point charges (point-charge calculation). The result of the point-charge calculation indicates that V_{zz} is mainly dependent on the thickness of the CoO_2 layers, because the effect of the neighboring O^{2-} ions is larger than that of oxonium ions and Na^+ ions that are distant to the Co ions. Due to the ionic charge contribution, the resonant frequency was found to increase with the compression of the CoO_2 layers.

Next, we discuss the contributions from the on-site $3d$ holes. These analyses were conducted on cuprate superconductors [77], and the local hole density was estimated from the NQR frequency [78]. Similar analyses were applied to the cobaltate, in order to evaluate the hole number at Co-$3d$ in the superconducting BLH compounds.

When we consider the a_{1g} and e_g' orbitals in the trigonal structure, the wave function for the Co-$3d$ electrons is described as a linear combination of these orbitals,

$$\psi_{3d} = \sqrt{1-w^2}|a_{1g} > + \frac{w}{\sqrt{2}}|e'_{g+} > + \frac{w}{\sqrt{2}}|e'_{g-} > . \tag{3}$$

Here, w^2 is the occupation of the e_g' orbitals. Using this wave function, the EFG $V_{\alpha\beta}$ originating from the on-site holes is derived as

$$V_{\alpha\beta} = \frac{e}{4\pi\epsilon_0}\int \psi_{3d}^* \frac{\partial^2}{\partial\alpha\partial\beta}\left(\frac{-1}{r}\right)\psi_{3d}\, d\tau \quad (\alpha, \beta = x, y \text{ and } z). \tag{4}$$

By diagonalizing the EFG tensor $V_{\alpha\beta}$, the eigen value $\lambda(\omega)$ is obtained, and is expressed as

$$\lambda(\omega) = \frac{e}{4\pi\epsilon_0} g_\alpha(\omega)\langle r^{-3}\rangle_{3d}, \tag{5}$$

where $\langle r^{-3}\rangle$ is the average volume of the $3d$ orbital. The values of $g_\alpha(\omega)$ are determined to be $g_z(\omega) > g_x(\omega) \geq g_y(\omega)$, and these values depend on the occupation of the e_g' orbitals. As a result, ν_{zz} from on-site hole contribution, is described as

$$\nu_{zz} = \frac{1}{14}\frac{e^2Q}{h}\frac{n_{3d}}{4\pi\epsilon_0} g_z(\omega)\langle r^{-3}\rangle_{3d}, \tag{6}$$

where $n_{3d} = n_{a_{1g}} + n_{e_g'}$ is the total hole number in the Co-$3d$ orbitals. In the simplest case, where one hole is in the a_{1g} orbital ($w = 0, n_{3d} = 1$), the values of $g_\alpha(0)$ are estimated to be $4/7$, $-2/7$, and $-2/7$ for $g_z(0)$, $g_x(0)$, and $g_y(0)$, respectively. Then ν_{zz} and η are evaluated to be 20.4 MHz and zero, respectively. Here, $\langle r^{-3}\rangle$ was set to be 6.7 a.u. and was multiplied by the reduction factor $\xi = 0.8$ [79, 80].

From the nuclear lattice relaxation rate $1/T_1$ measurement, we can obtain information concerning the low-energy dynamical susceptibility $\chi(q, \omega)$. Using the dynamical susceptibility, $1/T_1$ is expressed as

$$\frac{1}{T_1} = \frac{2\gamma_N^2 k_B T}{(\gamma_e \hbar)^2} \sum_q |A_q|^2 \frac{\chi(q, \omega_0)}{\omega_0}, \tag{7}$$

where γ_N and γ_e are the gyromagnetic ratios for the nuclei and electrons, respectively. It should be noted that the low-energy dynamical susceptibility summed up in q space is proportional to $1/T_1$ divided by T. The temperature dependence of the dynamical susceptibility is investigated on the basis of self-consistent renormalization (SCR) theory [81], and the temperature dependence of $1/T_1T$ near a magnetic instability is calculated using the results. In a two-dimensional system, $1/T_1T$ is proportional to $1/(T-\theta)$ near antiferromagnetic ordering and proportional to $1/(T-\theta)^{3/2}$ near ferromagnetic ordering. When the magnetic correlations possess a three-dimensional nature, these temperature dependencies change to $1/\sqrt{T-\theta}$ and $1/(T-\theta)$ for antiferromagnetic and ferromagnetic ordering, respectively. In these formulae, θ denotes the ordering temperature, if the system demonstrates magnetic ordering at finite temperature, and if the system has no ordering down to 0 K, then θ corresponds to the zero point fluctuation energy. In a special case, where θ is exactly 0 K, the magnetic fluctuations possess a quantum critical character, and the phase transition would occur at zero Kelvin.

Table 2. Na content, c-axis parameter c, transition temperature T_c, Co valence s and NQR frequency ν_3 arising from the transition between $\pm 5/2 \leftrightarrow \pm 7/2$.

No.	Na content	c (Å)	T_c (K)	Co^{+s}	ν_3 (MHz)	Ref.
1	0.348	19.684	4.7	—	12.30	[23]
2	0.339	19.720	4.6	—	12.30	[82]
3	0.348	19.569	2.8	—	12.07	[9]
4	0.35	19.603	4.6	—	12.32	[9]
5	0.331	19.751	< 1.5	3.42	12.54	[9]
6	0.351	19.714	4.6	3.37	12.45	[14, 83]
7	0.350	19.739	4.4	3.41	12.50	[84]
8	0.322	19.820	3.6	3.40	12.69	[84]
9	—	—	3.6	—	12.22	[84]
10	0.380	19.588	2.6	3.49	12.10	[84]
11	0.368	19.614	3.5	3.47	12.18	[84]
12	0.346	19.691	4.8	3.48	12.39	[84]

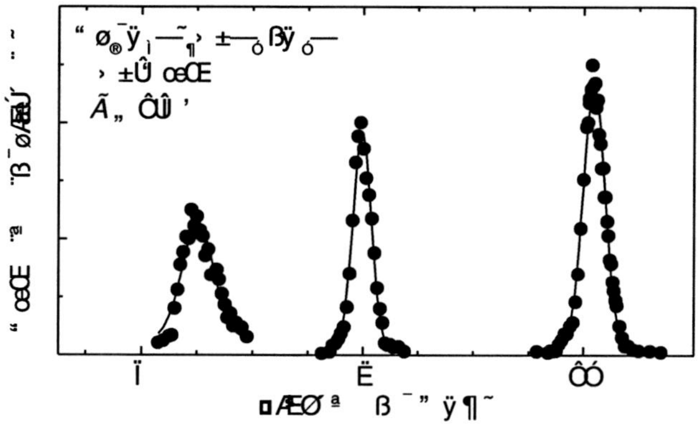

Figure 17. Typical Co-NQR spectra for the BLH compound. This representative spectrum was obtained for sample No. 11 at 1.5 K. From these peak frequencies, ν_{zz} and η were evaluated to be 4.08 MHz and 0.20, respectively. [84]

ÏŰÚ Õø ¥ –Æ ªÆª

NQR measurements were performed on twelve samples that were synthesized by the soft chemical procedure described in Ref. [1,7]. The crystallographic parameters and Na content x were precisely measured by XRD and ICP-AES, respectively. The results are listed in Table 2 together with various physical parameters. The Co valence s in several samples was determined by redox titration measurements. Using the values of s and x, the oxonium content z can be estimated from the relation $z = 4 - x - s$. The superconducting transition temperature T_c is defined as the onset temperature of superconducting diamagnetism, which was detected using an [illegible] NQR coil.

NQR spectra are obtained by recording the signal intensity with varying frequency. A typical Co-NQR spectrum for a superconducting cobaltate is shown in Fig. 17. Three NQR peaks, which arise from $m = \pm 1/2 \leftrightarrow \pm 3/2$, $m = \pm 3/2 \leftrightarrow \pm 5/2$, and $m = \pm 5/2 \leftrightarrow \pm 7/2$ transitions, are observed. Because the EFG is asymmetric in the ab plane, the separation between two peaks is not the same. Figure 18 (a) shows a close view of the NQR spectra arising from a $m = \pm 5/2 \leftrightarrow \pm 7/2$ transition in various samples with different H_2O, Na^+ and H_3O^+ contents. A single peak was observed for this transition, although a fractional content of Na ions randomly occupy the Na layer. The double hydrate layers screen the randomness of the Na layer, which is considered to be important for superconductivity. The resonant frequency of the $m = \pm 5/2 \leftrightarrow \pm 7/2$ transition, which is referred to as ν_3 in equation (2), is also presented in Table 2. From a comparison of ν_3 to various parameters in Table 2, it is revealed that ν_3 is strongly dependent on the physical properties of the samples, particularly that of the c-axis lattice parameter. The sample with $T_c \sim 4.6$ K (No. 6) has nearly the same ν_3 as those of previous samples with the same T_c (Nos. 2 and 4). This result confirms the reproducibility of ν_3 among those samples with the same properties.

Figure 18 (b) shows the temperature dependence of $1/T_1T$ for various BLH samples below 100 K, together with that for the MLH sample, in which superconductivity does not occur down to 1.5 K. The blue arrows in Fig. 18(b) indicate T_c of these samples. The values

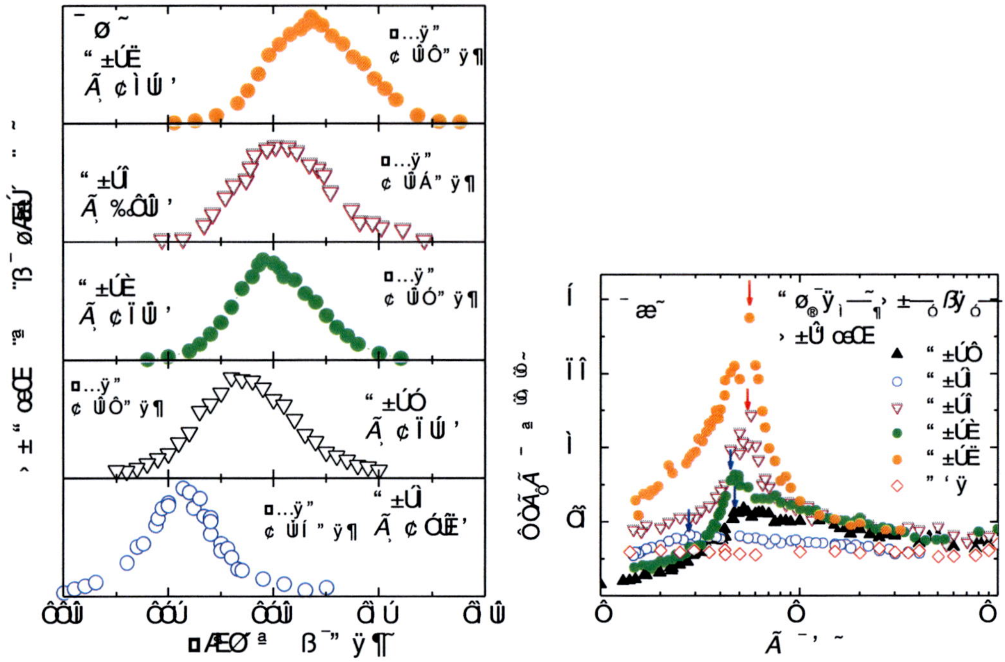

Figure 18. (a) Co-NQR spectra arising from the $\pm 5/2 \leftrightarrow \pm 7/2$ transitions in various samples. In magnetically ordered samples (No. 5 and No. 8), the spectra were measured at 8 K $> T_{\mathrm{M}}$. T_{c} and the full width at half maximum (FWHM) for each sample are represented together. (b) Temperature dependence of $1/T_1T$ in various samples. The blue (red) arrows indicate T_{c} (T_{M}). The sample legend colors are the same for both left and right figures. [84]

of $1/T_1T$ for all samples follow a unique temperature dependency from 200 K down to 70 K. The sample dependence of $1/T_1T$ is observed below 70 K. The non-superconducting MLH sample shows Korringa behavior down to 1.5 K, which is indicative of the absence of temperature-dependent magnetic fluctuations at low temperature, whereas the values of $1/T_1T$ in the superconducting samples (Nos. 1, 3 and 7) increase with decreasing temperature down to T_{c}. For the magnetically ordered samples (Nos. 5 and 8), $1/T_1T$ shows a prominent peak at T_{M}, which is indicated by red arrows, and an internal field appears at the Co nuclear site below T_{M}. It is noteworthy that the temperature dependence of $1/T_1T$ in the sample with $T_{\mathrm{c}} = 4.7$ K (No. 1) is the same as that for the magnetically ordered samples in the temperature range of $15 - 100$ K. This suggests that the magnetic fluctuations in superconducting samples with high T_{c} are similar to those in the magnetically ordered samples in the normal state. In the sample with $T_{\mathrm{c}} = 2.8$ K (No. 3), the enhancement of $1/T_1T$ below 70 K is moderate, and $1/T_1T$ gradually decreases below T_{c}. If T_{c} were suppressed by the inhomogeneity of the crystal structure, then the FWHM of the NQR spectrum for the lower-T_{c} sample would be broader than that for the higher-T_{c} sample. However, the FWHM for sample No. 3 is the narrowest among the samples. These results suggest that the inhomogeneity and/or randomness in samples does not suppress superconductivity, and that the sample dependence of $1/T_1T$ could be an indicator towards understanding the superconducting mechanisms in this compound.

It was found that the NQR frequency and the temperature variation of $1/T_1T$ are strongly sample dependent. From the comparison between Figs. 18(a) and (b), we indicated that below 70 K $1/T_1T$ is more significantly enhanced in the sample with higher NQR frequency, and that there is a relationship between low-temperature magnetic fluctuations and the NQR frequency. In the following sections, the origin of these sample dependencies is investigated, and a phase diagram is developed by employing some of sample dependent physical quantities.

4.3. Sample Dependence of the EFG

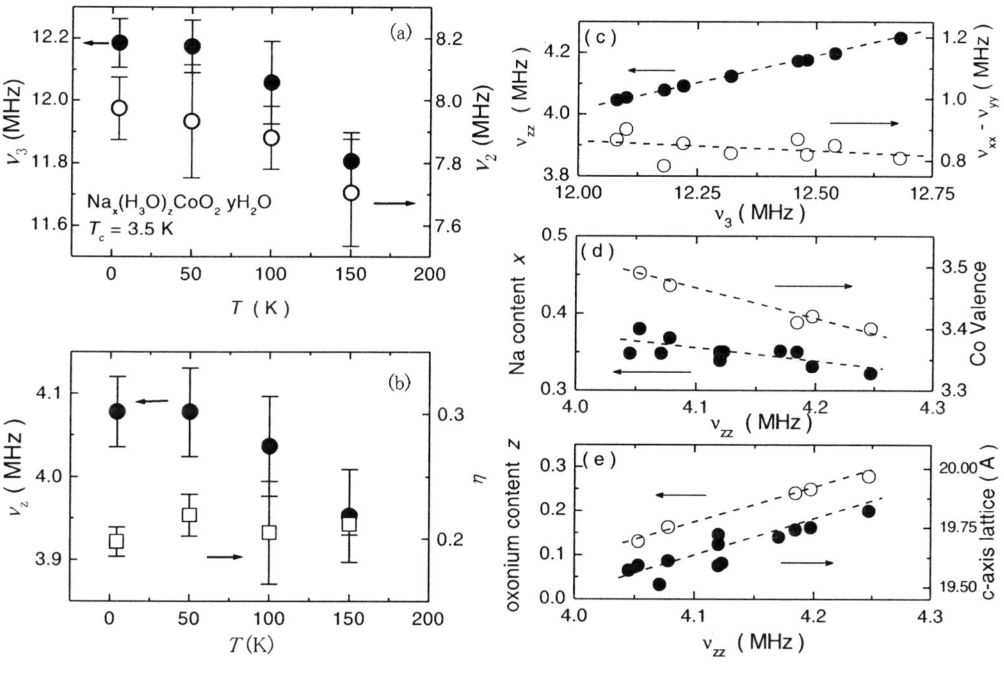

Figure 19. (a) Temperature dependency of ν_3 and ν_2 measured for sample No. 11. (b) Temperature dependency of ν_{zz} and η, evaluated from ν_2 and ν_3 displayed in (a). (c) ν_{zz} and $\nu_{xx} - \nu_{yy}$ as a function of ν_3. (d, e) Various physical parameters (Na content: x, Co valence: v, oxonium content: z, and c-axis lattice parameter) plotted against ν_{zz}. [84]

ν_2 and ν_3 were measured at various temperatures for sample No. 11, and at low temperature for several samples. As described in equation (2), ν_{zz} and η are obtained separately from the observation of two resonant frequencies. The temperature dependence of ν_{zz} and η was investigated up to 150 K for sample No. 11. As seen in Fig. 19 (a), both ν_2 and ν_3 are nearly constant below 50 K, and shift toward a lower frequency with increasing temperature. The temperature dependencies of ν_{zz} and η, which are derived from the temperature variations of ν_2 and ν_3, are shown in Fig. 19 (b). In contrast to the considerably strong temperature dependency of ν_{zz}, η is independent of temperature and is estimated to be $\sim$ 0.2. It was found that the temperature dependency of ν_{zz} is similar to that of the c-axis lattice parameter measured by the neutron diffraction method [85]. This suggests that the temperature dependency of ν_{zz} is mainly determined by that of the c-axis parameter.

The values of ν_{zz} and $\eta \equiv (\nu_{xx} - \nu_{yy})/\nu_{zz}$ were evaluated for all the samples at low temperature from the Co-NQR spectra of each sample. The values of ν_{zz} and $\nu_{xx} - \nu_{yy}(= \eta \times \nu_{zz})$ are plotted as a function of ν_3 in Fig. 19 (c). In contrast to the linear relationship between ν_{zz} and ν_3, the value of $\nu_{xx} - \nu_{yy}$ is nearly independent of ν_3. This indicates that the sample dependency of the NQR frequency ν_3 in BLH compounds originates mainly from the variation in ν_{zz}, which is caused by several factors, such as the hole density at the Co $3d$ orbitals and the crystal structure, particularly the distance between O^{2-} and Co^{s+} ions. To investigate the main factor that determines ν_{zz}, the values of ν_{zz} were plotted against various physical parameters listed in Table 2, and are shown in Figs. 19 (d) and (e). Among the physical quantities that seem to be related to ν_{zz}, it was pointed out that the c-axis parameter would be an important factor, because a relationship between the c-axis parameter and ν_{zz} was also observed from the temperature dependence of ν_{zz} for one sample. The oxonium content is considered to be the cause of the elongation of the c axis. Although the Co valence seems to be weakly correlated to ν_{zz}, it was considered that this relationship does not originate from the variation in hole density at the Co $3d$ orbitals from the following discussions.

In general, there are two sources for the EFG. One is the electrical charge of ions on the lattice surrounding Co sites, $V_{\rm lat}$, and the other is the intra-site contribution, $V_{\rm int}$, which is produced by the on-site Co-$3d$ holes. If we assume that the EFG is determined only by the on-site effect, ν_{zz} at the Co site, which is calculated on the basis of the on-site Co-$3d$ hole component, should decrease with the decrease in the on-site hole number, which is proportional to the Co valence s. This hypothesis is inconsistent with the experimental results displayed in Fig. 19 (d); ν_{zz} at the Co site increases with decreasing s. Alternatively, the sample dependence of ν_{zz} can be qualitatively explained by considering that ν_{zz} is mainly dependent on $V_{\rm lat}$. According to the point-charge calculations, $V_{\rm lat}$ was revealed to increase with compression of the CoO_2 block layer along the c axis, which is induced by water intercalation. It has been pointed out that the thickness of the CoO_2 block layer in the BLH compounds with large y values is thinner than that of the MLH compounds [46]. It seems that the absolute value of ν_{zz} is predominantly determined by the on-site hole contribution, and the change of ν_{zz} in samples is predominated by the lattice contribution.

It should be noted that η takes a non-zero value in the CoO_2 triangular lattice, because η is zero in a perfect triangular lattice, on which the EFG becomes isotropic along the x and y directions. The finite value of η suggests the presence of small distortions in the CoO_2 layer. However, a precise calculation of the EFG originating from on-site holes revealed that when holes occupy the e_g' orbital, η is finite without the presence of distortions. On the basis of the calculation, e_g' hole occupation $w^2 = 0.25$ and total hole number $n_{3d} = 0.31$ are derived from the experimental values of $\eta = 0.2$ and $\nu_{zz} = 4.1$ MHz. These values imply that the ratio between $n_{a_{1g}}$ and $n_{e_g'}$ is approximately 3:1, and the valence of Co is +3.31, which is close to that measured directly ($s \sim 3.45$). It is noted that the non-negligible η is not solely due to the Na-ordering effect or the distortion in the CoO_2 plane, but also to the presence of holes in the e_g' orbitals, to some degree.

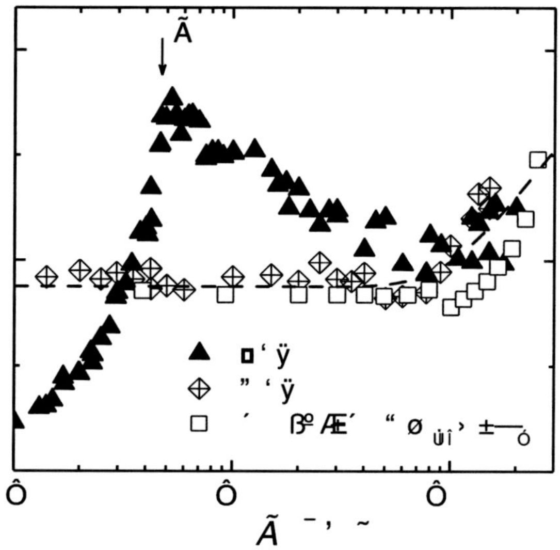

Figure 20. Temperature dependence of $1/T_1T$ in the superconducting BLH compound (No. 1), the MLH compound, and anhydrous cobaltate $Na_{0.35}CoO_2$. The experimental data for the anhydrous cobaltate was reported by Ning and Imai [86]. The dashed line is the fitted curve for the temperature dependence of $1/T_1T$ in the MLH and anhydrous compounds. [84]

ÏŰÚ ” ø a·· ◻¥ ···´ø·· ± Õ´ ªÆ± º´ ·· ø º ” ø a·· ø¥ß —Æª Æº Õø ¥

With the NQR technique, low energy magnetic fluctuations are investigated from the measurement of $1/T_1T$, because $1/T_1T$ is proportional to the integral of the dynamical susceptibility in q space, which is related to the low energy magnetic fluctuations. In this section, we study the magnetic fluctuations in BLH compounds and discuss the relationship between magnetic fluctuations and superconductivity.

Firstly, $1/T_1T$ for the superconducting BLH sample (No. 1) is compared with that for the non-superconducting MLH sample [23] and anhydrous $Na_{0.35}CoO_2$ [86]. Figure 20 shows $1/T_1T$ for these samples. $1/T_1T$ for the MLH sample is almost the same as that for the anhydrous sample, both of which show Korringa behavior, which is characterized by constant $1/T_1T$ in the low-temperature region. In contrast, $1/T_1T$ for the superconducting BLH compound is enhanced below 70 K down to T_c, although above 100 K $1/T_1T$ is nearly the same as that for the other two samples. The results from the comparison of the BLH compound with the MLH and anhydrous compounds suggest that low-temperature enhancement of $1/T_1T$ plays an important role in the occurrence of superconductivity. In the temperature region greater than 100 K, $1/T_1T$ increases for all the samples as the temperature increases, which is reminiscent of the spin-gap behavior in cuprate superconductors. It is considered that the spin-gap behavior of $1/T_1T$ in the high-temperature region is related to the pseudogap revealed by a photoemission experiment [87]. A similar spin-gap behavior was reported for anhydrous Na_xCoO_2 with $x \leq 0.7$ [88, 89]. The spin-gap behavior at

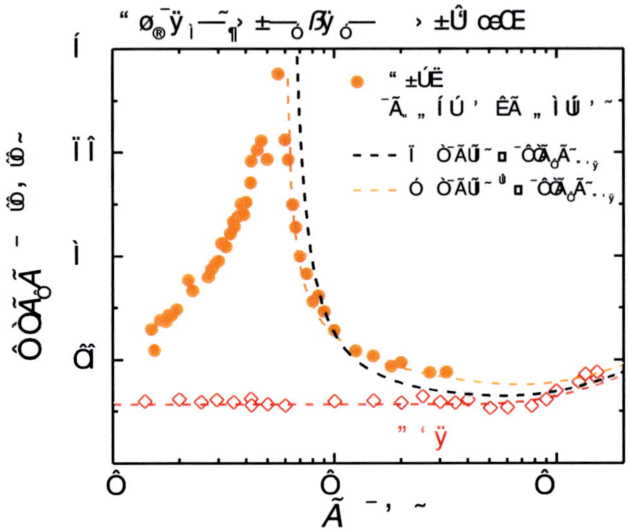

Figure 21. Temperature dependence of $1/T_1T$ in a magnetically ordered sample (No. 8). A prominent peak is observed at $T_M = 6$ K. The orange and black dashed lines show the fitting curves for the temperature dependence of $1/T_1T$ above T_M. [84]

high temperature and the Korringa behavior at low temperature of the MLH and anhydrous samples, shown in Fig. 20, are consistently fitted by

$$\left(\frac{1}{T_1T}\right)_{\mathrm{MLH}} = 8.75 + 15\exp\left(-\frac{\Delta}{T}\right) \quad (\mathrm{sec}^{-1}\mathrm{K}^{-1}) \tag{8}$$

with $\Delta = 250$ K. The value of Δ is in good agreement with a pseudogap having an energy scale of $\sim$ 20 meV, which was reported in Ref. [87]. It seems that the spin-gap behavior is unrelated to superconductivity, because the behavior is commonly observed in both superconducting and non-superconducting cobaltate compounds. Rather, we suggest that the low-temperature enhancement of $1/T_1T$ is strongly related to superconductivity. The properties of the low-temperature spin dynamics will be discussed quantitatively.

In order to characterize the magnetic fluctuations in the BLH compounds, we first analyze the temperature dependence of $1/T_1T$ in a magnetically ordered sample (No. 8), as shown in Fig. 21. $1/T_1T$ for the MLH compound is shown together in the figure. It is noteworthy that the most prominent divergence of $1/T_1T$ at $T_M \sim 6$ K is observed in sample No. 8, which has the highest ν_{zz} value among the twelve samples. This result confirms the relationship between the NQR frequency and the magnetic fluctuations in BLH compounds. It was found that a fairly good fit to the temperature dependence of $1/T_1T$ could be obtained by a function consisting of two contributions, expressed as

$$\left(\frac{1}{T_1T}\right)_{\mathrm{No.8}} = \left(\frac{1}{T_1T}\right)_{\mathrm{MLH}} + \frac{20}{\sqrt{T-6}} \quad (\mathrm{sec}^{-1}\mathrm{K}^{-1}). \tag{9}$$

The first term on the right-hand side is the contribution from the spin dynamics with weak temperature dependence, which is observed even in the MLH and anhydrous compounds. The second term represents the magnetic contribution, which gives rise to magnetic

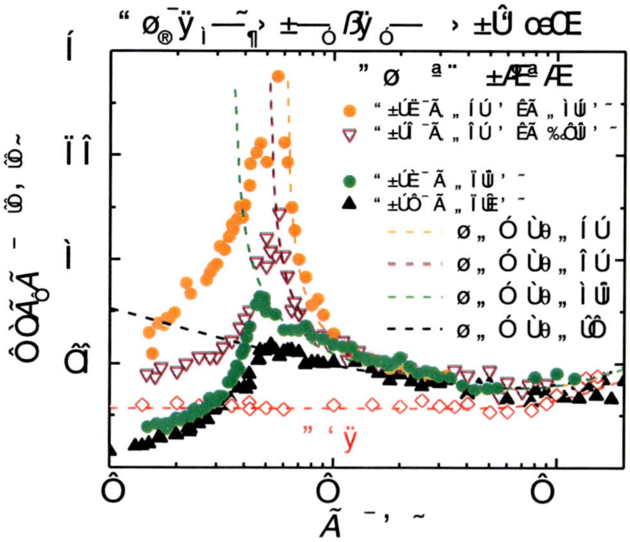

Figure 22. Sample dependence of $1/T_1T$ for the samples with ν_3 higher than 12.3 MHz. Magnetic ordering is observed in samples No. 5 and No. 8. The dashed lines are the fitting curves, and the fitting parameters for each sample are shown in the figure. [84]

ordering. The functional form of the magnetic contribution, $1/\sqrt{T - T_{\rm M}}$, is the temperature dependence anticipated in a three-dimensional itinerant antiferromagnet, using the framework of the self-consistent renormalization (SCR) theory [81]. As explained in §4..1, the functional form for a two-dimensional (2-D) antiferromagnet, $1/(T - T_{\rm M})$, would be more appropriate if the two-dimensional crystal structure of the BLH compound were taken into account. However, the experimental data does not fit well with the temperature dependence of a 2-D antiferromagnet, as shown in Fig. 21. Considering the exponent of the fitting function, the magnetic correlations might be determined as three-dimensional just above $T_{\rm M}$, due to the magnetic ordering, even if the dominant fluctuations are of a 2-D nature. Therefore, a fitting function with two fitting parameters, a and θ was adopted to understand the systematic change of the magnetic fluctuations,

$$\left(\frac{1}{T_1T}\right) = \left(\frac{1}{T_1T}\right)_{\rm MLH} + \frac{a}{\sqrt{T - \theta}}. \tag{10}$$

Figure 22 shows the temperature dependence of $1/T_1T$ for sample Nos. 5, 7, and 8, together with that for sample No. 1 and the MLH compound. The NQR frequencies ν_3 of sample Nos. 5, 7 and 8 are higher than 12.3 MHz, which is the ν_3 of sample No. 1. As shown in Fig. 22, the systematic change in the temperature dependence of $1/T_1T$ is consistently explained by adjusting the ordering temperature θ without changing the coefficient a. It should be noted that θ for sample No. 1 with $T_{\rm c} \sim 4.7$ K is evaluated to be -1 K, which indicates that the system is very close to the quantum critical point $(\theta = 0)$.

Next, the temperature dependence of $1/T_1T$ in the superconducting samples was analyzed using the same fitting function expressed in equation (10) for consistency. Figure 23 shows the temperature dependency of $1/T_1T$ for superconducting samples Nos. 1, 3

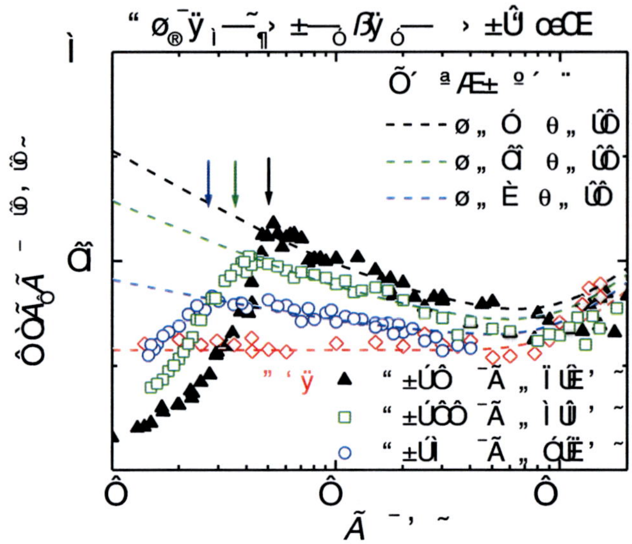

Figure 23. Sample dependence of $1/T_1T$ for the samples with ν_3 lower than 12.3 MHz. All the samples are superconducting. The parameters for the fitting curves of each sample are represented in the figure. [84]

and 11, together with that for the MLH sample. The temperature dependency of $1/T_1T$ for these samples shows a satisfactory fit using the same function as that described for the magnetically ordered samples, and using the parameters displayed in the figure. Even for the samples with various values of T_c, the experimental results are explained by the same $\theta = -1$ K. This result is in contrast to that for the magnetically ordered samples shown in Fig. 22, in which a was invariant.

It was noticed that among the superconducting samples, the coefficient a is larger for samples with higher T_c. To explore the relationship between a and T_c, the normalized spin term $F(t)$ is derived from the following formula

$$F(t) = \frac{1}{T_c}\left[\left(\frac{1}{T_1T}\right)_{\mathrm{BLH}} - \left(\frac{1}{T_1T}\right)_{\mathrm{MLH}}\right],$$

where $t = T/T_c$ is the reduced temperature [90]. The spin term was divided by T_c, in order to normalize the absolute value. In Fig. 24, $F(t)$ is plotted against the reduced temperature. All the $F(t)$ curves for the five different superconducting samples are scaled to a single line which is proportional to $1/\sqrt{t+1}$, when the spin term is normalized using T_c. It was found that $F(t)$ follows a universal function, which depends only on the reduced temperature t. Even though the temperature dependency of the spin fluctuations is strongly sample dependent, it is scaled to the universal function, which is characterized by T_c for each sample. This result directly indicates that superconductivity in this compound is related to spin fluctuations. A similar scaling was performed on a high-T_c cuprate superconductor on the over-dope side [91]. The scaling plot shown here is different from that performed in Ref. [91] at the point where the absolute value of the spin term is normalized by T_c. Generally, when the energy distribution of the spin fluctuations is assumed to be of a Lorentzian

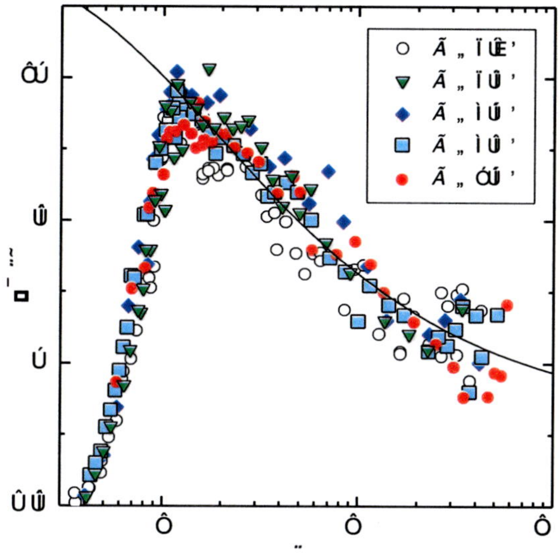

Figure 24. Scaling plot with respect to T_c for five different samples. The definition of the universal function $F(t)$ and reduced temperature t are given in the text. The temperature dependence of five samples is scaled to a single line. [90]

form, $1/T_1T$ is proportional to the value of $\chi_{\tilde{q}}/\Gamma_{\tilde{q}}\xi^3$, where $\chi_{\tilde{q}}$, $\Gamma_{\tilde{q}}$ and ξ are the weight of the spin fluctuations at dominant q vector, $\tilde{q}$, the characteristic energy of the fluctuations, and the magnetic correlation length, respectively. The temperature dependence of $F(t)$ discussed above is accounted for by the temperature dependence of $\Gamma(t)$, which depends only on t. For the BLH cobaltate compounds, the spin term has to be normalized, because $\chi_{\tilde{q}}$ is also sample dependent, due to the difference in the Fermi surfaces that contribute to magnetic fluctuations. Because the absolute value is normalized by T_c, we suggest that $\chi_{\tilde{q}}$ is also related to T_c. It is considered that the superconductivity is stabilized when the spin fluctuations possess a large weight in the low energy region.

Co-NMR measurements were also performed in strong magnetic fields, in order to study the low-temperature magnetic fluctuations below T_c. External magnetic fields were applied precisely parallel to the ab plane. The temperature dependence of $1/T_1T$ under various magnetic field strengths is shown in Fig. 25. Above 5 K, the temperature variation of $1/T_1T$ is independent of the external magnetic field. Equation (10) was employed to fit the temperature dependency of $1/T_1T$ obtained from the Co-NMR measurements, and the resulting fitting parameters are $a = 30$ and $\theta = -1$ K. The difference of coefficient a from the NQR measurements is due to the angle dependence of the hyperfine coupling constant. The principal axis of the Co-NMR measurement is in the ab plane, because of the Zeeman interaction, while that for Co-NQR measurements is along the c axis. The good fit for NMR results confirms the validity of equation (10) for the study of systematical changes of magnetic fluctuation in various BLH samples.

As shown in Fig. 25, $1/T_1T$ increases toward low temperatures in the normal state, and the increase is interrupted by the occurrence of superconductivity. In the field-induced normal state, $1/T_1T$ continues to increase, and follows the fitting curve defined above 5

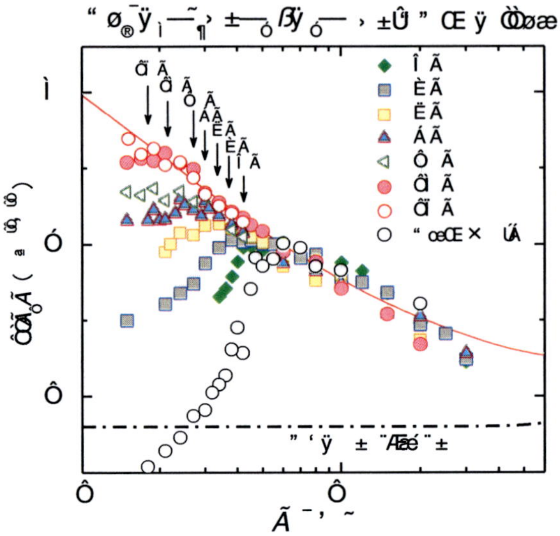

Figure 25. The temperature dependency of $1/T_1T$ in various magnetic fields. The red solid curve is the fitted curve for the temperature dependency of $1/T_1T$ in the normal state. The arrows indicate the temperature at which $1/T_1T$ deviates from the normal-state temperature variation.

K. This is a significant contrast to Fermi-liquid (T_1T = const.) behavior. The continuous increase of $1/T_1T$ indicates that magnetic fluctuations continue to be enhanced down to $T_c(H)$ without displaying Fermi-liquid behavior. This non-Fermi liquid behavior, which is understood by equation (10) with $\theta = -1$ K, strongly suggests that magnetic fluctuations possess a quantum critical character. This behavior is reminiscent of the specific-heat results for $CeCu_2Si_2$ [92] and $CeCoIn_5$ [93] in magnetic fields, both of which are situated near a quantum critical point.

Now we will briefly comment on the structure of magnetic fluctuations in q space. Although $1/T_1T$ is the integral of low energy magnetic fluctuations in q space, and the magnitude of $1/T_1T$ is insensitive to the q structure of magnetic fluctuations, we can approximately deduce the q structure from the $1/T_1T$ measurement at a ligand site, such as the O site, and from the Knight shift measurement. It was suggested that the q-dependence of the magnetic fluctuations has a peak in the small q-vector region, since $1/T_1T$ at the O site is scaled to that at the Co site [14, 88]. If there were magnetic correlations with $Q = (\pi, \pi)$ far from the Brillouin-zone center, the magnetic correlations would be filtered out at the O site due to the symmetry of the O position. In addition, the temperature dependence of the spin part of the Knight shift shows a gradual increase with decreasing temperature but is not proportional to the temperature dependence of $1/T_1T$ below 30 K; therefore, the low-temperature magnetic fluctuations below 70 K are considered to have incommensurate structure with the peak at $q \sim 0$ other than $q = 0$ [14].

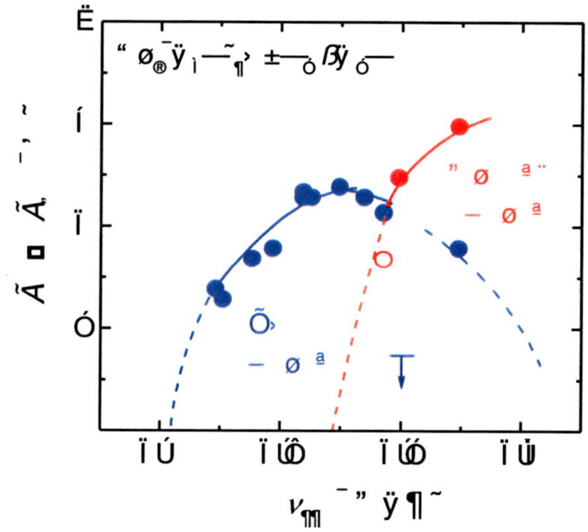

Figure 26. Phase diagram for $Na_x(H_3O)_zCoO_2 \cdot yH_2O$, in which T_c and T_M are plotted against ν_{zz}. The sample with the highest ν_{zz} shows a magnetic anomaly at 6 K and superconductivity at 3.5 K. The fitting parameter θ for sample No. 7 is also shown by the red open circle. [84]

ÏÛÚ – ø ª ‹ ø ÆB ±Æ' ø$_x$ ÿ 3—$_z$› ±—2 · yÿ 2—

On the basis of the studies using NQR measurements, it was found that the NQR frequency of the Co site is very sample sensitive, and reflects the important change of the Co-$3d$ electronic state. In a previous paper, we plotted T_c for each sample against the NQR frequencies of the $m = \pm 5/2 \leftrightarrow \pm 7/2$ transitions ν_3, and developed a phase diagram for $Na_x(H_3O)_zCoO_2 \cdot yH_2O$ [82, 94].

Figure 26 shows the revised phase diagram, which was constructed by using the experimental data from twelve samples, where the horizontal axis is ν_{zz}. It was found that the superconducting phase forms a "dome" shape around $\nu_{zz} \sim 4.15$ MHz, and that the highest T_c seems to occur at the point where T_M becomes zero, ÚÚ at the quantum critical point. The phase diagram we have developed is reminiscent of that for Ce-based heavy-Fermion superconductors, where antiferromagnets change to superconductors by application of pressure [95, 96]. We suggest that the superconductivity of BLH cobaltate is induced by the quantum critical fluctuations, similar to that for heavy-Fermion superconductivity [95].

In heavy-Fermion superconductors, the coupling between f electrons and s electrons is controlled by the application of pressure. As a result, the bandwidth of the conduction electrons increases with the application of pressure. We discuss the physical meaning of the horizontal axis used for the phase diagram of the BLH cobaltate superconductor. Although ν_{zz} is determined by both V_{lat} and V_{int} contributions, the sample dependence of ν_{zz} is mainly due to the variation of V_{lat}, which is related to the subtle distortion of the CoO_2 block layer along the c axis. This crystal distortion splits the t_{2g} levels of the Co-$3d$ orbitals into doublet e'_g states and a singlet a_{1g} state. Therefore, ν_{zz} is regarded as the physical parameter related to the crystal-field splitting between a_{1g} and e'_g states. According to the band calculation, these two different bands cross the Fermi energy and form two kinds of Fermi surfaces (FSs); six hole pockets around the K points (e'_g-FS) and a large hole Fermi surface around the Γ point (a_{1g}-FS) [67]. The a_{1g}-FS was observed from angle-resolved photoemission spectroscopy (ARPES) experiments [97,98], but e'_g-FS has not been detected yet. When e'_g levels shift upward and reach the Fermi energy to form e'_g-FSs due to the increase of crystal-field splitting between e'_g and a_{1g} bands originating from trigonal crystal distortions, it is considered that hole-like e'_g-FSs appear and increase the volume in BLH compounds.

The above scenario explains the evolution of the magnetic fluctuations from anhydrous $Na_{0.35}CoO_2$ and the MLH compound to the superconducting BLH compounds and further to magnetically ordered BLH compounds. We pointed out that the magnetic fluctuations in the BLH compounds are interpreted in terms of multiband properties, because two different temperature dependencies are observed for $1/T_1T$. The temperature dependences of $1/T_1T$ in the anhydrate and MLH compounds, which is expressed as $(1/T_1T)_{\mathrm{MLH}}$ in equation (10), are nearly sample independent, and are governed by the robust a_{1g}-FS. The Korringa behavior and the pseudogap behavior of $(1/T_1T)_{\mathrm{MLH}}$ are supported by a theoretical study concerning the properties of the magnetic fluctuations originating from the a_{1g}-FS [70]. The low-temperature behavior of $1/T_1T$ in the BLH compounds, which is expressed as $a/\sqrt{T-\theta}$, is considered to be due to the e'_g-FSs. From the scaling plot displayed in Fig. 24, the increase in the spin term is accounted for by the appearance and the increase of a new Fermi surface, which is anticipated to be e'_g-FSs. Therefore, when the area of the e'_g-FSs is increased by crystal-field splitting, $1/T_1T$ is enhanced at low temperature due to the dominance of the spin term. Superconductivity is considered to occur with the appearance of the e'_g-FSs and the transition temperature increases with the dominance of magnetic fluctuations originating from the e'_g-FSs. Although the magnetic fluctuations for sample Nos. 5, 7, and 8, with ν_{zz} higher than 4.18 MHz are stronger than those for sample No. 1 with $T_{\mathrm{c}} = 4.7$ K, the superconducting transition temperature of these samples are lower than 4.7 K, and superconductivity is not observed down to 1.5 K for sample No. 5. In these samples, magnetic ordering is observed at T_{M}. (Magnetic ordering is expected in No. 7 when superconductivity is suppressed by a magnetic field.) When magnetic ordering occurs in these samples, the volume of the FS responsible for superconductivity decreases due to the opening of the magnetic gap associated with the magnetic ordering. The suppression of T_{c} in the higher-ν_{zz} samples is interpreted in this way.

The experimental results given here seem to be consistently explained by the theoretical scenario developed by Yanase ª¨ ø¥Úand Mochizuki ª¨ ø¥Úin which magnetic fluctuations with a small characteristic wave vector q are induced by the nesting of the e'_g-FSs [69,99]. Although ferromagnetic fluctuations with the peak at $q = 0$ were excluded from the comparison between bulk susceptibility χ_{bulk} and $1/T_1T$ in the superconducting compounds, the magnetic fluctuations with $q \sim 0$ seem to be consistent with the experimental results. Detailed inelastic neutron-scattering measurements are required, in order to reveal the character of the magnetic fluctuations in the superconducting samples. In addition, a detailed ARPES study on superconducting $Na_x(H_3O)_zCoO_2 \cdot yH_2O$ is also desired, although the absence of the e'_g-FS has been reported for anhydrous $Na_{0.35}CoO_2$. We suggest, on the basis of the $1/T_1T$ results shown in Fig. 20, that the FSs of the BLH compounds are quite different from those of the anhydrate and MLH compounds.

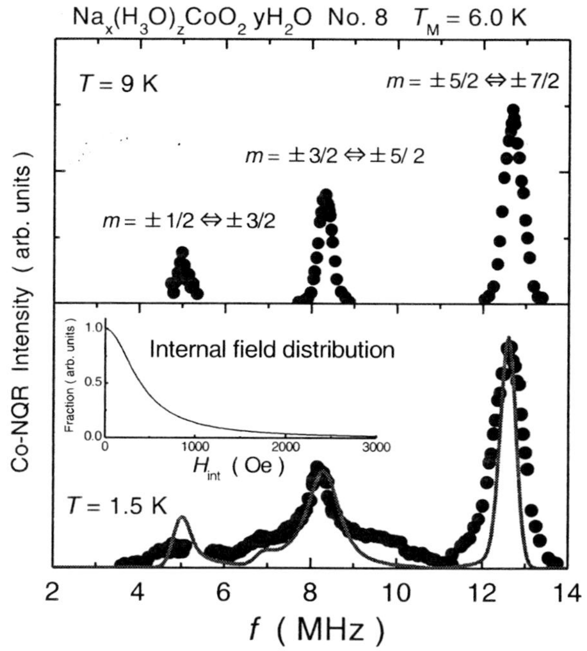

Figure 27. Co-NQR spectra for sample No. 8. The upper figure shows the spectrum for the normal state, and the lower figure shows that for the magnetically ordered state. The red line is the calculated NQR spectrum, assuming the internal field distribution shown in the inset. [84]

4.6. Relationship between Superconductivity and Magnetism

Magnetic ordering was found under zero magnetic field for sample Nos. 5 and 8, whose NQR frequency was higher than the superconducting sample with $T_c = 4.7$ K (No. 1) [9]. As shown in Fig. 21, a prominent peak of $1/T_1T$ was observed at T_M in these samples. Below T_M, the Co-NQR spectra become broad due to the appearance of an internal field at the Co site, which originates from the Co-$3d$ spins. The Co-NQR spectra above and below T_M are displayed in Fig. 27. The magnitude of the internal field is considered to be distributed, because of the broad spectral structure below T_M. In the case where a well-defined internal field appears in the magnetically ordered state, the NQR spectra would split or shift, rather than broaden. The broad NQR spectra are approximately explained assuming a certain type of internal field distribution.

When the Co nuclear spins, having electric quadrupole moment, are in a small magnetic field, the Zeeman interaction is added to the nuclear Hamiltonian described by equation (1). The Zeeman interaction is expressed as

$$\mathcal{H}_Z = -\gamma_n \hbar I \cdot H_{int},$$

where H_{int} is the internal field at the Co nuclear site. Firstly, we pointed out that the line width of the NQR peak arising from the highest-frequency transition $m = \pm 7/2 \leftrightarrow \pm 5/2$ is not largely changed by the appearance of the internal fields, while the line width of the

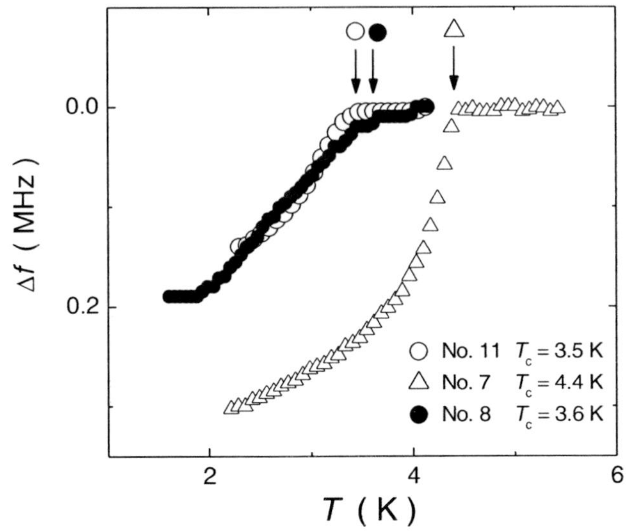

Figure 28. The results of ac-susceptibility measurements for samples No. 7, 8, and 11, using the *in situ* NQR coil. The superconducting transition in sample No. 8 is broad, but the Meissner signal of sample No. 8 is comparable to that of sample No. 11, which has a similar T_c. [84]

middle transition $m = \pm 5/2 \leftrightarrow \pm 3/2$ below T_M becomes more than twice as broad as that in the normal state. This is because the internal fields are directed perpendicular to the *c* axis, which is the principal axis of the EFG. In the case where the internal fields are parallel to the c axis, the highest-frequency transition should be strongly affected, because the magnetic quantum numbers m of the nuclear states concerning this transition are the largest. We calculated the energy levels of each nuclear state for internal fields which are directed perpendicular to the c axis by solving the total Hamiltonian $\mathcal{H} = \mathcal{H}_Q + \mathcal{H}_Z$, and revealed that the middle transition becomes the broadest under this condition. We considered an internal field distribution, shown in the inset of Fig. 27, and simulated the NQR spectra below T_M using the distribution. A good fit was achieved for the simulated spectra, shown by the red line in Fig. 27, to the experimental results. This indicates the appearance of a distributed internal field in the magnetically ordered state. From the internal field distribution, the magnetic structure is anticipated to be of incommensurate spin density wave type, which is consistent with the results of $1/T_1T$ measurements.

In addition to the magnetic ordering, the Meissner shielding effect was observed in sample No. 8 below 3.5 K. The superconducting transition of sample No. 8 is broader than that of superconducting samples with $T_c \sim 4.4$ K. It is important to reveal whether the magnetic and superconducting phases coexist, by microscopic or macroscopic levels. From the observation of Meissner shielding using an *in situ* NQR coil, a major part of sample No. 8 seems to be superconducting because the magnitude of the Meissner shielding is comparable to that in other samples, as shown in Fig. 28. However, from the $1/T_1T$ measurement at 12.69 MHz, the clear anomaly associated with the superconducting transition cannot be detected at 3.5 K. Taking into account that the static moments are distributed in the magnetically ordered state, we consider that superconductivity and magnetism coexist in an inhomogeneous magnetic structure; superconductivity occurs in the smaller magnetic moment region and magnetism occurs in the larger moment region. Such coexisting states are observed in $CeCu_2(Si_{1-x}Ge_x)_2$ with small Ge concentration ($0 < x < 0.05$) [100]. We suggest that the inhomogeneous coexistence is characteristic of compounds whose T_M is higher than T_c. To investigate the relationship between superconductivity and magnetism in $Na_x(H_3O)_zCoO_2 \cdot yH_2O$, further experiments are still required. In particular, pressure experiments on sample No. 8 are considered to be very important.

5. Conclusion

We have constructed phase diagrams of $Na_x(H_3O)_zCoO_2 \cdot yH_2O$ compound with the BLH structure from chemical composition analyses and microscopic NQR and NMR measurements. These phase diagrams confirmed that a magnetically ordered state exists near the superconducting phase.

In the chemical phase diagram shown in Fig. 12, superconducting and magnetic phases are described in the $x - s$ plane. The superconducting transition temperatures of the BLH samples were found to depend on two control parameters, *i.e.* Na content x and the Co valence s. Previously, there have been many tuning parameters used to synthesize strictly controlled samples due to the complex sample-preparation conditions, which are characteristic of soft chemical procedures. We extracted two important parameters, x and s, for the control of this compound, and explained that Na content can be manipulated by using the advanced soft chemical procedure that was introduced in §3.. With this method, it becomes possible to carry out measurements on a properly controlled sample, although it is still difficult to control the Co valence. For the measurement of physical properties, attention should be given not only to the Na content, but also to the Co valence of the sample.

The phase diagram constructed using the results of NQR measurements (Fig. 26) also explains the variation of superconducting transition temperatures in samples. The NQR measurements concluded that the phase diagram is interpreted with a single parameter ν_{zz}, which is related to the subtle distortions of the CoO_2 layers. This phase diagram does not contain information regarding the chemical composition of each sample, especially the composition of H_2O-Na-H_2O block layers. However, the superconducting and magnetic properties are described in the NQR phase diagram, because the NQR frequency is the parameter which reflects the electronic state at the Co site, and thus reflects the strength of magnetic fluctuations. The relationship between the NQR frequency and magnetic fluctuations was investigated using $1/T_1T$ measurements of each sample. From the fact that superconducting properties in $Na_x(H_3O)_zCoO_2 \cdot yH_2O$ are clearly related to the strength of magnetic fluctuations, we consider that the superconductivity is induced by magnetic fluctuations that possess a quantum critical character.

NMR measurements under strong magnetic fields revealed that the quantum critical point seems to exist in the sample with an NQR frequency $\nu_{zz} \sim 4.15$ MHz. The sample with ν_{zz} higher than this frequency is in a more strongly correlated state, where magnetic ordering occurs below 6 K, and the sample with ν_{zz} lower than this frequency is in a less correlated state, where the Fermi-liquid state is realized.

Comparing the NQR phase diagram with the chemical phase diagram, the Fermi-liquid state with low NQR frequency corresponds to a large Na content and a large Co valence region. The magnetically ordered state with high NQR frequency corresponds to low Na content and a low Co valence region. We consider that the ν_{zz} axis in the phase diagram constructed from NQR measurements corresponds to the axis perpendicular to the phase boundary between superconductivity and magnetism in the $x - s$ phase diagram shown in Fig. 12.

In this article, we investigated BLH $Na_x(H_3O)_zCoO_2 \cdot yH_2O$ from chemical and microscopic viewpoints, and discussed the relationship between magnetism and superconductivity observed in BLH compounds with various chemical composition. The phase diagram we constructed is reminiscent of those for unconventional superconductors, such as the cuprate and heavy-Fermion superconductors. From the experimental results, we strongly suggest that the superconductivity in $Na_x(H_3O)_zCoO_2 \cdot yH_2O$ is realized near the quantum critical point, and that the superconductivity is induced by magnetic fluctuations.

Acknowledgments

We would like to extend considerable gratitude to Dr. Takada and Dr. Sasaki (NIMS) for their experimental support, and to Mr. Takenouchi and Mr. Kosuda for the chemical and physical analyses of the compositions. We thank Prof. Shimoyama (University of Tokyo) for allowing us to observe and refer to unpublished TG data. We also thank K. Yoshimura, C. Michioka, Y. Itho H. Ohta, and Y. Maeno for support in the experiments and for valuable discussions, and thank S. Fujimoto, K. Yamada, Y. Yanase, M. Mochizuki, and M. Ogata for useful discussions. This study is partially supported by Grants-in-Aid for Scientific Research from the Japan Society for the Promotion of Science (JSPS) (Nos. 16340111 and 18340102) and the Ministry of Education, Culture, Sports, Science and Technology (MEXT) of Japan (No. 16076209), and by CREST of the Japan Science and Technology Agency (JST) and the 21COE program on "Center for Diversity and Universaliti in Physics" form MEXT.

References

[1] Takada, K.; Sakurai, H.; Takayama-Muromachi, E.; Izumi, F.; Dilanian, R. A.; Sasaki, T. *Nature* 2003, 422. 53.

[2] Li, S. Y.; Taillefer, L.; Hawthorn, D. G.; Tanatar, M. A.; Paglione, J.; Sutherland, M.; Hill, R. W.; Wand, C. H.; Chen, X. H. *Phy. Rev. Lett.* 2004, 93. 056401.

[3] Bednorz, J. G.; Muller, K. A. *Z. Physik B* 1986, 64. 189.

[4] Maeno, Y.; Hashimoto, H.; Yoshida, Y.; Nishizaki, S.; Fujita, T.; Bednorz, J. G.; Lichtenberg, F. *Nature* 1994, 372. 532.

[5] Schaak, R. E.; Klimczuk, T.; Foo, M. L.; Cava, R. J. *Nature* 2003, 424. 527.

[6] Milne, C. J.; Argyriou, D. N.; Chemseddine, A.; Aliouane, N.; Veira, J.; Landsgesell, S.; Alber, D. *Phys. Rev. Lett.* 2004, 93. 247007.

[7] Sakurai, H.; Takada, K.; Sasaki, T.; Takayama-Muromachi, E. *J. Phys. Soc. Jpn.* 2005, 74. 2909.

[8] Sakurai, H.; Tsujii, N.; Suzuki, O.; Kitazawa, H.; Kido, G.; Takada, K.; Sasaki, T.; Takayama-Muromachi, E. *Phys. Rev. B* 2006, 74. 092502.

[9] Ihara, Y.; Ishida, K.; Michioka, C.; Kato, M.; Yoshimura, K.; Takada, T.; Sasaki, T.; Sakurai, H.; Takayama-Muromachi, E. *J. Phys. Soc. Jpn.* 2005, 74. 867.

[10] Ohta, H.; Michioka, C.; Itoh, Y.; Yoshimura, K. *J. Phys. Soc. Jpn.* 2005, 74. 3147.

[11] Barnes, P. W.; Andeev, M.; Jorgensen, J. D.; Hinks, D. G.; Claus, H.; Short, S. *Phys. Rev. B* 2005, 72. 134515.

[12] Sakurai, H.; Takda, K.; Yoshii, S.; Sasaki, T.; Kindo, K.; Takayama-Muromachi, E. *Phys. Rev. B* 2003, 68. 132507.

[13] Chou, F. C.; Cho, J. H.; Lee, Y. S. *Phys. Rev. B* 2004, 70. 144526.

[14] Ihara, Y.; Ishida, K.; Yoshimura, K.; Takada, K.; Sasaki, T.; Sakurai, H.; Takayama-Muromachi, E. *J. Phys. Soc. Jpn.* 2005, 74. 2177.

[15] Higemoto, W.; Ohishi, K.; Koda, A.; Ranjan, S.; kadono, R.; Ishida, K.; Takada, K.; Sakurai, H.; Takayama-Muromachi, E.; Sasaki, T. *Phys. Rev. B* 2004, 70. 134508.

[16] Chou, F. C.; Cho, J. H.; Lee, P. A.; Abel, E. T.; Matan, K.; Lee, Y. S.*Phys. Rev. Lett.* 2004, 92. 157004.

[17] Cao, G.; Feng, C.; Xu, Y.; Lu, W.; Shen, J.; Fang, M.; Xu, Z. *J. Phys. Condens. Matter* 2003, 15. 519.

[18] Jin, R.; Sales, B. C.; Li, S.; Mandrus, D. *Phys. Rev. B* 2005, 72. 060512.

[19] Lorenz, B.; Cmaidalka, J.; Meng, R. L.; Chu, C. W. *Physica C* 2004, 402. 106.

[20] Oeschler, N.; Fisher, R. A.; Phillips, N. E.; Gordon, J. E.; Foo, M. L.; Cava, R. J. *Chin. J. Phys.* 2005, 43. 574.

[21] Ueland, B. G.; Schiffer, P.; Schaak, R. E.; Foo, M. L.; Miller, V. L.; Cava, R. J. *Physica C* 2004, 402. 27.

[22] Yang, H. D.; Lin, J.-Y.; Sun, C. P.; Kang, Y. C.; Huang, C. L.; Takada, K.; Sasaki, T.; Sakurai, H.; Takayama-Muromachi, E. *Phys. Rev. B* 2005, 71. 020504.

[23] Ishida, K.; Ihara, Y.; Maeno, Y.; Michioka, C.; Kato, M.; Yoshimura, K.; Takada, K.; Sasaki, T.; Sakurai, H.; Takayama-Muromashi, E. *J. Phys. Soc. Jpn.* 2003, 72. 3041.

[24] Fujimoto, T.; Zheng, q. G.; Kitaoka, T.; Meng, R. L.; Cmaidalka, J.; Chu, C. W. *Phys. Rev. Lett.* 2004, 92. 047004.

[25] Sasaki, T.; Badica, P.; Yoneyama, N.; Yamada, K.; Togano, K.; Kobayashi, N. *J. Phys. Soc. Jpn.* 2004, 73. 1131.

[26] Badica, P.; Kondo, T.; Togano, K.; Yamada, K. *J. Phys. Chem. Soli.* 2006, 67. 590.

[27] Motohashi, T.; Ueda, R.; Naujalis, E.; Tojo, T.; Terasaki, I.; Atake, T.; Karppinen, M.; Yamauchi, H. *Phys. Rev. B* 2003, 67. 064406.

[28] Higemoto, W.; Ohishi, K.; Koda, A.; Kadono, R.; Sakurai, H.; Takada, K.; Takayama-Muromachi, E.; Sasaki, T. *Physica B* 2006, 374–375. 274.

[29] Takada, K.; Sakurai, H.; Takayama-Muromachi, E.; Izumi, F.; Dilanian, R. A.; Sasaki, T. *Adv. Mater.* 2004, 16. 1901.

[30] Takada, K.; Osada, M.; Izumi, F.; Sakurai, H.; Takayama-Muromachi, E.; Sasaki, T. *Chem. Mater.* 2005, 17. 2034.

[31] Sakurai, H.; Takada, K.; Sasaki, T.; Izumi, F.; Dilanian, R. A.; Takayama-Muromachi, E. *J. Phys. Soc. Jpn.* 2004, 73. 2590.

[32] Foo, M. L.; Klimczuk, T.; Li, Lu; Ong, N. P.; Cava, R. J. *Soli. Sta. Commun.* 2005, 133. 407.

[33] Mistry, S.; Arnold, D. C.; Nuttall, C. J.; Lappas, A.; Green, M. A. *Chem. Comm.* 2002, 21. 2440.

[34] Motohashi, T.; Naujalis, E.; Ueda, R.; Isawa, K.; Karppinen, M.; Yamauchi, H. *Appl. Phys. Lett.* 2001, 79. 1480.

[35] Sakurai, H.; Takenouchi, S.; Tsujii, N.; Takayama-Muromachi, E. *J. Phys. Soc. Jpn.* 2004, 73.

[36] Park, S.; Lee, Y.; Moodenbaugh, A.; Vogt, T. *Phys. Rev. B* 2003, 68. 180505.

[37] Liu, C.-J.; Liao, C.-Y.; Wang, L.-S.; Su, C.-H.; Neeleshwar, S.; Chen, Y.-Y. *Physica C* 2004, 416. 43.

[38] Liu, C.-J.; Hung, W.-C.; Wang, J.-S.; Liu, C.-J. *Am. Chem. Soc.* 2005, 127. 830.

[39] Braconnier, J.-J.; Delmas, C.; Fouassier, C.; Hagenmuller, P. *Mater. Rev. Bull.* 1980, 15. 1797.

[40] Takada, K.; Fukuda, K.; Osada, M.; Nakai, I.; Izumi, F.; Dilanian, R. A.; Kato, K.; Takata, M.; Sakurai, H.; Takayama-Muromachi, E.; Sasaki, T. *J. Mater. Chem.* 2004, 14. 1448.

[41] Ohta, H.; Michioka, C.; Itho, Y.; Yoshimura, K. *Private Communication* 2006,.

[42] Foo, M. L.; Klimczuk, T.; Cava, R. J. *Mat. Res. Bull.* 2005, 40. 665.

[43] Karppinen, M.; Asako, I.; Motohashi, T.; Yamauchi, H. *Chem. Mater.* 2004, 16. 1693.

[44] Chen, D. P.; Chen, H. C.; Maljuk, A.; Kulakov, A.; Zhang, H.; Lemmens, P.; Lin, C. T. *Phys. Rev. B* 2004, 70. 024506.

[45] Sakurai, H.; Takada, K.; Sasaki, T.; Takayama-Muromachi, E. *E. Chinese J. Phys.* 2005, 43. 566, (cond-mat/0408426).

[46] Sakurai, H.; Takada, K.; Takayama-Muromachi, E. *In Progress in Supreconductivity Research*; Columbus F.; Ed.; Nova Publishers Inc.; in press.

[47] Takada, K.; Sakurai, H.; Takayama-Muromachi, E.; Izumi, F.; Dilanian, R. A.; Sasaki, T. *J. Solid State Chem.* 2004, 177. 372.

[48] Foo, M. L.; Schaak, R. E.; Miller, V. L.; Klimczuk, T.; Rogado, N. S.; Wang, Y.; Lau, G. C.; Craley, C.; Zandbergen, H. W.; Ong, N. P.; Cava, R. J. *Soli. Sta. Commun.* 2003, 127. 33.

[49] Sakurai, H.; Osada, M.; Takayama-Muromachi, E. *Private Communication*.

[50] Iliev, M. N.; Litvinchuk, A. P.; Meng, R. L.; Sun, Y. Y.; Cmaidalka, J.; Chu, C. W. *Physica C* 2004, 402. 239.

[51] Fujita, K.; Mochida, T.; Nakamura, K. *Jpn. J. Appl. Phys.* 2001, 40. 4644.

[52] Wang, H.; Chen, Z.; Jin, Q. *Mater. Lett.* 2005, 59. 3917.

[53] Molenda, J.; Delmas, C.; Dordor, P.; Stoklosa, A. *Solid State Ionics* 1969, 12. 473.

[54] Chou, F. C.; Abel, E. T.; Cho, J. H.; Lee, Y. S. *J. Phys. Chem. Soli.* 2005, 66. 155.

[55] Shimoyama, J. *Private Communication*.

[56] Viciu, L.; Huang, Q.; Cava, R. J. *Phys. Rev. B* 2006, 73. 212107.

[57] Sakurai, H.; Takada, K.; Sasaki, T.; Takayama-Muromachi, E. *J. Mag. Mag. Mat.* 2007, 310 e136.

[58] Nakatsuji, S.; Maeno, Y. *Phys. Rev. Lett.* 2000, 84. 2666.

[59] Leligny, H.; Grebille, D.; Perez, O.; Masset, A. C.; Hervieu, M.; Raveau, B. *Acta Cryst.* 2000, 56B. 173.

[60] Masset, A. C.; Michael, C.; Maignan, A.; Hervieu, M.; Toulemonde, O.; Studer, F.; Raveau, B.; Hejtmanek, J. *Phys. Rev. B* 2000, 62. 166.

[61] Miyazaki, Y.; Onoda, M.; Oku, T.; Kikuchi, M.; Ishii, Y.; Ono, Y.; Morii, Y.; Kajitani, T. *J. Phys. Soc. Jpn.* 2002, 71. 491.

[62] Yamauchi, H.; Sakai, K.; Nagai, T.; Matsui, Y.; Karppinen, M. *Chem. Mater.* 2006, 18. 155.

[63] Ishiwata, S.; Terasaki, I.; Kusano, Y.; Takano, M. *J. Phys. Soc. Jpn.* 2006, 75. 104716.

[64] Shizuya, M.; Isobe, M.; Baba, Y.; Nagai, T.; Matsui, Y.; Takayama-Muromachi, E. *J. Solid. State. Chem* 2006, 179. 3974.

[65] Shizuya, M.; Isobe, M.; Baba, Y.; Nagai, T.; Osada, M.; Kusuda, K.; Takenouchi, S.; Matsui, Y.; Takayama-Muromachi, E. *J. Soli. Sta. Chem* in press,.

[66] Isobe, M.; Onoda, M.; Shizuya, M.; Tanaka, M.; Takayama-Muromachi, E. *J. Phys. Soc. Jpn.* 2007, 76. 014602.

[67] Singh, D. J. *Phys. Rev. B* 2000, 61. 13397.

[68] Kuroki, K.; Onari, S.; Tanaka, Y.; Arita, R.; Nojima, T. *Phys. Rev. B* 2006, 73. 184503.

[69] Mochizuki, M.; Yanase, Y.; Ogata, M. *J. Phys. Soc. Jpn.* 2005, 74. 1670.

[70] Yada, K.; Kontani, H. *J. Phys. Soc. Jpn.* 2005, 74. 2161.

[71] Sakurai, H.; Takada, K.; Sasaki, T.; Takayama-Muromachi, E. *Physica C* 2006, 445-448. 31.

[72] Clogston, A. M. *Phys. Rev. Lett* 1962, 9. 266.

[73] Chandrasekhar, B. S. *Appl. Phys. Lett.* 1962, 1. 7.

[74] Maśka, M. M.; Mierzejewski, M.; Andrzejewski, B.; Foo, M. L.; Cava, R. J.; Klimczuk, T. *Phys. Rev. B* 2004, 70. 144516.

[75] Werthamer, N. R.; Helfand, E.; Hohenberg, P. C. *Phys. Rev.* 1966, 147. 295.

[76] Sun, C. P.; Lin, J.-Y.; Kang, Y. C.; Huang, C. L.; Takada, K.; Sasaki, T.; Sakurai, H.; Takayama-Muromachi, E.; Yang, H. D. *Physica B* 2006, 378-380. 865.

[77] Hanzawa, K.; Komatsu, F.; Yosida, K. *J. Phys. Soc. Jpn.* 1990, 59. 3345.

[78] Zheng, q. G.; Kitaoka, Y.; Ishida, K.; Asayama, K. *J. Phys. Soc. Jpn.* 1995, 64. 2524.

[79] Ray, R.; Ghoshray, A.; Ghoshray, K.; Nakamura, S. *Phys. Rev. B* 1999, 59. 9454.

[80] Mukhamedshin, I. R.; Alloul, H.; Collin, G.; Blanchard, N. *Phys. Rev. Lett.* 2005, 94. 247602.

[81] Moriya, T. *J. Mag. Mag. Mat.* 1991, 100. 261.

[82] Ihara, Y.; Ishida, K.; Michoka, C.; Kato, M.; Yoshimura, K.; Takada, K.; Sasaki, T.; Sakurai, H.; Takayama-Muromachi, E. *J. Phys. Soc. Jpn.* 2004, 73. 2069.

[83] Ihara, Y.; Ishida, K.; Takeya, H.; Michioka, C.; Kato, M.; Itho, Y.; Yoshimura, K.; Takada, K.; Sasaki, T.; Sakurai, H.; Takayama-Muromachi, E. *J. Phys. Soc. Jpn.* 2006, 75. 013708.

[84] Ihara, Y.; Takeya, H.; Ishida, K.; Ikeda, H.; Michioka, C.; Yoshimura, K.; Takada, K.; Sasaki, T.; Sakurai, H.; Takayama-Muromachi, E. *J. Phys. Soc. Jpn.* 2006, 75. 124714.

[85] Lynn, J. W.; Huang, Q.; Brown, C. M.; Miller, V. L.; Foo, M. .; Schaak, R. E.; Jones, C. Y.; Mackey, E. A.; Cava, R. J. *Phys. Rev. B* 2003, 68. 214516.

[86] Ning, F. L.; Imai, T.; Statt, B. W.; Chou, F. C. *Phys. Rev. Lett.* 2004, 93. 237201.

[87] Shimojima, T.; Yokoya, T.; Kiss, T.; Chainani, A.; Shin, S.; Togano, T.; Zhang, C.; Chen, C.; Watanabe, S.; Takada, K.; Sasaki, T.; Sakurai, H.; Takayama-Muromachi, E. *J. Phys. Chem. Soli.* 2006, 67. 282.

[88] Ning, F. L.; Imai, T. *Phys. Rev. Lett.* 2005, 94. 227004.

[89] Yokoi, M.; Moyoshi, T.; Kobayashi, Y.; Soda, M.; Yasui, Y.; Sato, M.; Kakurai, K. *J. Phys. Soc. Jpn.* 2005, 74. 3046.

[90] Ihara, I.; Takaya, H.; Ishida, K.; Michioka, C.; Yoshimura, K.; Takada, K.; Sasaki, T.; Sakurai, H.; Takayama-Muromachi, E. *J. Phys. Chem. Soli.* in press.

[91] Tokunaga, Y.; Ishida, K.; Kitaoka, Y.; Asayama, K. *Soli. Sta. Commun.* 1997, 103. 43.

[92] Gegenwart, P.; Langhammer, C.; Geibel, C.; Helfrich, R.; Lang, M.; Spam, G.; Steglich, F.; Hom, R.; Donnevert, L.; Link, A.; Assmus, W. *Phys. Rev. Lett.* 1998, 81. 1501.

[93] Ikeda, S.; Shishido, H.; Nakashima, M.; Settai, R.; Aoki, D.; Haga, Y.; Harima, H.; Aoki, Y.; Namiki, T.; Sato, H.; Onuki, Y. *J. Phys. Soc. Jpn.* 2001, 70. 2248.

[94] Ishida, K.; Ihara, Y.; Takeya, H.; Michioka, C.; Yoshimura, K.; Takada, K.; Sasaki, T.; Sakurai, H.; Takayama-Muromachi, E. *Physica B* 2006, 378-380. 869.

[95] Mathur, N. D.; Grosche, F. M.; Julian, S. R.; Walker, I. R.; Freye, D. M.; Haselwimmer, R. K. W.; Lonzarich, G. G. *Nature* 1998, 394. 39.

[96] Kawasaki, S.; Mito, T.; Zheng, q. G.; Thessieu, C.; Kawasaki, Y.; Ishida, K.; Kitaoka, Y.; Muramatsu, T.; Kobayashi, T. C.; Aoki, D.; Haga, Y.; Settai, R.; Onuki, Y. *Phys. Rev. B* 2002, 65. 20504.

[97] Hasan, M. Z.; Chuang, H.-D.; Qian, D.; Li, Y. W.; Kong, Y.; Kuprin, A.; Fedorov, A. V.; Kimmerling, R.; Rotenberg, E.; Rossnagel, K.; Hussain, Z.; Koh, H.; Rogado, N. S.; Foo, M. L.; Cava, R. J. *Phys. Rev. Lett.* 2004, 92. 246402.

[98] Yang, H.-B.; Wang, S.-C.; Sekharan, A. K. P.; Matsui, H.; Souma, S.; Sato, T.; Takahashi, T.; Takeuchi, T.; Campuzano, J. C.; Jin, R.; Sales, B. C.; Mandrus, D.; Wang, Z.; Ding, H. *Phys. Rev. Lett.* 2004, 92. 246503.

[99] Yanase, Y.; Mochizuki, M.; Ogata, M. *J. Phys. Soc. Jpn.* 2005, 74. 430.

[100] Kawasaki, Y.; Ishida, K.; Obinata, K.; Tabuchi, K.; Kashima, K.; Kitaoka, Y.; Trovarelli, O.; Geibel, C.; Steglich, F. *Phys. Rev. B* 2002, 66. 224502.

In: New Research on Superconductivity
Editor: Barry P. Martins, pp. 105-141
ISBN: 978-1-59454-197-1

Chapter 3

CONTRIBUTION OF THE ELECTRON-PHONON INTERACTION TO HIGH-T_C SUPERCONDUCTIVITY: TUNNELING STUDY OF $Bi_2Sr_2CaCu_2O_8$

Nobuo Tsuda[1], Daisuke Shimada[2] and Nobuaki Miyakawa[1]
[1]Department of Applied Physics, Tokyo University of Science, 1-3 Kagurazaka, Shinjuku-ku, Tokyo 162-8601, Japan
[2]Department of Physics, Shizuoka University, 836 Ohya, Suruga-ku, Shizuoka 422-8529, Japan

Abstract

The tunneling studies by us are reviewed on the following matters: Determination of the spectral function $\alpha^2F(\omega)$ for slightly overdoped $Bi_2Sr_2CaCu_2O_8$ (Bi2212), good correspondence of $\alpha^2F(\omega)$ with the phonon structures, corresponding phonon modes calculated by Paltzer and De Wette, energy gap $2\Delta(0)$ and critical temperature T_c estimated from $\alpha^2F(\omega)$, layer number dependence of T_c, and unusually large $2\Delta(0)/k_BT_c$ for underdoped Bi2212. It is emphasized that the large $2\Delta(0)/k_BT_c$ for underdoped Bi2212 cannot be explained by a usual electron-phonon (e-ph) interaction alone and that this is as substantial as a high value of T_c for high-T_c superconductivity.

$\alpha^2F(\omega)$ was applied for calculating dc resistivity, spin-lattice relaxation rate, isotope effect on T_c, dispersion relation and damping rate of a quasi-particle, and optical damping rate. They are compared with corresponding experimental ones. The comparison determines what is explained and what is not explained by the e-ph interaction.

Furtherrmore, we discuss a role of spin in high-T_c superconductivity. For explaining the large $2\Delta(0)/k_BT_c$ for underdoped Bi2212, a pair-break model by Cu magnetic moments is proposed. Finally, we propose an experiment for clarifying whether a spin enhancement of the e-ph interaction really occurs or not.

1. Introduction

Since the discovery of high critical temperature (T_c) superconductivity by Bednorz and Müller [1], great efforts have being poured into a clarification of a mechanism of high T_c, and vast amount of experimental results have been accumulated. They indeed explored a very complicated property of electrons (holes), but still there are quite few experiments which directly indicate what the attractive medium between electrons is.

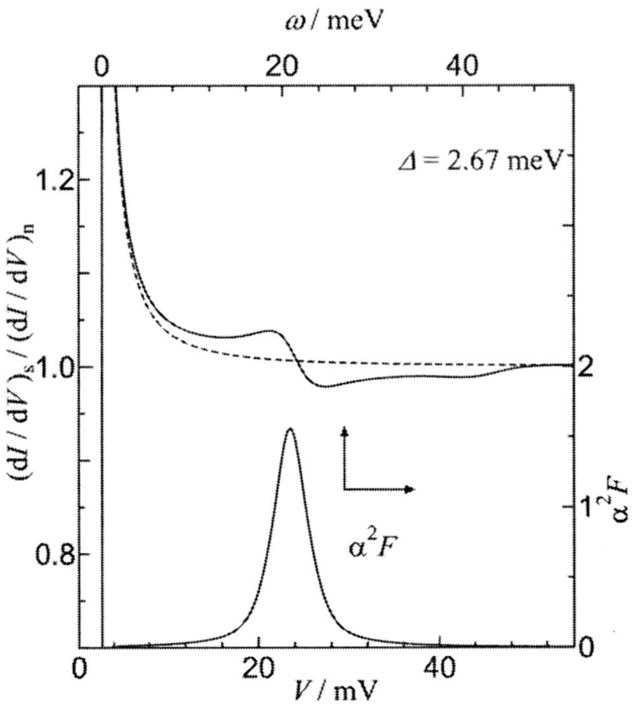

Figure 1. Phonon structure in dI/dV. Calculation was done for a model spectral function at the bottom.

A direct method is to ask to electrons themselves by what they are glued, and it is a tunneling method. As indicated by Scalapino et al. [2], we can know what glue is by analyzing tunneling conductance of a superconductor. McMillan and Rowell firstly executed such an analysis for Pb, and determined a spectral function of the electron-phonon (e-ph) interaction [3]. Figure 1 shows how the phonon structure appears in the tunneling differential conductance.

There is another experimental method for clarifying the pairing mechanism. It is an angle-resolved photoelectron spectroscopy (ARPES). Investigations of a dispersion relation of an electron in a superconducting state by ARPES may also indicate glue, but the spectroscopic resolution is, even at present, not so fine as in a tunneling experiment.

Therefore, we applied a tunneling method to $Bi_2Sr_2CaCu_2O_8$ (Bi2212), which is known to have a rather stable surface among various cuprates, and discovered fine phonon structures in the tunneling conductance [4].

The composition of this paper is as follows: In Section 2, we review our tunneling experiments and explain a characteristic feature of the spectral function $\alpha^2F(\omega)$. It indicates a substantial contribution of phonons to high-T_c superconductivity. In Section 3, we apply thus determined $\alpha^2F(\omega)$ to various properties, such as isotope shift of T_c, dc resistivity, relaxation rate of a nuclear magnetic moment, dispersion relation and a damping rate of a quasi-particle, optical damping rate, etc. for estimating to how much extent the e-ph interaction is effective in these properties. In Section 4, we discuss about a spin contribution to high-T_c superconductivity since it is believed by many researchers that a spin plays an important role for enhancing T_c together with phonons. In this section, we concentrate our discussions on a pair destruction by a magnetic moment and an experiment which may clarify a spin contribution to the pairing, if it really exists. Finally, in Section 5, conclusion is briefly stated.

2. Tunneling Spectral Function of the Electron-Phonon Interaction

2.1. Tunneling Conductance and the Spectral Function

The tunneling junction was a mechanical contact between a slightly overdoped (OD) Bi2212 crystal, whose T_c, $2\Delta(0)$, and $2\Delta(0)/k_BT_c$ were 86 ± 2 K, 48 ± 2 meV, and 6.5 ± 0.4, respectively, and a degenerate GaAs doped with Zn, whose resistivity was about 0.003 Ωcm at 300 K [4]. The junction is schematically shown in an inset of Fig. 2 [4]. It is noted that T_c was determined from the tunneling conductance itself and not from dc resistance of the bulk crystal. The latter was 89.3 K at mid-resistance with a transition width of 3 K. A thin Bi2212 crystal of 2 mm × 5 mm × 0.1 mm was strongly pressed against the GaAs crystal in air at 300

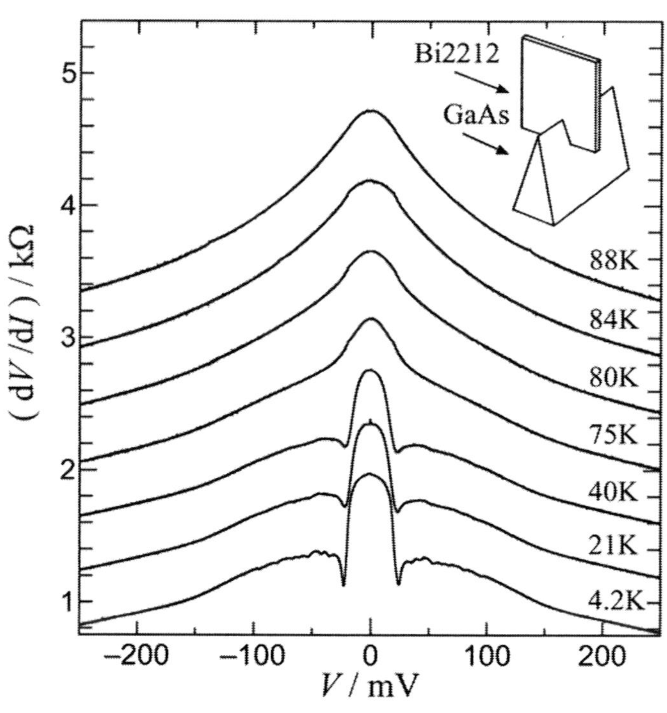

Figure 2. Differential resistance was plotted against voltage [4]. Tunneling electrons flowed into Bi2212 under a positive bias. The ordinate is for the bottom line, and it was shifted upward by 0.4 kΩ each. Inset is a schematic of the contact. (Permitted by the American Physical Society, Copyright (1995).)

K just before cooling to lower temperature. For reinforcing the thin Bi2212 crystal, it was coated with epoxy resin. The current flowed macroscopically in parallel with the Cu–Cu (not Cu–O–Cu) direction but it will not be an actual atomic-scale flow direction because of a deformation at the contact. In reality, the edge of GaAs penetrated into the Bi2212 crystal. Thus, the contact area was much wider than that of an atomic-scale point contact. The characteristic feature of the junction is in use of degenerate GaAs as a counter electrode. Since a tunneling barrier is a Shottky barrier in GaAs, the surface of Bi2212 must be superconducting for observing a superconducting tunneling phenomenon. When the surface is insulating as is for a usual crystal kept in air, such a tunneling feature is not observed. In other words, the GaAs needle automatically searches for a good portion of the Bi2212 surface. This will be one of the reasons why the well-resolved spectral function was observed only for our junction. This semiconductor method was firstly applied to Pb-In alloys by one of the authors (N. T.), and is especially useful for mechanically soft materials [5]. For instance, phonon structures are easily observed for Pb. However, there is a deficit of large leak current due to a weak Shottky barrier. The first symptom of the e-ph interaction appeared in our tunneling study in 1989, but a giant background resistance peak hindered a precise determination of $\alpha^2F(\omega)$ [6].

Before explaining about $\alpha^2F(\omega)$, it must be noted that a so-called current peak often appears at or near the phonon position and masks an intrinsic phonon structure especially for a mechanical contact [7]. However, this current peak appears as a large and sharp resistance peak while the phonon structure appears as a lower-voltage conductance peak and a higher-voltage conductance dip in a dI/dV curve as shown in Fig. 1. An inelastic tunneling process also produces a signal at the same energy with the elastic tunneling process. In the inelastic process, electrons emit phonons in the barrier, and thus an additional tunneling path opens up at the phonon energy. Therefore, a conductance peak may appear in the differential conductance curve. Thus we can distinguish the two signals at the same energy. Moreover, for our GaAs junction, the inelastic phonons must be those of GaAs. Other structures may also appear in a tunneling conductance, and it is very difficult observing an intrinsic phonon structure for a transition metal oxide whose surface is, in general, not stable and brittle. After trials of more than hundreds times over more than five years, the present clear phonon signals were almost accidentally observed. At present, it will be understandable why the observation was as difficult as an accidental occasion. Tunneling experiments were also repeated many times using a Au tip instead of a GaAs tip, but they were not so successful as the GaAs experiment though very faint structures were scarcely detected [7].

Returning to $\alpha^2F(\omega)$, it was determined by Shimada et al. from a normalized tunneling conductance $(dI/dV)_s/(dI/dV)_n$ at 4.2 K with the McMillan-Rowell inversion procedure where temperature T was put to 0 K [3, 4]. Before the calculation, normalization of conductance must be done. At first, 0.507 mS was subtracted from the observed conductance, which was an average for both bias voltages, at 4.2 K to reduce the conductance zero at zero bias voltage. Next, 0.340 mS was subtracted from the (averaged) observed conductance at 88 K to make the normalized conductance unity at 200 mV. In the calculation, the tunneling matrix element was assumed to be constant and the voltage dependence was attributed to the effective density of states. The intraband Coulomb repulsion μ^* was assumed to be 0.11 since it gave a good convergence in the McMillan-Rowell procedure. It is as small as for a normal metal but the larger values did not give better coincidence [4].

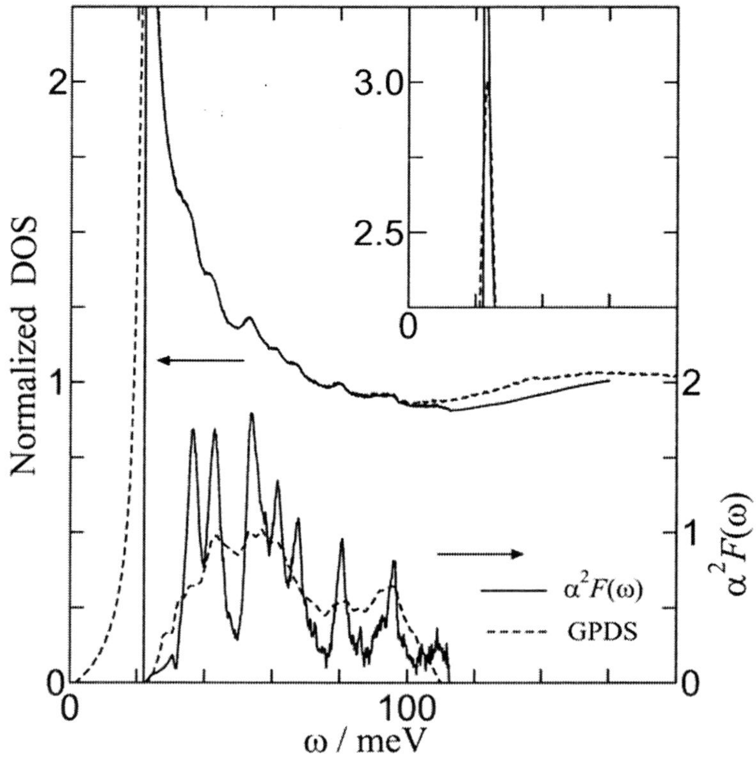

Figure 3. The normalized conductance at 4.2 K which was averaged for both biases (upper broken line), and the calculated normalized density of states at 0 K (upper solid line) [4]. Sharp edge peaks outside the frame were replaced to the upper right panel. The left-hand ordinate is for the normalized quantities. The spectral function and the generalized phonon density of states [8] are shown at the bottom. (Permitted by the American Physical Society, Copyright (1995).)

Figure 3 presents $\alpha^2F(\omega)$ together with the experimental (normalized) conductance (broken line) and the calculated electron density of states (solid line) [4]. The generalized phonon density of states (GPDS) of Bi2212, which was determined by Renker et al., is also shown [8]. 'Generalized' means that corrections to a neutron scattering cross section are not done. In Fig. 3, thus calculated solid line and the experimental broken line show to what extent the self-consistency is satisfied. As is seen, there is a slight discrepancy between the two in an energy range 100 meV to 160 meV. This discrepancy may be due to the normalization procedure and/or a pseudogap phenomenon which may grow at lower temperatures in this energy range. Below the gap, there is also a slight discrepancy, and it might be due to the normalization procedure, d-wave feature, if it is true, and/or a magnetic pair-break effect which will be explained in Section 4 of this chapter. Except these small discrepancies, the over-whole fitting is satisfactory.

Farther, higher temperature conductance was calculated by applying thus determined $\alpha^2F(\omega)$ to the Marsiglio-Schossmann-Carbotte (MSC)-Eliashberg equations [2, 9–12]. Shiina wrote a calculation program of subsequent equations referring to Marsiglio et al.'s theory and the imaginary-axis Eliashberg equations (atomic unit) [9–11].

$$\tilde{\omega}(\omega) = \omega + \int_0^\infty d\Omega \alpha^2 F(\Omega)[K_+(\Omega-\omega) - K_+(\Omega+\omega)] + i\pi \int_0^\infty d\Omega \alpha^2 F(\Omega)$$
$$\times \left[\frac{\tilde{\omega}(\omega-\Omega)}{\left[\tilde{\omega}^2(\omega-\Omega) - \tilde{\Delta}^2(\omega-\Omega)\right]^{1/2}} \{n(\Omega) + f(\Omega-\omega)\} + \frac{\tilde{\omega}(\omega+\Omega)}{\left[\tilde{\omega}^2(\omega+\Omega) - \tilde{\Delta}^2(\omega+\Omega)\right]^{1/2}} \{n(\Omega) + f(\Omega+\omega)\} \right], \tag{1}$$

$$\tilde{\Delta}(\omega) = \int_0^\infty d\Omega \alpha^2 F(\Omega)[K_-(\Omega-\omega) + K_-(\Omega+\omega)] - 2\pi T\mu^* \sum_{m=0}^{\infty} \theta(\omega_c - |\omega_m|) \frac{\tilde{\Delta}_m}{\left[\tilde{\omega}_m^2 + \tilde{\Delta}_m^2\right]^{1 2}} + i\pi \int_0^\infty d\Omega \alpha^2 F(\Omega)$$
$$\times \left[\frac{\tilde{\Delta}(\omega-\Omega)}{\left[\tilde{\omega}^2(\omega-\Omega) - \tilde{\Delta}^2(\omega-\Omega)\right]^{1 2}} \{n(\Omega) + f(\Omega-\omega)\} + \frac{\tilde{\Delta}(\omega+\Omega)}{\left[\tilde{\omega}^2(\omega+\Omega) - \tilde{\Delta}^2(\omega+\Delta)\right]^{1 2}} \{n(\Omega) + f(\Omega+\omega)\} \right], \tag{2}$$

where Ω and ω are phonon and electron energy, respectively. $n(\Omega)$ and $f(\omega)$ are the Bose-Einstein and the Fermi-Dirac distribution function, respectively. μ^* is an intra-band Coulomb repulsion.

$$K_+(\omega) = 2\pi T \sum_{m=0}^{\infty} \frac{\tilde{\omega}_m}{\left[\tilde{\omega}_m^2 + \tilde{\Delta}_m^2\right]^{1/2}} \frac{\omega_m}{\omega_m^2 + \omega^2}, \tag{3}$$

$$K_-(\omega) = 2\pi T \sum_{m=0}^{\infty} \frac{\tilde{\Delta}_m}{\left[\tilde{\omega}_m^2 + \tilde{\Delta}_m^2\right]^{1/2}} \frac{\omega}{\omega_m^2 + \omega^2}. \tag{4}$$

$$\tilde{\omega}_n = \omega_n + \pi T \sum_{m=-\infty}^{\infty} \lambda(n-m) \frac{\tilde{\omega}_m}{\left[\tilde{\omega}_m^2 + \tilde{\Delta}_m^2\right]^{1/2}}, \tag{5}$$

$$\tilde{\Delta}_n = \pi T \sum_{m=-\infty}^{\infty} \left[\lambda(n-m) - \mu^* \theta(\omega_c - |\omega_m|)\right] \frac{\tilde{\Delta}_m}{\left[\tilde{\omega}_m^2 + \tilde{\Delta}_m^2\right]^{1/2}}, \tag{6}$$

$$\lambda(n-m) = \int_0^\infty d\Omega \frac{2\Omega \alpha^2 F(\Omega)}{(\omega_n - \omega_m)^2 + \Omega^2}, \tag{7}$$

$$\omega_n = (2n+1)\pi T. \tag{8}$$

Solving the MSC-Eliashberg equations, the density of states and the normalized conductance $G = dI/dV$ are given by

$$\frac{N_S(\omega)}{N_N(0)} = \mathrm{Re}\left(\frac{\tilde{\omega}(\omega)}{\left[\tilde{\omega}^2 - \tilde{\Delta}^2(\omega)\right]^{1/2}}\right), \tag{9}$$

$$\frac{G_S(\omega)}{G_N(\omega)} = \int_{-\infty}^{\infty} \mathrm{d}\omega' \left(-\frac{\partial f(\omega' - \omega)}{\partial \omega'}\right) \frac{N_S(\omega')}{N_N(0)}, \tag{10}$$

where S and N stand for a superconducting and a normal state, respectively. Tunneling matrix element was assumed to be constant. The gap edge is a real part of Δ_0 for which [10]

$$\tilde{\omega}(\Delta_0) = \tilde{\Delta}(\Delta_0) \tag{11}.$$

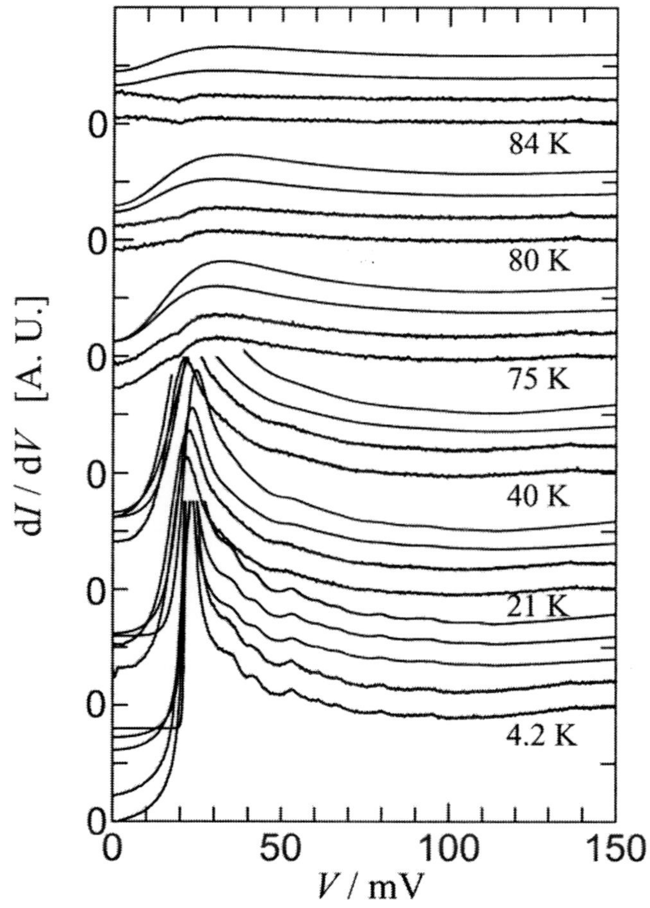

Figure 4. Temperature variation of normalized conductance [9]. The uppermost line; calulated one of eq. (10), the second; calculated one following the BTK theory [13] with the barrier strength Z = 1.5 except for 4.2 K where Z = 2 line was also added, the third; positive-bias experimental line, the fourth; inverted negative-bias experimental line. A small jog at about 140 mV is due to the normalization. The origin of the ordinate was shifted each. (Permitted by the Physical Society of Japan, copyright (1995).)

The calculated higher-temperature $\mathrm{d}I/\mathrm{d}V$ and $\mathrm{d}^2I/\mathrm{d}V^2$ are shown in Figs. 4 and 5 [9]. Again, the experimental conductance seems to be well reproduced by the temperature independent $\alpha^2F(\omega)$, and it represents certain reliability of $\alpha^2F(\omega)$. In Fig. 4, calculated results using the BTK theory [13] were also shown where the density of states of eq. (9) was used with the barrier strength $Z = 1.5$ or 2. When Fig. 5 is scrutinized, one may note a slight discrepancy in the position of a dip in the $\mathrm{d}^2I/\mathrm{d}V^2$ between the calculated and the experimental line. The dip appears at a phonon energy $\hbar\Omega$ from the gap edge, $\Delta + \hbar\Omega$ in case of an e-ph interaction. As seen from Fig. 5, the calculated dip displaces a little toward a lower voltage at higher temperatures [9]. At present, physical meaning of the displacement is not clear, but we discuss about this quite small discrepancy in conjunction with a magnetic pair-break model in Section 4 of this chapter.

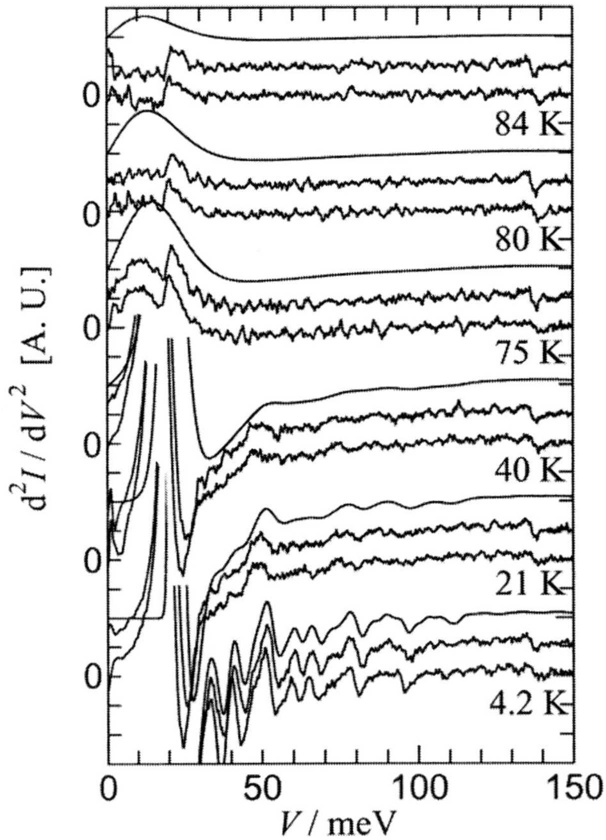

Figure 5. Second derivative conductance $\mathrm{d}^2I/\mathrm{d}V^2$ [9]. Details are the same with those of Figure 4 except that the BTK lines were omitted. (Permitted by the Physical Society of Japan, copyright (1995).)

Another feature to be noted is the structures in $\mathrm{d}^2I/\mathrm{d}V^2$ at the two-phonon frequency range. As seen from Fig. 6, the experimental structures at 4.2 K are as small as the calculated ones [14]. It indicates that the Eliashberg equations are useful even at the two-phonon frequency range, and thus the Migdal approximation will be useful either for the slightly overdoped Bi2212 whose energy gap 2Δ is 48 meV at 4.2 K.

The present $\alpha^2F(\omega)$ is the only one example of well-resolved spectral function. However, similar but less-resolved spectral functions were reported by Vedeneev et al. [15], and Gonnelli et al. [16] as shown in Fig. 7. These results indicate certain reproducibility of $\alpha^2F(\omega)$. However, the poor reproducibility of tunneling phonon structures necessitated a certification by another experimental method, and, finally, the phonon structures appeared in a dispersion relation of a quasi-particle in $La_{2-x}Sr_xCuO_4$ (LSCO) as will be explained in Section 3 of this chapter [17].

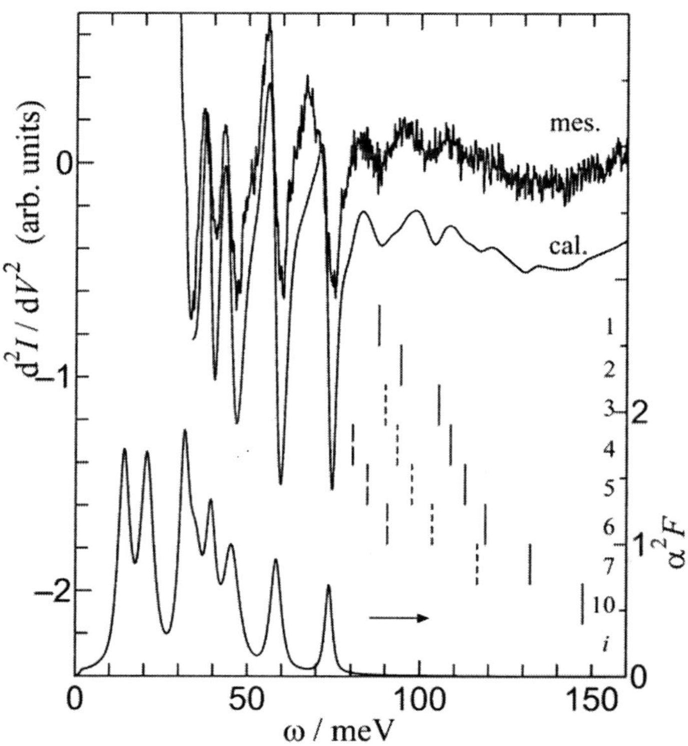

Figure 6. The calculated and the experimental d^2I/dV^2 [14]. The short lines indicate the composite phonon frequencies $\omega_i + \omega_j$ above 80 meV; broken, dotted, and solid lines stand for the peak number j = 6, 7, and 10, respectively. The number i is shown on the right. A dip in d^2I/dV^2 is at the phonon peak position 1–7, and 10 of the bottom $\alpha^2F(\omega)$. Origin of the abscissa is at Δ. From this result, we consider that the Migdal approximation is good at least for the slightly overdoped Bi2212. (Permitted by Elsevier, copyright (1998).)

We calculated $2\Delta(0)$ and T_c for $\alpha^2F(\omega)$ using the isotropic Eliashberg equations and they were 44.4 meV and 87.3 K, respectively. Then, $2\Delta(0)/k_BT_c$ was 5.9. On the other hand, the experimental values were 48 ± 2 meV and 86 ± 2 K in good coincidence with the calculated values. $2\Delta(0)/k_BT_c$ was 6.5 ± 0.4. The coincidences among the calculated and the experimental $2\Delta(0)$ and T_c are surprising when we remember that the usual band theory is widely believed to be inapplicable to high-T_c superconductivity, and they cast vague doubt on such a belief at least about the quasi-particles near the Fermi level.

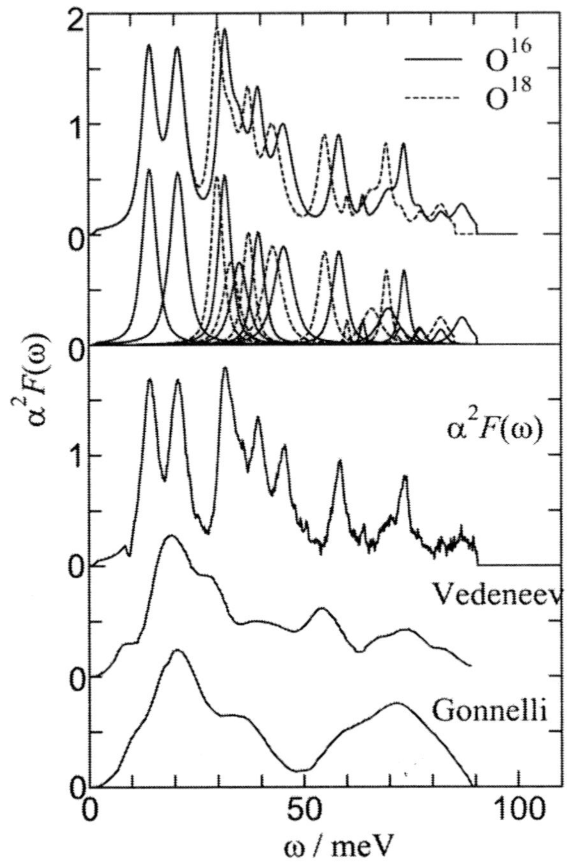

Figure 7. Upper panel: Isotope shifted model $\alpha^2F(\omega)$ in which only the frequency was shifted for mostly oxygen vibration. Lower panel: $\alpha^2F(\omega)$ by Vedeneev et al. [15] and Gonnelli et al. [16]. (Adapted by permission from Elsevier, copyright (1994, 1997).)

2.2. Electron-Phonon Interaction and the Spectral Function

The spectral function of an e-ph interaction $\alpha^2F(\omega)$ is given for a simple metal by [11]

$$\alpha^2F(\omega)=\left(\frac{\Omega_a}{(2\pi)^3}\right)^2\frac{1}{N(0)\hbar}\sum_j\int d^3\boldsymbol{k}\,\delta(\varepsilon_{\boldsymbol{k}})\int d^3\boldsymbol{k}'\delta(\varepsilon_{\boldsymbol{k}'})\left|g_{\boldsymbol{kk}'j}\right|^2\delta(\omega-\omega_j(\boldsymbol{k}'-\boldsymbol{k})), \tag{12}$$

$$g_{\boldsymbol{kk}'j}=\left\langle\Psi_{\boldsymbol{k}}\left|\varepsilon^j(\boldsymbol{k}-\boldsymbol{k}')\cdot\nabla V\right|\Psi_{\boldsymbol{k}'}\right\rangle\left(\frac{\hbar}{2M\omega_j(\boldsymbol{k}-\boldsymbol{k}')}\right)^{1/2}, \tag{13}$$

where Ω_a is the atomic volume of metal whose mass is M, and V is the crystal potential. $\boldsymbol{\varepsilon}^j(\boldsymbol{k})$ is the polarization vector for the $(j\boldsymbol{k})$ th phonon mode. Therefore, we can decide an origin of the interaction by comparing $\alpha^2F(\omega)$ with a phonon density of states of Bi2212. The GPDS was experimentally determined by Renker et al. by a neutron scattering method [8], and a phonon density of states (PDOS) was theoretically given by Paltzer and de Wette [14]. In the latter work, they calculated a phonon dispersion relation fitting to the Raman phonon frequencies of Bi2212, and thus it is independent of the tunneling work. In Figs. 3 and 8, $\alpha^2F(\omega)$, GPDS, and PDOS are shown. As seen from Fig. 8, correspondences among peaks in $\alpha^2F(\omega)$ and the ones in PDOS are quite good, and there is no other peak in $\alpha^2F(\omega)$. Therefore, we conclude that the attractive interaction is principally due to the e-ph interaction. However, we cannot deny a possibility that the e-ph interaction is enhanced in a structure-less way by other interaction such as a spin interaction until the intensity of each peak in $\alpha^2F(\omega)$ is theoretically explained by an e-ph interaction alone. It is noted that Zeyher once calculated a high enough T_c assuming a cylindrical Fermi surface for $YBa_2Cu_3O_7$ (YBCO) [18].

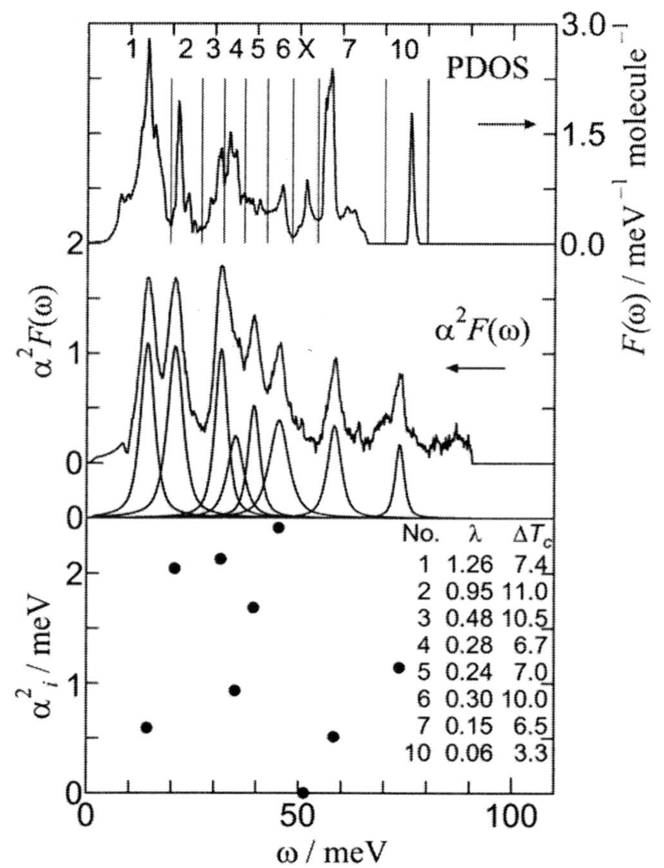

Figure 8. Phonon density of states calculated by Paltzer and de Wette (PDOS) [14], decomposition of $\alpha^2F(\omega)$ into eight peaks, α_i^2, λ_i, and ΔT_c [14]. Note that a dispersion-less mode X is absent in $\alpha^2F(\omega)$. As for the definition of each quantity, see the text. (Permitted by Elsevier, copyright (1998).)

We estimated how much each peak in $\alpha^2F(\omega)$ contributes to superconductivity by recalculating T_c after eliminating the corresponding peak from the entire $\alpha^2F(\omega)$. Thus calculated decrease in T_c, ΔT_c, is shown in Fig. 8. Summation of ΔT_c is not equal to T_c because of nonlinearity of the equations. We also estimated $\lambda_i \equiv 2\int_{\text{peak } i} d\omega\, \alpha^2F(\omega)/\omega$ and $\alpha_i \equiv$

$\int_{\text{peak } i} d\omega\, \alpha^2F(\omega) \,/\, \int_{\text{peak } i} d\omega\, F(\omega)$, and they are also shown in Fig. 8. The total interaction strength $\int_0^\infty d\omega\, \alpha^2F(\omega)$ was 52.5 meV, and the total density of phonon modes $\int_0^\infty d\omega\, \alpha^2F(\omega)$ was put to 45 per molecule. The mass enhancement factor $\lambda \equiv 2\int_0^\infty d\omega\, \alpha^2F(\omega)/\omega$ was 3.9. As seen from these results, the lower frequency peaks P1 and P2 contribute much to λ but not so much to T_c. This might be strange at a first glance when one were used to calculate T_c from λ for a single-phonon mode. However, it is not strange when numerous phonon modes distribute in a wide energy range such as in the present case. For such $\alpha^2F(\omega)$, it is erroneous estimating T_c from λ using the well known McMillan formula [19]. The more proper quantity for estimating T_c than λ is a total interaction strength $\int_0^\infty d\omega\, \alpha^2F(\omega)$. In fact, it was numerically found that T_c is approximately proportional to the total strength when it is as large as 50 meV [11, 20]. This approximate relation is later used for discussing a layer number dependence of T_c.

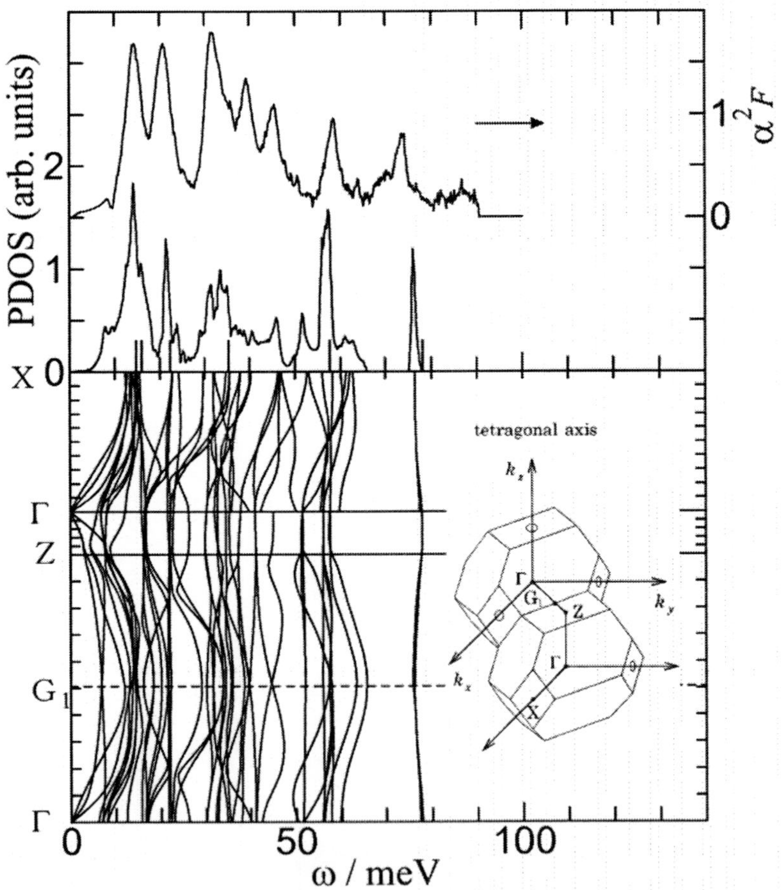

Figure 9. Spectral function $\alpha^2F(\omega)$ is compared with the phonon density of states. The phonon dispersion relations were calculated by Paltzer and de Wette fitting to the Raman phonon frequencies which are shown by short solid lines on the abscissa [14]. A large dip at about 25 meV separates the mostly heavy metal vibrations and the mostly light oxygen vibrations. Γ–X is along the Cu–O–Cu direction. (Permitted by Elsevier, copyright (1998).)

In Fig. 8, the interaction strength per mode α_i is also shown. It is an averaged interaction strength of each phonon mode which is included in the peak *i*. There are many modes within each peak as shown in Fig. 9. To our regret, we cannot decide interaction strength of each phonon mode because of an insufficient resolution of a tunneling method. The averaged strength α_i distributes from 0.5 meV to 2.5 meV per one mode. In comparison, α_i is 1.5 meV/mode for Pb [11]. These analyses indicate that there is no particularly strong peak in $\alpha^2F(\omega)$ which solely gives more than half of T_c.

When the e-ph interaction is investigated from a phonon side, the strength will be determined for each phonon mode. In fact, Chung et al. discovered a particular phonon whose interaction strength with electrons and something else is particularly large by a neutron study of $YBa_2Cu_3O_{6.95}$ [21]. Egami called it a 'superphonon'. If certain peak in $\alpha^2F(\omega)$ is composed of only one particular phonon mode, whose interaction strength must be extraordinarily large since the denominator is one. However, even in such a case, high T_c is not solely given by such a particular interaction because of a reason stated above. Davis et al. also found one structure in a vacuum contact tunneling conductance curve and assigned it a phonon structure [22]. This structure might be a 'superphonon' in consistent with Egami group's result. At the same time, it will be worth noting that there appears only one structure either for our tunneling conductance curve at 21 K as shown in Fig. 4 [9]. This is due to a thermal smearing of fine structures. Therefore, we must wait for a better resolution of other tunneling method for evaluating an idea of 'superphonon'. At least, there is no giant peak in $\alpha^2F(\omega)$ for slightly overdoped Bi2212. In this connection, it is noted that frequencies of some phonons were observed to change at T_c of RBCO (R; rare metal, $T_c \geq 90$ K) and their amounts were explained using the Eliashberg equations by Thomsen et al. and Altendorf et al.[23, 24].

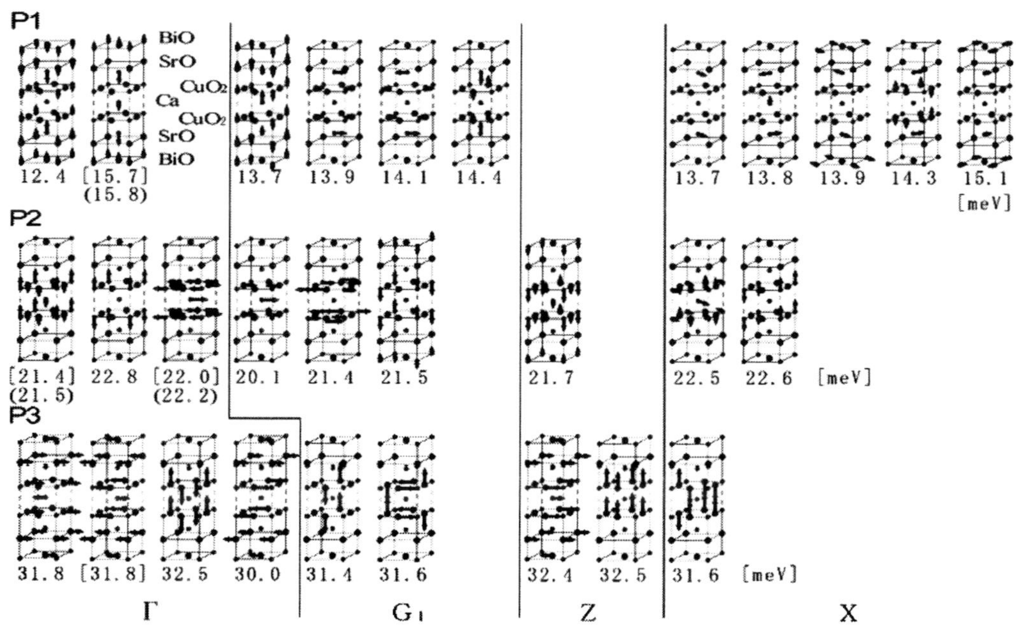

Figure 10. Continued on next page.

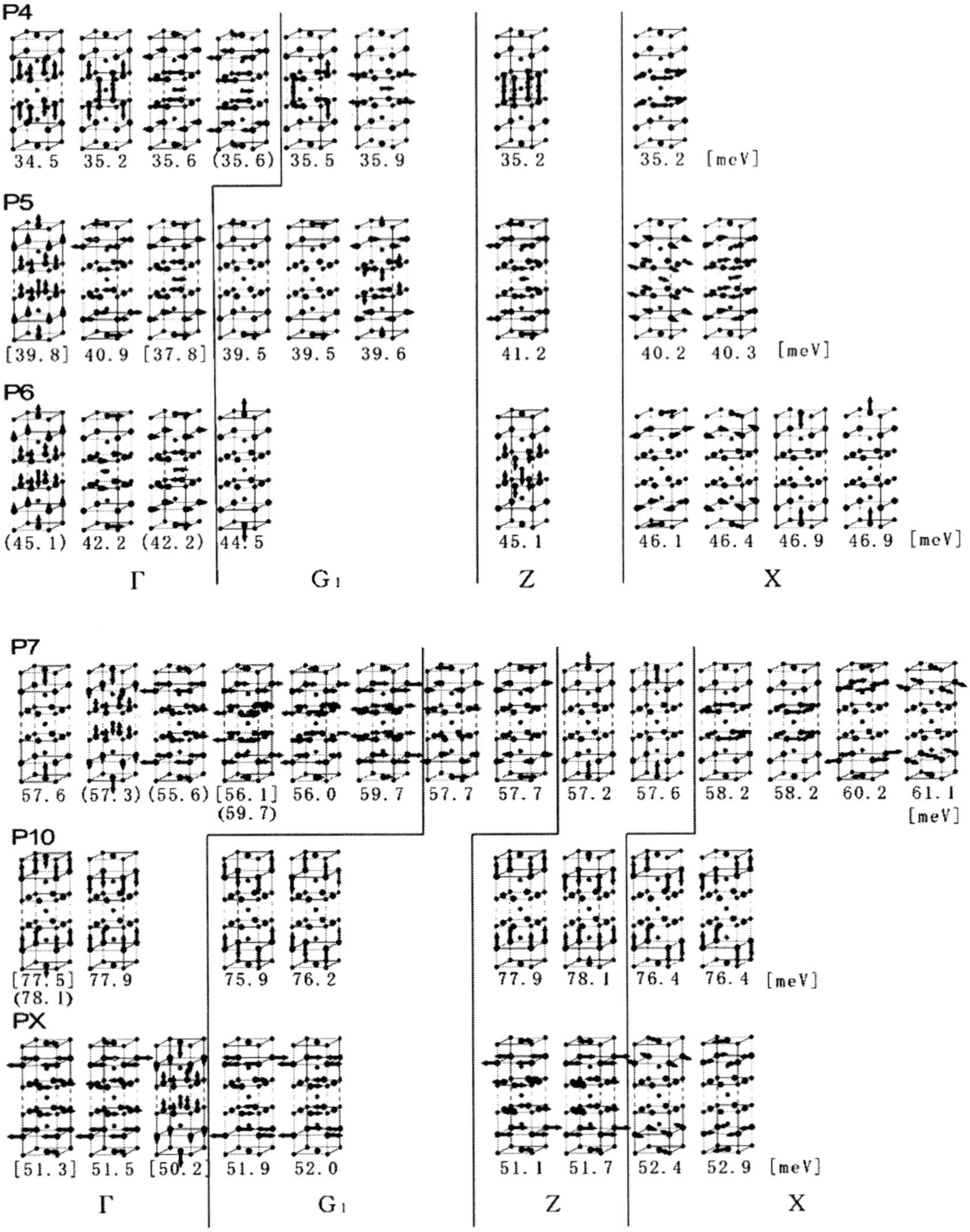

Figure 10. Atomic polarization vectors and their frequencies at special points in the Brilloin zone after Paltzer and de Wette [14]. Arrows qualitatively indicate the displacements, and the frequency in square and round brackets stand for the transverse and longitudinal optical modes, respectively. The largest circles are oxygen atoms. (Permitted by Elsevier, copyright (1998).)

There was no extraordinarily large shift. According to Zeyher's calculation [18], high-T_c is not strange when one believes a usual simple band theory. However, it is not agreed upon in a community because of the observations of very complicated phenomena such as a pseudogap and an inhomogeneity which are not the properties of simple metals. Kamimura et al. proposed a model in which a strong intra-atomic exchange interaction is avoided by assuming two 2p orbitals corresponding to the Cu spin direction [25].

Now, we explain about the phonon modes which may constitute each peak in $\alpha^2F(\omega)$. Paltzer and de Wett calculated dispersion relations of phonons in Bi2212 as shown in Fig. 9 [14]. The atomic displacements at some symmetric points in a reciprocal lattice are shown in Fig. 10 [14]. As seen from Figs. 9 and 10, there are numerous modes within an energy range of each peak in $\alpha^2F(\omega)$. This numerousness is one of the characteristic features of oxides which are composed of many atoms. When current carriers prevail over the entire lattice, this numerousness will not be beneficial to enhancing an e-ph interaction. However, when electrons are confined in the CuO_2 planes as in high-T_c superconductors, atomic vibrations outside the CuO_2 planes may interact with electrons in the planes and the numerousness may contribute to the enhancement. This is a beneficial point of the layer structure of cuprates for achieving a high T_c. Zeyher emphasized an importance of such a long-range e-ph interaction [18]. Cohen et al. also noticed the long-range feature in their frozen-phonon calculation of the interaction [26–28]. We may recognize this long-range nature in the layer number dependence of T_c [29]. As is well known, T_c increases with the number of CuO_2 plane in a molecular unit. It increases until three-layer substance and then saturates or degrades in any series of cuprates. Sometimes, a local structure around Cu such as a coordination number or a bond length is claimed to be a cause of this phenomenon. However, the layer number dependence is quite a general dependence which is common to high-T_c cuprates. Therefore, the physical meaning of the layer number dependence must be quite simple and clear in a mechanism of high T_c. It is simple in case of the e-ph interaction when we become aware of an approximate relation [11, 20],

$$T_c \propto \int \mathrm{d}\omega\alpha^2 F(\omega) \tag{14}$$

Further, when the interaction strength is approximately the same for all the peaks in $\alpha^2F(\omega)$ as shown in Fig. 8,

$$\int \mathrm{d}\omega\alpha^2 F(\omega) \propto \int \mathrm{d}\omega F(\omega) \tag{15}$$

The last quantity is a number of phonon modes in a primitive cell. The mode density increases with the number of constituent ions in the cell, and thus the number of CuO_2 layers. Saturation occurs due to the screening of vibration field by a conductive CuO_2 layer. This is of course very rough estimation of the effective number density of active modes, and corrections for interaction strength of each phonon mode and charge distribution etc. are necessary [29, 30]. Nevertheless, it indicates very clearly a physical meaning of the layer number in high-T_c problem. In the above discussion, poor screening of the Coulomb interaction along the c-axis of cuprates is recognized to be substantial. Importance of such poor screening for raising T_c was noted many years ago by Salchow et al. for Na_xWO_3 which shows superconductivity near the metal to insulator transition [31].

Atomic polarization vectors of the body-centered tetragonal lattice are shown for some special points in the Brillouin zone in Fig. 10 following Paltzer and de Wette [14]. The cell is not a primitive cell but a molecular cell of Bi2212, and the displacement amount is qualitative. The displacement of an ion on an edge of the cell was shown in accordance with the symmetry of the reciprocal lattice point. These modes indicate that the lower frequency peaks P1 and P2 are mostly heavy metal ion vibrations and the other higher frequency modes are mostly oxygen vibrations. A breezing-like mode is found at P7 and an apical oxygen vibration along the c-axis is found at P10. These modes were emphasized for their strong interaction with electrons by Tachiki et al, and Bishop et al. [32, 33]. At the same time, we note that there are many other modes which may significantly contribute to the interaction, and they should not be overlooked. Thus, high-T_c will not be due to some special phonon modes. A detailed study of the frequency change of all these modes at T_c will elucidates out each contribution even though the changes will be very small. Cohen et al. and Rodriguez et al. calculated the interaction strength of various modes of YBCO, and found that the interaction strength sensitively depends on the wave vector of a phonon [26, 34]. According them, vibrations of Ba, Cu, out-of-phase O (in plane), in-phase O (in plane) and apical O mainly contribute to the interaction.

2.3. Anomalously Large Gap Ratio $2\Delta(0)/k_BT_c$ for Underdoped Bi2212

Oda et al. and Miyakawa et al. discovered that $2\Delta(0)/k_BT_c$ became much larger than 10 when Bi2212 was less doped as shown in Fig. 11, while the experimental ratio for the present slightly overdoped Bi2212 is 6.5 and the calculated ratio is 5.9 [35–37]. Some people doubt that the gap is not a superconducting gap but a pseudogap because the ratio is extraordinarily large. However, Miyakawa et al. showed that the gap is a superconducting gap by observing a Josephson current whose intensity is well explained by the gap amplitude [36]. Moreover, if the observed sharp gap edge were that of a pseudogap, there should have been a much sharper gap edge of superconductive coherent state inside the pseudogap. There was none. Therefore, the large $2\Delta(0)/k_BT_c$ for underdoped (UD)Bi2212 is due to superconductivity. The large ratio is serious against the phonon mechanism since the ratio does not exceed about 10 in use of the Eliashberg equations as shown in Fig. 12. In Fig. 12, the ratio was calculated by multiplying a certain constant (>1) on all the peaks of $\alpha^2F(\omega)$ (open circle) or a constant (>1) on P1 and P2 leaving the other peaks as they are. Enhancing $\alpha^2F(\omega)$ of a lower peak i enlarges the ratio as is easily understood. Thus, the ratio much larger than about 10 cannot be explained by the Eliashberg equations. Marsiglio et al. also noted a saturation of the ratio (<13) against a coupling constant [11, 38]. Apart from the numerical calculation, the doping dependence of the ratio is intuitively strange because the gap increases while T_c decreases as the carrier density decreases. Such a reverse tendency was also noted in a dispersion relation of a quasi-particle as will be discussed in Section 3 of this chapter.

Reminiscence of the large ratio seems to be found even for the present $2\Delta(0)/k_BT_c$. The experimental one is 6.5 ± 0.4 while the calculated one is 5.9. Further, a small mismatch between the experiment and the calculation was also noted in a position of dips in the d^2I/dV^2–V lines at higher temperatures as was described in Section 2.1 of this chapter. As shown in Fig. 5, the experimental dip position decreases a little faster than the calculated one as temperature increases. For fitting the calculated dip to the experimental dip, we must assume a small reduction of $\alpha^2F(\omega)$ as temperature increases [39]. This phenomenon is similar to the anomalous $2\Delta(0)/k_BT_c$ because both phenomena suggest a more rapid decrease of $\Delta(T)$ with increasing temperature than that calculated for the temperature independent $\alpha^2F(\omega)$. Thus, these phenomena suggest that a mechanism which makes $2\Delta(0)/k_BT_c$ larger for UDBi2212 still weakly survives for the slightly overdoped Bi2212.

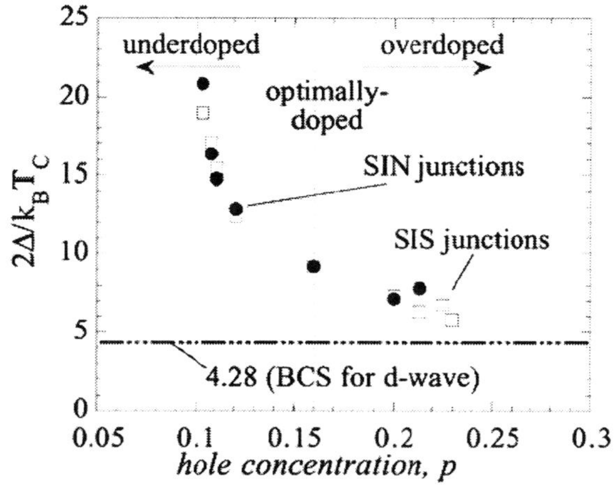

Figure 11. $2\Delta/k_BT_c$ of Bi2212 [37]. (Permitted by Research Trends, Copyright (2002).)

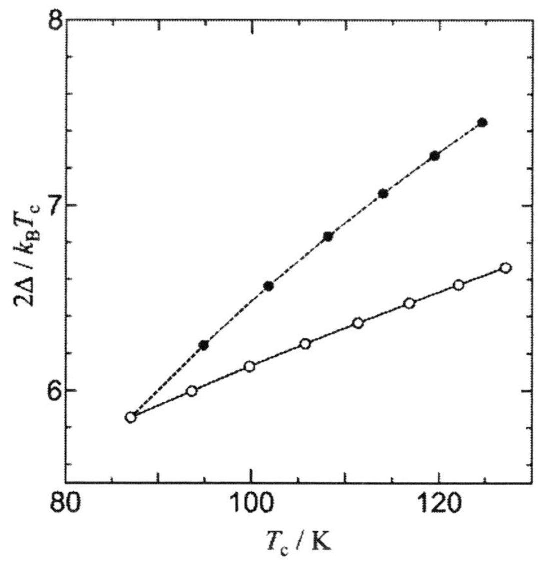

Figure 12. $2\Delta(0)/k_BT_c$ and T_c were calculated for enhanced $\alpha^2F(\omega)$. Open circle; some constant was multiplied on $\alpha^2F(\omega)$ for getting higher T_c, closed circle; some constant was multiplied on the first two peaks of $\alpha^2F(\omega)$ leaving the others unchanged.

Thus, the anomalously large $2\Delta(0)/k_BT_c$ is as substantial as high T_c for high-T_c superconductivity because even if high T_c is explained by the phonon mechanism the large

$2\Delta(0)/k_BT_c$ cannot be explained only by the use of $\alpha^2F(\omega)$ and the Eliashberg equations. A d-wave will not be a cause of the large ratio since the weak coupling BCS $2\Delta(0)/k_BT_c$ is only increased from 3.52 to 4.27 by a d-wave order parameter [40]. Moreover, the larger gap and the lower T_c are completely contradictory. Therefore, some mechanism is necessary for reducing $2\Delta(0)$ at higher temperatures. In this respect, reduction of T_c by a substitution of Zn for Cu is interesting because a break of short-range magnetic ordering results in a break of Cooper pair. When a magnetic correlation length is shorter than a superconducting coherence length, a resultant effective magnetic moment will destroy a pairing state. Since a magnetic correlation length becomes shorter and a superconducting coherence length becomes longer with increasing temperature, the pair break effect becomes stronger as temperature increases. Such a magnetic effect may become more evident as the carrier density is decreased.

The problem is, then, whether such a pair break is as large as bringing about the large gap ratio, and we estimated the strength of the pair break as follows. The pair break term was introduced into the Eliashberg equations as in Ref. [41], and the rate was assumed to be

$$\tau_p^{-1}(T) = \tau_{p0}^{-1} \exp\left(-\frac{E_a}{k_BT}\right) \tag{16}$$

The pre-exponential factor was put to 230 meV which is an extrapolated value of the pair break rate 2 meV at $Co_{0.02}$ (nominal, $Bi_2Sr_2CaCu_{1.98}Cu_{0.02}O_8$) and 7 meV at $Co_{0.06}$ (nominal $Bi_2Sr_2CaCu_{1.94}Cu_{0.06}O_8$) to a Co 100% concentration [42]. The activation energy was fitted to the experimental T_c of slightly underdoped Sample Q of Ref. [43]. The energy gap $2\Delta(0)$, T_c, and $2\Delta(0)/k_BT_c$ were 67 meV, 82 K, and 9.4, respectively [43]. For getting a value of 67 meV, we multipled 1.54 on the present $\alpha^2F(\omega)$, and then the activation energy was determined for getting a value of 82 K. It was 16.4 meV which is comparable with a spin gap of 16 meV for YBCO whose T_c was 59 K and 28 meV for YBCO whose T_c was 91 K [44]. Since the values of the pre-exponential factor and the activation energy are not extraordinary, we (except N. M.) consider that the pair break is a cause of the large $2\Delta(0)/k_BT_c$ for UDBi2212.

Another completely different way for explaining the large $2\Delta(0)/k_BT_c$ is considering that T_c is not a temperature at which the superconducting gap amplitude becomes zero but T_c is a temperature at which the coherence of superconducting state is lost [45, 46]. Renner et al. reported that the superconducting gap smoothly continues to a psudogap, and a pseudogap state is a precursor of a superconducting state [47]. As will be explained in Section 3.8 of this chapter, we (except N. M.) do not agree with this idea.

2.4. Anisotropy and Inhomogeneity

There are many experimental reports which indicate an anisotropic superconducting gap at least on the surface. If it is a d-wave, there may be a giant zero-bias conductance peak ($G_s(0)/G_n(0)>2$) along a nodal direction [48]. However, there was no evident d-wave feature in our tunneling experiments. Therefore, if the order parameter were a d-wave either in our experimental situation, the tunneling current might have flown out from a very limited portion of the crystal momentum space. However, the situation is yet opaque.

Another important feature of the electronic state is inhomogeneity which was forecasted by Phillips [49] and revealed at least on the surface by Davis's group [22, 50, 51]. Then, to estimate its effect on the tunneling phonon structure, we smeared the conductance of Fig. 3 assuming a rectangular or a Gaussian distribution in $\Delta(0)$. Thus smeared conductance is shown in Fig. 13, where numbers beside each line stand for half a width of the rectangle or a standard deviation [7]. As seen from Fig. 13, even the distribution of 3 meV around the averaged value of $\Delta(0)$, the phonon structures are smeared out leaving the largest one. This shows that such a smearing did not occur for our tunneling experiment. It is not yet clear whether the interior of a sample is inhomogeneous as the surface is (see inset of Fig. 2), or whether it is needed a new theoretical frame work for analyzing a space- and momentum-dependent atomic-scale phenomenon [22, 49] Superficially, our tunneling conductance was satisfactorily analyzed within a usual frame work of the band theory.

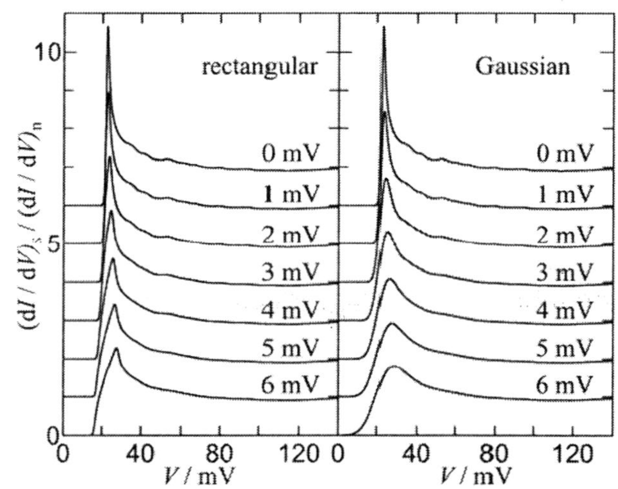

Figure 13. The phonon structures are smeared when the gap distributes in a rectangular or a Gaussian way [7]. A number beside each line is half a distribution width or a standard deviation in Δ/e. (Permitted by SPIE, copyright (2002).)

3. Application of $\alpha^2 F(\omega)$ to Various Properties Obtained by Experiments Other than Tunneling

The present $\alpha^2 F(\omega)$ decisively indicates a substantial contribution of phonons to high-T_c superconductivity. However, it is only one example of well-resolved spectral function. Therefore, it is desired to test its validity by analyzing various experimental properties obtained by other experimental methods such as ARPES.

3.1. Isotope Effect on T_c

Firstly, isotope effect on T_c is calculated for the substitution of ^{18}O for ^{16}O [52]. The experimental isotope effect parameter α_O is not unusually smaller than the BCS value 0.5 for underdoped cuprates but it becomes as small as 0.025 for optimally doped cuprates such as $YBa_{1-x}La_xCu_3O_{7-\delta}$ [53]. Such carrier density dependence is also observed for other cuprates [54–57]. Since T_c of UD cuprates cannot be explained by $\alpha^2F(\omega)$ in the meaning of Section 2.3 of this chapter, its shift should neither be explained by use of $\alpha^2F(\omega)$ alone. However, it will be worth estimating an amount of the shift for slightly overdoped Bi2212 by modifying $\alpha^2F(\omega)$ because whose $2\Delta(0)/k_BT_c$ is approximately explainable by use of $\alpha^2F(\omega)$.

Since the peaks other than P1 and P2 of $\alpha^2F(\omega)$ are mostly oxygen vibrations, ω of oxygen peaks is shifted by an amount $(16/18)^{1/2}\omega$ for the ^{18}O substitution. Since it is not well known how the height of each peak changes by the substitution, we kept the height constant for the substitution and only shifted the frequency. Then, T_c was calculated for the shifted $\alpha^2F(\omega)$, and it was 87.0 K. Thus, the shift in T_c was −3.2 K and α_O was 0.30 which is much larger than the experimental one. The shift can be reduced to −0.32 K ($\alpha_O = 0.03$) when the height is increased by $(18/16)^{1/2}$. However, it is not certain whether such a simple scenario is applicable to the very complicated isotope effect. For instance, Inada and Nasu numerically showed that the isotope shift may become almost zero when a crystal potential becomes steeper for larger amplitude of vibration [58].

3.2. Spin-Lattice Relaxation Rate T_1^{-1}

Disappearance of a coherent peak in a spin-lattice relaxation rate T_1^{-1} of nuclear magnetic resonance at T_c once attracted attention for suggesting exotic superconductivity. However, Allen and Rainer soon showed that it is not strange when the e-ph interaction is strong [59]. We also calculated T_1^{-1} following Allen and Rainer using $\alpha^2F(\omega)$ as shown in Fig. 14 [60]. It was normalized at T_c. The coherent peak really disappeared. Since the relaxation occurs either through a magnetic pass, the experimental one [61] increases even above T_c but it is another problem.

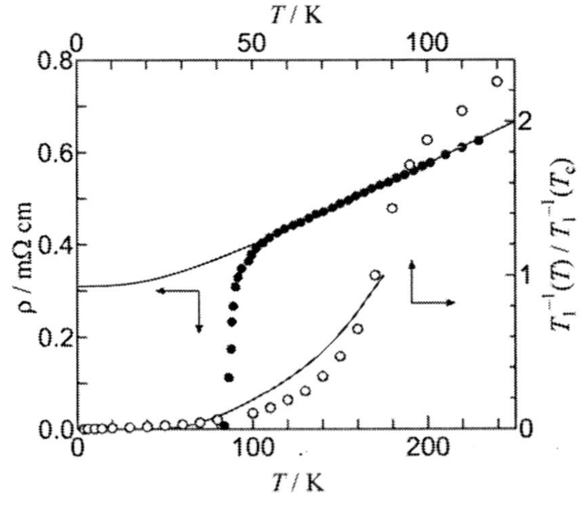

Figure 14. Resistivity along the a-axis [63] and a spin-lattice relaxation rate of Cu nucleus T_1^{-1} [61] of Bi2212 are compared with the calculated ones [60]. (Permitted by Elsevier, copyright (1996).)

3.3. DC Resistivity

When the attractive interaction is expressed by $\alpha^2F(\omega)$, the in-plane dc resistivity ρ will either be determined by $\alpha^2F(\omega)$. Tralshawala et al. did such a trial for $Nd_{2-x}Ce_xCuO_4$, and found a good consistency [62]. We also calculated ρ following them as shown in Fig. 14 where transport $\alpha^2F(\omega)_t$ was approximated by the present tunneling $\alpha^2F(\omega)$. The experimental ρ is due to Honma et al. [63]. They measured ρ along an a-axis and a b-axis of Bi2212, and the a-axis resistivity is shown in Fig. 14. The fitting parameters were residual resistivity and a plasma frequency ω_p which was 3.07 eV for the a-axis ρ and 1.94 eV for the b-axis ρ, while Bozovic reported a value of 2.4 ± 0.3 eV for Bi2212 whose resistivity at 300 K was about 100 μΩcm [64]. Then $n/(m^*/m_0) = n/(1+\lambda) = n/4.9$ is 6.8×10^{21} cm^{-3} and 5.4×10^{21} cm^{-3} for respective directions, where n is the carrier density, and m_0 is the free electron mass. The band mass was assumed to be m_0. For reference, the number density of Cu is 9.0×10^{21} cm^{-3}. Thus, the fitted n is larger than the Cu density, and it will be much larger when the band mass of an electron is larger than a free electron mass. Therefore, explaining the absolute value of dc resistivity is not a simple task, and such a difficulty is also encountered for optical conductivity. Zeyher and Kulić claim that the Coulomb repulsion dominates a forward scattering and farther that $\alpha^2F(\omega)_t$ is different from $\alpha^2F(\omega)$ depending on a band structure [65, 66]. Concerning to the temperature dependence of ρ, however, the approximately linear temperature dependence was well reproduced as shown in Fig. 14. Nevertheless, we do not attribute all the linear temperature dependences to the e-ph interaction because the calculated ρ saturated at about 15 K while Ando et al. discovered a linear temperature dependence down to 0.66 K for $Bi_2Sr_{1.95}La_{0.05}CuO_x$ [67]. Farther, Watanabe et al. demonstrated that another scattering mechanism other than the phonon's strongly occurs for underdoped Bi2212 [68].

3.4. Dispersion Relation of a Quasi-particle: ARPES

Shen's group and Lanzara's group studied a dispersion relation and a damping rate of a quasi-particle by ARPES for various cuprates and for the isotope substituent [17, 69–72]. They arrived at a conclusion that quasi-particles dominantly interact with phonons. Then, it now becomes possible to test the tunneling $\alpha^2F(\omega)$ for its validity. Verga et al. already tried to deduce a spectral function from the ARPES results [73, 74]. We also calculated a dispersion relation following them (atomic unit) [73, 75].

$$E_k = \varepsilon_k + \mathrm{Re}\,\Sigma(E_k), \tag{17}$$

$$\Sigma(\omega) = \int_0^\infty \mathrm{d}\Omega\, \alpha^2F(\Omega)\left[-2\pi i\left\{n(\Omega)+\frac{1}{2}\right\}+\psi\left(\frac{1}{2}+i\frac{\Omega-\omega}{2\pi T}\right)-\psi\left(\frac{1}{2}-i\frac{\Omega+\omega}{2\pi T}\right)\right], \tag{18}$$

$$\psi(1+z) = -\gamma + \sum_{m=1}^{\infty}\frac{z}{m(m+z)}, \quad \gamma;\ \text{Euler's constant}, \tag{19}$$

where E_k, ε_k, Σ are the energy, bare energy, and a self energy of a quasi-particle, and $\psi(z)$ is the digamma function. A liner relation was assumed for the bare state, and the calculated line was fitted to the experimental one at higher energy. Multiphonon corrections to the self energy [76] and/or the corrections beyond the Migdal theorem [77, 78] were not taken into account because even the tunneling structures in d^2I/dV^2 at the two-phonon frequency range were well explained within the Migdal approximation as shown in Fig. 6. This is important because discussions are hot on a role of various interactions at energy which is higher than the phonon energy.

Figures 15 and 16 compare thus calculated dispersion relations with experimental ones [70, 75]. In Fig. 15, the calculated normal state E–k lines at 20 K, 100 K and 300 K are shown, and the phonon structures are observable at 20 K. The structures are smeared out at higher temperatures. Such phonon structures were in fact observed by Zhou et al. for UD$La_{2-x}Sr_xCuO_4$ at about 20 K [17]. The experiment was an energy scan. Their work verifies that many phonon modes contribute to the interaction as the tunneling $\alpha^2F(\omega)$ indicated. Some people say that the contribution of phonons is a matter of course whatever the mechanism of high-T_c is, and the observation of phonons is useless to the clarification of the mechanism. It will be true when only the phonons whose energy is larger than superconducting gap energy are observed because a purely electronic mechanism of attraction may cut-off the spectral function below the gap energy [79]. However, many phonon modes, whose energy is smaller than the gap energy, do appear in $\alpha^2F(\omega)$, and it puts aside an electronic mechanism.

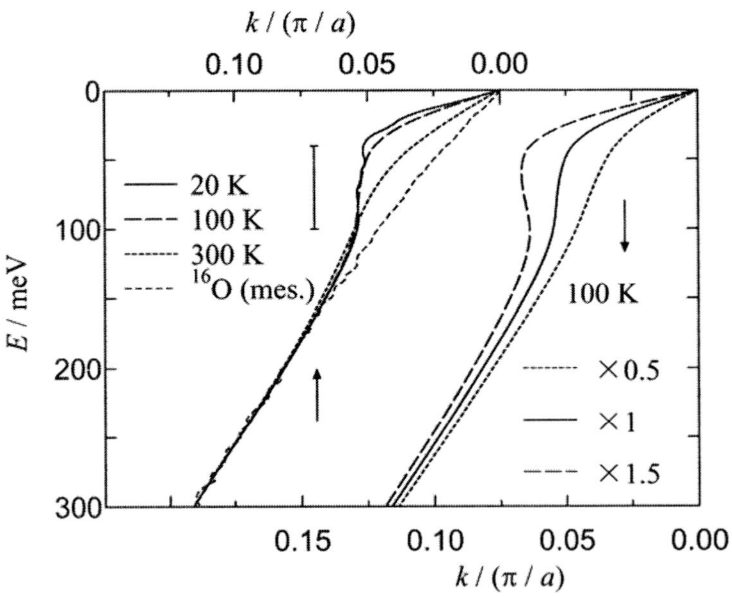

Figure 15. Calculated dispersion relations at 20 K, 100 K, and 300 K (left), and for 0.5 ×, 1×, and 1.5 × $\alpha^2F(\omega)$ at 100 K (right) [75]. Calculation was done for hypothetical normal states. The phonon structures are seen at 20 K. The ARPES relation is along the nodal direction at 100 K [70]. A solid bar indicates a length of 60 meV. (Adapted by permissions from Macmillan Publishers Ltd. Nature, copyright (2004) and Elsevier, copyright (2006).)

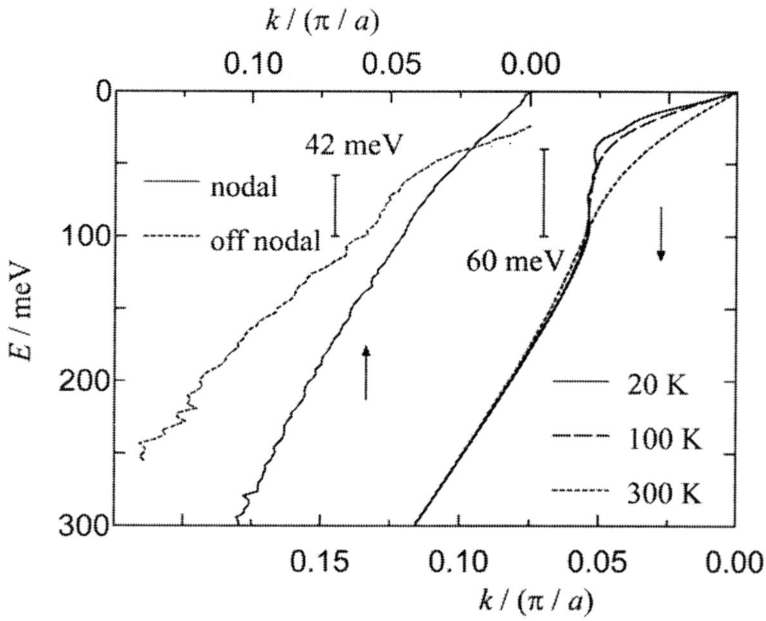

Figure 16. Calculated dispersion relations at 20 K, 100 k, and 300 K for $\alpha^2F(\omega)$ [75]. Calculation was done for hypothetical normal states. The ARPES ones are along the nodal and off-nodal direction at 25 K [70]. Solid bars indicate 42 meV and 60 meV, respectively. (Adapted by permissions from Macmillan Publishers Ltd. Nature, copyright (2004) and Elsevier, copyright (2006).)

Now, interaction strength is discussed. It is estimated from a shift of quasi-particle energy E_k at the maximum phonon energy as seen from Figs. 15 and 16. The tunneling interaction strength $\int_0^\infty d\omega\, \alpha^2F(\omega)$ was 52.5 meV, and such a shift may appear when the interaction within the phonon energy decisively contributes to the attractive interaction. In Fig. 15 such a shift is not clear in the experimental line along the nodal direction (Cu–Cu direction), but a certain shift is observable along the off-nodal direction (not exactly the Cu–O–Cu direction) as seen from Fig. 16. The shift seems to be as large as 42 meV. It is smaller than 52.5 meV, but the experimental situation will not be enough for discussing the amount of this difference. At least, we may consider that a fairly large interaction occurs in the phonon energy. However, it is strange that such a shift seems to be absent along the nodal direction in Fig. 16. It is quite strange since the bending of E–k lines, $(dE/dk)_{E=0\text{meV}}/(dE/dk)_{E>100\text{meV}}$, is almost isotropic at 100 K [70]. The difference in ratio was less than 13 % in any direction. It shows that the interaction is almost isotropic at least at 100 K. Therefore, the absence cannot be understood unless; (1) some unknown interaction grows only along the off-nodal direction below 100 K, and the e-ph interaction itself is as weak as unobservable within the single-phonon energy in cuprates or (2) a steeper declination of the higher energy E–k line along the nodal direction made the existing shift unclear. As for (1), Sato et al. claim that a spin interaction anisotropically grows below T_c from their ARPES study [80]. Perhaps, the raw ARPES data may be similar among the various groups, and the problem will be how to analyze the data. The present $\alpha^2F(\omega)$ is not of spin origin, and the spin contribution will be discussed in Section 4 of this chapter. As for (2), it is noted that the experimental E–k line at the step along the off-nodal direction is nearly parallel with the higher energy E–k line along the nodal direction. Therefore, there is a possibility that the absence of the shift along the nodal direction is merely superficial. If the absence is substantial, $\alpha^2F(\omega)$ must be much smaller than 0.5 times the present value along the nodal direction as seen from Fig. 15 where the dispersion relation at 100 K was calculated for 0.5 times the present value. It is hardly believable considering the almost isotropic interaction strength at 100 K.

3.5. Isotope Effect on the Dispersion Relation

Gweon et al. observed an isotope effect of ^{18}O substitution on the dispersion relation for optimally doped Bi2212 [70]. It strongly suggests that the observed shift and the bending are due to the e-ph interaction. At the same time, the observed isotope effect is strange in two respects: One is a large effect in an energy range much larger than the single-phonon maximum energy, and the other is that the shift amount in the single-phonon energy range is anisotropic and it becomes even reversal along some direction [70].

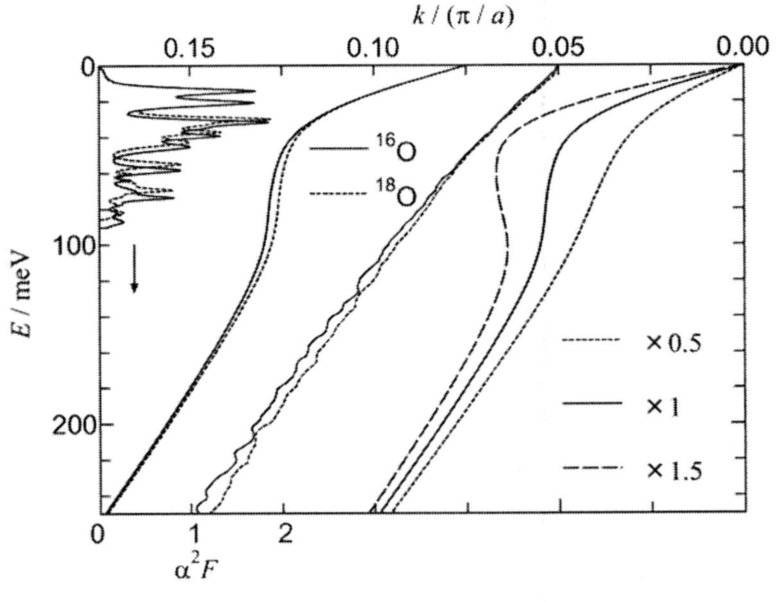

Figure 17. Isotope shifted dispersion relations at 100 K. ARPES one (mid) is along the nodal direction [70]. Calculated ones are for isotope shifted $\alpha^2F(\omega)$'s (left), and for 0.5 ×, 1×, and 1.5 × $\alpha^2F(\omega)$ (right) [75]. The shifted $\alpha^2F(\omega)$'s are also shown. (Adapted by permissions from Macmillan Publishers Ltd. Nature, copyright (2004) and Elsevier, copyright (2006).)

We calculated E_k for the isotope shifted $\alpha^2F(\omega)$ as shown in Fig. 17 [70, 75]. The isotope shifted $\alpha^2F(\omega)$ was obtained by merely shifting each peak of mainly oxygen vibrations as was explained in Section 3.1 of this chapter. As explained there, the calculated amount will be larger than an actual one because each peak was shifted keeping its height the same. It is noted that the present $\alpha^2F(\omega)$ is a function of energy alone and is not directional. Now, the experimental and the calculated E–k lines differ greatly as seen from Fig. 17. The most distinct difference appears at higher energy. Gweon et al. attributed the higher energy isotope shift to a multi-phonon processes beyond the Migdal approximation [70]. However, as was explained in Section 2.1 of this chapter, our tunneling study indicates that the Migdal approximation is not bad at least for slightly overdoped Bi2212 whose mid-resistance T_c was 89 K while T_c was 92 K for Gweon et al.'s Bi2212. Then, it is difficult to suppose a substantially different e-ph interaction process between the two Bi2212 crystals. Instead, we suppose that the oxygen lattice positions may differ between ^{16}O and ^{18}O, and the large isotope effect at higher energy might be a structural effect. For clarifying this important problem about the Migdal approximation, a precise structural study on the oxygen positions is waited for. We once experienced very large amplitude of the vibration of O in ReO_3, and then, the position of O in cuprates is guessed to be amplitude dependent and thus isotope dependent.

In an earlier experiment, we observed small structures at the highest two- and three-phonon frequencies in the conductance curve, and proposed multi-phonon superconductivity [81]. However, such small structures were not observed in a later clearer conductance curve of Ref. [4], and we suspect that the earlier multi-phonon structures were current peaks originated in a critical current at somewhere along a conduction path. As we demonstrated in Ref. [7], such a current peak tends to appear at a phonon position. Now, at a single-phonon energy range the isotope effect is not so clear in the experiment at 100 K along the nodal direction [70], and it will be due to an apparent absence of the step in the *E*–*k* line. At lower 25 K, the isotope effect becomes clearer, and the sense of the shift is consistent with the calculation along the nodal direction [70]. However, along the off-nodal direction, the shift seems to be in a reverse sense. Such a reverse shift is hardly believable for an ordinary e-ph interaction.

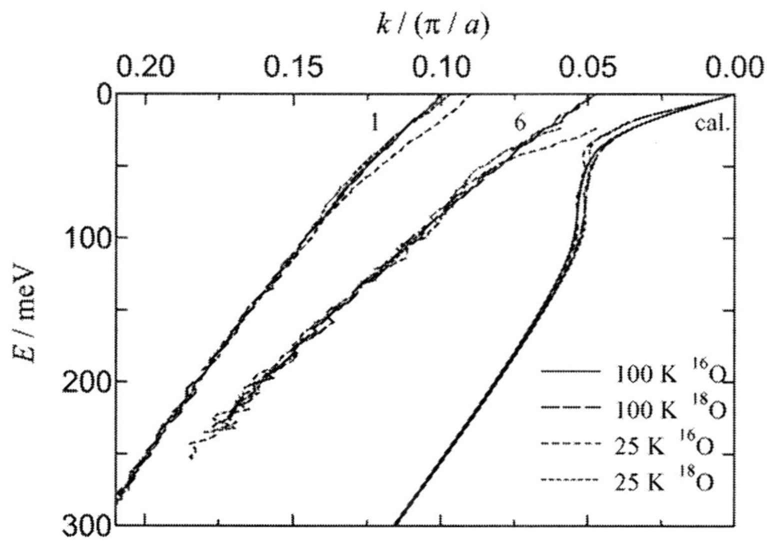

Figure 18. The ARPES dispersion relations at 25 K and 100 K [70] were normalized by the higher energy relations as explained in the text. The calculated one at 25 K is for a hypothetical normal state [75]. (Permitted by Elsevier, copyright (2006).)

The strange anisotropy and the absence of the shift will not be free from the strange higher-energy dispersion relation. Therefore, we normalized the dispersion relation of ARPES [70] by the higher energy relation as shown in Fig. 18 [75]. Certain constant was multiplied on the crystal momentum so that each E–k line has the same declination at higher energy. After the normalization each line was shifted parallel on the abscissa so that they fell on a single line at higher energy. Note that the origin of the abscissa ($k = 0$) is different for each line, and the ^{18}O line runs above a ^{16}O line due to the parallel displacement. The directions 1 and 6 stand for the nodal and off-nodal direction, respectively, according to Ref. [70]. The similar behavior of thus normalized lines suggests that the shift is of the same origin, that is, a usual isotope effect, and the various strange phenomena come from the higher energy dispersion relations.

Oxygen isotope substitution has a large effect on some magnetic phenomena [82]. Therefore, one might consider that even the T_c shift of high-T_c cuprates occurs through a magnetic path and T_c itself is of a magnetic origin. However, it is not true as far as the tunneling result is correct.

Finally, it is noted that a frequency shift and a change in the height of the spectral function are theoretically quite different matter but experimentally they give quite similar dispersion relations as seen from Fig. 17.

3.6. Damping Rate of a Quasi-particle: ARPES

We calculated a single-particle damping rate $1/\tau_s(\omega, T) = -2\,\mathrm{Im}\,\Sigma(\omega, T)\,/\,[1 + \lambda(0, T)]$, where Σ and λ are an electron self-energy and a mass enhancement factor as follows (atomic unit) [83–85].

$$\frac{1}{\tau_s(\omega,T)} = \frac{\pi}{1+\lambda(0,T)} \int_0^\infty \mathrm{d}\Omega\,\alpha^2 F(\Omega)\left[2\coth\left(\frac{\Omega}{2T}\right) - \tanh\left(\frac{\omega+\Omega}{2T}\right) + \tanh\left(\frac{\omega-\Omega}{2T}\right)\right], \quad (20)$$

$$\lambda(0,T) = 2\int_0^\infty \mathrm{d}\Omega \frac{\alpha^2 F(\Omega)}{\Omega} G(\Omega,T),$$

$$G(\Omega,T) = \frac{1}{2}\left(\frac{\Omega}{2T}\right)^2 \int_{-\infty}^{\infty} \mathrm{d}x \frac{1}{\cosh^2 x} \frac{1}{\left(\frac{\Omega}{2T}\right)^2 - x^2}. \quad (21)$$

$$\left(\omega > \tau_s^{-1}\right)$$

This is a low-energy approximation. In Fig. 19, we compared the calculated τ_s^{-1} at 100 K with the ARPES damping rate at 100 K by Bogdanov et al. [85, 86]. The photoelectron damping rate, an energy width of a quasi-particle, is plotted with □ which includes an impurity scattering rate. The ARPES sample was ODBi2212 with T_c of 91K [86]. As seen from Fig. 19, the tunneling $\alpha^2F(\omega)$ approximately gives the same damping rate (τ_s^{-1}) as the ARPES damping rate at least near the Fermi level. However, since the observed rate does not show any temperature dependence and the calculation almost violates the approximation limit

$\omega\tau > 1$, it will be too early to conclude that the single-particle damping rate near the Fermi level is entirely explainable by use of the tunneling $\alpha^2F(\omega)$.

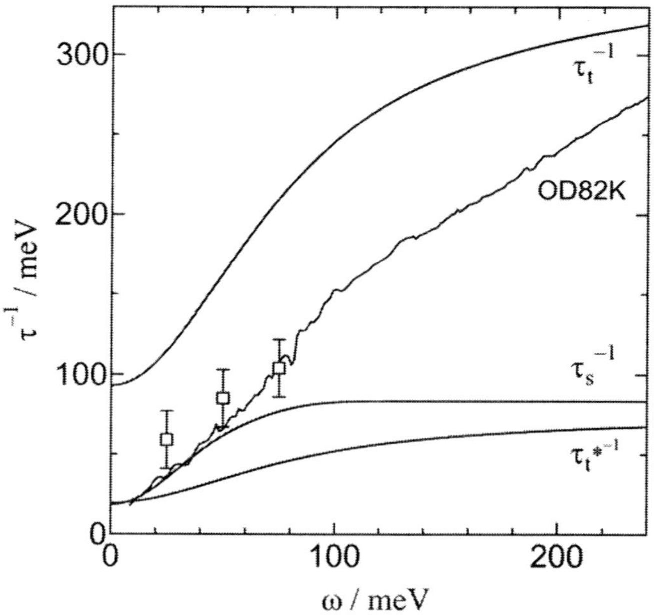

Figure 19. Calculated $1/\tau_s(\omega, 100\text{ K})$, $1/\tau_t(\omega, 90\text{ K})$, $1/\tau_t^*(\omega, 90\text{ K})$, experimental (unrenormalized) optical damping rate at 90 K (solid line) [87], and ARPES width □ at 100 K [86] were plotted [85]. Experiments were for overdoped Bi2212. Impurity scattering rate was not subtracted. See the text. (Adapted by permissions from the American Physical Society, copyright (1996) and Elsevier, copyright (2007).)

3.7. Optical Damping Rate

The quasi-particle damping rate may also be determined by an optical method. In Fig. 19, the optical damping rate determined by Puchkov et al. at 90 K is also shown by a solid line [87]. The sample was ODBi2212 whose T_c was 82K. As seen from Fig. 19, the experimental optical damping rate is approximately the same with the calculated single-particle damping rate in an energy range within about 60 meV from the Fermi level. This result is consistent with the fact that the tunneling conductance was satisfactorily analyzed by the Eliashberg equations at least for ODBi2212.

However, things will not be so simple. For instance, Kulić emphasized that the transport $\alpha^2F(\omega)_t$ is different from $\alpha^2F(\omega)$, and that the damping process of optically excited electron-hole pair is different from that in the ARPES experiment for which a single-particle damping is a good approximation [66]. Then, we calculated the optical unrenormalized damping rate $1/\tau(\omega, T)_t$ following Shulga et al. [88] though $\alpha^2F(\omega)_t$ was again approximated by $\alpha^2F(\omega)$. (atomic unit)

$$\frac{1}{\tau(\omega,\ T)_{\mathrm{t}}} = \frac{\pi}{\omega}\int_0^{\infty} \mathrm{d}\Omega\ \alpha^2 F(\Omega)\left[2\omega\coth\left(\frac{\Omega}{2T}\right) - (\omega+\Omega)\coth\left(\frac{\omega+\Omega}{2T}\right) + (\omega-\Omega)\coth\left(\frac{\omega-\Omega}{2T}\right)\right] \tag{22}$$

$$(\omega > 1/\tau_{\mathrm{t}})$$

In which an impurity scattering term was omitted. In Fig. 19, thus calculated $1/\tau(\omega, T)_{\mathrm{t}}$ is also shown. As seen from Fig. 19, the calculated rate is much larger than the experimental rate of Puchkov et al. One of the reasons will be that the calculated rate exceeds the approximation condition in a wide energy range. The second will be the approximation of $\alpha^2F(\omega)_{\mathrm{t}}$ by $\alpha^2F(\omega)$. Moreover, Zeyher and Kulić claim that the strong Coulomb repulsion suppresses the transport resistivity which is estimated from the e-ph interaction alone [65]. Farther, it is noted that the renormalization factor $1/(1+\lambda)$ is absent in eq. (22) and the experimental rate by Puchkov et al. is an unrenormalized result [87]. Such a disagreement between the calculated and the observed value was also noted for dc resistivity as explained in Section 3.3 of this chapter. In Fig. 19, we also plotted renormalized $1/\tau_{\mathrm{t}}^*(\omega, T) = 1/\tau_{\mathrm{t}}(\omega, T)[1+\lambda(0, T)]$, since necessity of such a correction had been suggested by Allen for a strong e-ph coupling [89]. The renormalized rate is smaller than the experimental one. Thus, explaining the absolute value of the transport coefficients is still theoretically problematic.

Finally, we wish to note again that the APRES width of ODBi2212 with T_{c} of 91K and the optical damping rate of ODBi2212 with T_{c} of 82K do not differ so much.

3.8. Pseudogap

The optical study showed an evolution of the pseudogap for UD cuprates [87]. Such pseudogap phenomena have been observed in almost all the electronic and lattice properties of cuprates, and the pseudogap state has been disputed whether it is a precursor of a superconducting state or not. Whether it is a precursor or not will be judged by observing superconducting gap amplitude at T_{c}. If the amplitude is finite at T_{c}, the pseudogap state will be a phase-incoherent superconducting state and thus a precursor of superconductivity, while if the amplitude of a superconducting gap is zero at T_{c}, the pseudogap state is not a phase-incoherent superconducting state and thus not a precursor, but a back ground state on which superconductivity arises.

Determination of a superconducting gap from tunneling conductance at higher temperatures is not a simple task because the observed gap-edge-peaks keep their separation almost constant against temperature even if the gap closes at T_{c} as seen from Fig. 20. Therefore, estimation of the energy gap at T_{c} needs a help of theory, and thus the estimated gap value depends on the analysis way. There are many tunneling results which demonstrate that the superconducting gap seems to continue smoothly to the pseudogap, but it does not mean as it seems.

Therefore, it is useful if there is a method in which a superconducting gap can be purely experimentally estimated. A mesa junction is such a junction in which many s(superconductor)–i(insulator)–s–i–s–i–s layers are accumulated as if a mesa. Suzuki et al. prepared a mesa junction of Bi2212 ($2\Delta(0)/k_{\mathrm{B}}T_{\mathrm{c}}$ = 6.6) and suggested that the superconducting gap diminishes toward T_{c} while the pseudogap structure exists up to 150 K

[90]. Later, Krasnov et al. also prepared a mesa junction of Bi2212 ($2\Delta(0)/k_BT_c$ = 6.7 and 8.3), and found that the amplitude of a superconducting gap becomes zero at T_c while the pseudogap structure still exists at higher temperatures, and thus concluded that the two phenomena are coexisting different phenomena [91]. However, there might be a disputation whether such a conclusion is correct either for an extremely underdoped cuprate.

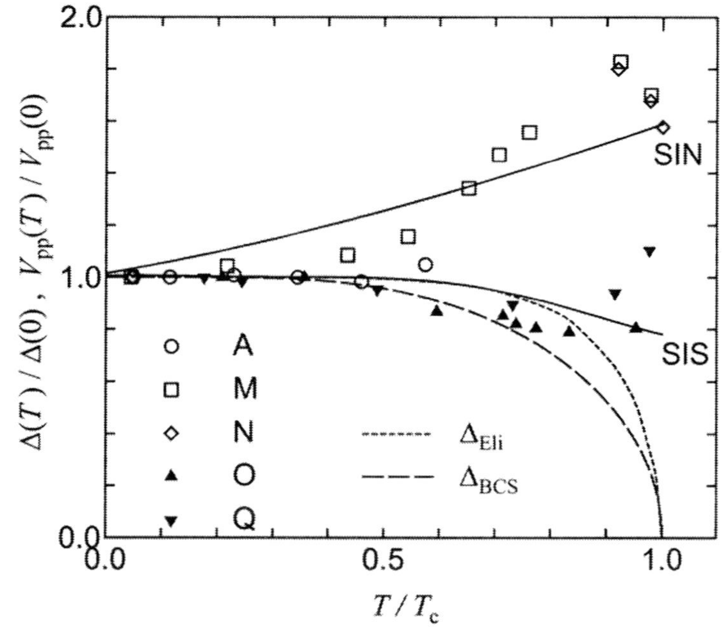

Figure 20. Normalized gap edge separation $V_{pp}(T)/V_{pp}(0)$, normalized gap $\Delta(T)/\Delta(0)$, experimental $V_{pp}(T)/V_{pp}(0)$ for superconductor-insulator-normal metal (SIN) junctions (symbols) and for superconductor-insulator-superconductor junctions (SIS) (symbols) were plotted for demonstrating how $V_{pp}(T)$ varies with temperature while $\Delta(T)$ becomes zero at T_c [43]. Calculations of $V_{pp}(T)/V_{pp}(0)$ (solid line) and $\Delta(T)/\Delta(0)$ (dotted line Δ_{Eli}) were done using the Eliashberg equations and $\alpha^2F(\omega)$. In the present text, sample O (▲) is noted. (Permitted by Elsevier, copyright (2001).)

The closing of a superconducting gap at T_c is also supported by our tunneling study as follows. In this study, a Au tip was used instead of a GaAs tip [43]. The junction was a SIS junction of slightly underdoped Bi2212 whose $2\Delta(0)$, T_c, and $2\Delta(0)/k_BT_c$ were 64 ± 1 meV, 84 ± 4 K, and 8.8 ± 0.6, respectively (sample O of Ref. [43]). In Fig. 20, the gap edge separation V_{pp} was plotted against temperature (▲), and it was compared with the calculated separation (line SIS). The calculation was done as explained in Section 2.1 of this chapter. Even though the gap edge separation seems to continue smoothly to a pseudogap edge separation, the superconducting gap $\Delta_{Eli}(T)$ becomes zero at T_c. Fig. 21 shows more directly the conductance curves [43]. This is, however, not a direct proof of disappearance of an amplitude of superconducting order parameter at T_c because we used a theory.

From these results, we (except N. M.) consider that the pseudogap state will not be an incoherent Cooper pair state at least for Bi2212 whose $2\Delta(0)/k_BT_c$ is smaller than about 9.

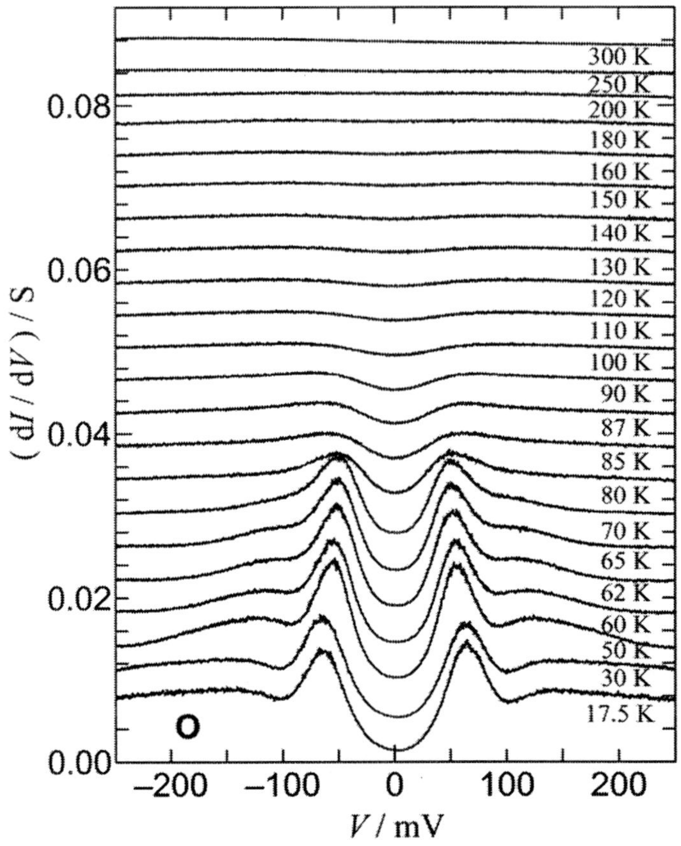

Figure 21. Differential tunneling conductance curve of sample O of Figure 20 [43]. The lowest two lines have larger gap edge separation which is demonstrated with ▲ in Figure 20. (Permitted by Elsevier, copyright (2001).)

4. Spin Contribution (?) to the Electron-Phonon Interaction

Phonon contribution to high-T_c superconductivity is thus clear. At the same time, it is also clear that the presence of magnetic moments influences various properties of cuprates as having been discussed by many authors. What is unclear is a spin contribution to the attractive pairing interaction. Since the spins give effects on various properties especially for underdoped cuprates, their contribution to the attractive interaction will also be large for less-doped cuprates if it really exists.

In this section, we propose an experimental method for clarifying a spin contribution to the attractive interaction especially referring to the underdoped cuprates.

4.1. Pair Break by the Magnetic Moment

Magnetic impurities effectively destruct the Cooper-pair state. For cuprates, even a nonmagnetic impurity such as Zn effectively reduces T_c. Since a nonmagnetic element destructs the short-range antiferromagnetic ordering of Cu spins, a nonmagnetic element may also act as an effective magnetic moment. Farther, even an antiferromagnetically ordering moment becomes an effective magnetic moment when the magnetic correlation length becomes shorter than the superconducting coherence length. The former becomes shorter while the latter becomes longer as temperature increases, and the pair break effect will become stronger at higher temperatures even if no magnetic impurities are doped. Substitution of Cu by any other element also shortens the magnetic correlation length but, in this case, the shortening occurs even at the lowest temperature. Thus, non-substituted cuprates and substituted cuprates will show different temperature dependence of the pair breaking. In fact, substitution of Co for Cu reduces both $\Delta(0)$ and T_c, and the ratio $2\Delta(0)/k_BT_c$ slightly decreases as Co concentration increases [42, 92]. This is quite different from the increase of $2\Delta(0)/k_BT_c$ for underdoped cuprates. For underdoped cuprates, the increase in the hyperfine field at the Cu nucleus [93] indicates a growing of Cu magnetic moment by less doping, and thus the pair break effect will become more effective.

Then, we proposed a pair break model by the effective magnetic moment for explaining the anomalously large $2\Delta(0)/k_BT_c$ of UDBi2212 as described in Section 2.3 of this chapter. Farther, we wish to note that the antiparallel spin ordering along the gapped direction (Cu–O–Cu) simultaneously produces a parallel spin ordering along the nodal direction (Cu–Cu) which will make the Cooper pair state unstable along the nodal direction.

4.2. Spin Enhancement of the Electron-Phonon Interaction if Any (?)

Tunneling $\alpha^2F(\omega)$ and ARPES dispersion relation indicate a substantial contribution of the e-ph interaction to high-T_c superconductivity. However, spin contribution to the attractive interaction is still disputed since various properties are greatly influenced by Cu spins especially for less-doped cuprates. Since each peak in $\alpha^2F(\omega)$ for slightly overdoped Bi2212 corresponds well to phonon density of states, the spins will merely enhance each e-ph interaction at least for overdoped cuprates even if any spin enhancement is true. Detection of such an enhancement might be less difficult for underdoped cuprates than for overdoped cuprates though the large damping rate [87] will smear any spectroscopic fine structure.

In this connection, we note on a very large $2\Delta(0)/k_BT_c$ of UDBi2212. In our view, the large ratio is as characteristic as high T_c itself for high-T_c superconductivity as explained in Section 2.3 of this chapter. Then, if there were a spin enhancement, it would appear in this substantial phenomenon. Reduction of T_c by spins was already explained. However, there is another mechanism for explaining the large ratio. It is the enhancement of $\Delta(0)$ larger than the value which is expected from T_c by the Eliashberg equations. Our proposal is an experimental detection of this enhancement of $\Delta(0)$ if it really exists.

The idea is as follow. Even if the spin enhancement occurs at lower temperatures, it will diminish when the antiferromagnetic correlation length becomes shorter than the superconducting coherence length whatever a mechanism of the enhancement is. On the other hand, the e-ph interaction strength is essentially independent of temperature since the force constants of the lattice little change. Then, if the weakening of the enhancement really occurs, it may be observed by a tunneling through the narrowing of the gap-edge separation with increasing temperature.

As shown in Fig. 20, the gap edge separation does not become narrower (SIN junction) or shows a slight narrowing (SIS junction) at elevated temperatures even though the superconducting gap itself is greatly reduced, when $\alpha^2F(\omega)$ does not change. On the other hand, when $\alpha^2F(\omega)$ becomes smaller, the gap-edge separation may become narrower at any temperature. Thus, even if the fine structures of $\alpha^2F(\omega)$ are smeared out by a strong damping of quasi-particles in underdoped cuprates, we may observe a weakening of $\int\alpha^2F(\omega)\mathrm{d}\omega$ through the narrowing of the gap-edge peak to peak separation.

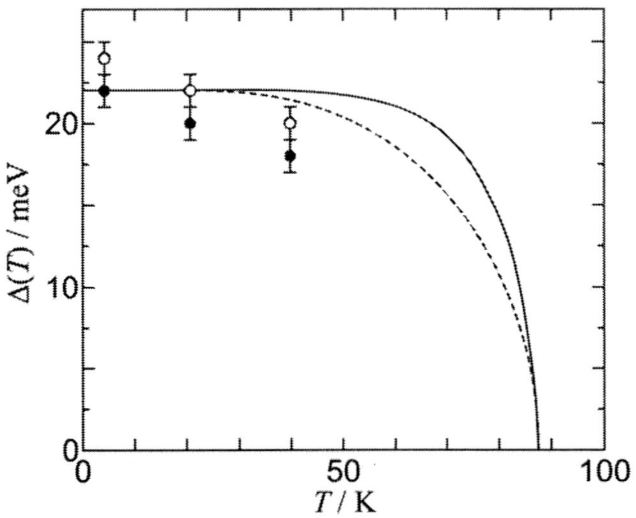

Figure 22. Rapid decrease in the experimental gap compared with the calculated gap (solid line) for the present slightly overdoped Bi2212 [9]. The experimental gap (open circle) was determined from the zero-energy position of the phonon density of states whose structures were adjusted to those of the experimental conductance curve. Thus, it is essentially free from theory. The other gap (solid circle) was determined from the amount of deviation of the experimental dip position in d^2I/dV^2 from the calculated one which is shown in Figure 5. The broken line represents the BCS relation. (Permitted by the Physical Society of Japan, copyright (1995).)

We searched for such a narrowing in the published data. One is our datum. As explained in Section 2.1 of this chapter and as shown in Fig. 5, there seems to be a slight decrease of $\alpha^2F(\omega)$ as temperature increases even for the slightly overdoped Bi2212. It is shown more clearly in Fig. 22 [9, 39]. However, it is not clear whether the apparent decrease in $\Delta(T)$ is a decrease of the enhancement in question or a reminiscence of the magnetic-pair break effect. The second is also our datum. As seen from Figs. 20 and 21, the gap-edge separation is almost the same at 17.5 K and 30 K and decreases suddenly at 50 K for the sample O (▲) whose T_c was 84 ± 4 K and $2\Delta(0)/k_BT_c$ was 8.8 ± 0.6. However, it is neither certain whether the observed narrowing is due to the spin enhancement effect for the following reasons. Firstly, such a narrowing is explainable by the magnetic pair break mechanism as demonstrated in Ref. [43]. Secondly, the junction was a mechanical contact between the sample O and a Au tip, and the SIS junction was accidentally formed. Such an accidental formation of an SIS junction was often experienced during making a mechanical contact. At that time, the gap value greatly varied when we contacted on and off the Au tip at macroscopically the same point (about 0.1 × 0.1 mm^2) of the sample [7, 43]. It is due to inhomogeneity of Bi2212. Therefore, the gap might change during the heating up from 4.2 K to 60 K due to thermal stress, and the observed change in the gap-edge separation might have been due to the thermal shock. The third example is seen in Dipasupil et al.'s data [94]. The separation at 6 K, where $2\Delta(6\text{ K})/k_BT_c = 10.0$, became narrower at 40 K, where $2\Delta(40\text{ K})/k_BT_c = 7.5$. T_c was 82 K [94]. There is a clear narrowing in the gap edge separation similar to our sample O, but there is no datum between 6 K and 40 K, and thus we cannot conclude that the spin enhancement appeared in their sample for the same reasons as for our sample O. The fourth example is again our datum. One of the authors (N. M.) measured the tunneling conductance of $Ba_2Ca_3Cu_4O_8(O_{0.2}F_{0.8})_2$, whose T_c was 80 K and $2\Delta(7\text{K})/k_BT_c \approx 13$, from 7.1 K up at about 10 K separation [95]. There, we found no narrowing at all at lower temperatures.

As for the ARPES result of Sato et al., which suggested the growing of the attractive interaction below T_c as mentioned at Section 3.4 of this chapter, we wait for a convergence of disputations among ARPES people.

Thus, at present, we have no clear indication of the spin enhancement of the e-ph interaction for underdoped cuprates.

5. Conclusion

The critical temperature of slightly overdoped Bi2212, which is as high as 86 K, will be due to the strong electron-phonon interaction. The spectral function of the interaction was determined using a Shottky barrier tunneling junction between Bi2212 and degenerate GaAs. The structure of the function coincided with the structure of the phonon density of states, and the calculated $2\Delta(0)$ and T_c agreed with the observed values. However, the function is yet only one well resolved example.

For complementing poor reproducibility of the function, it was applied for calculating a dispersion relation and a damping rate of a quasi-particle, and the calculated results were compared with the recent ARPES results. For us, there seems to be no substantial disagreement between the tunneling and the ARPES results in the single-phonon energy range. However, at higher energy range, the damping rate is much larger than what is expected from the spectral function. Moreover, the large isotope effect of ARPES study in this higher energy range is strange, and we speculated that it would be an isotope effect on the atomic position of oxygen whose amplitude of vibration is very large.

It was pointed out that $2\Delta(0)/k_BT_c$, which is larger than about 10 for underdoped cuprates, cannot be explained by the similar spectral function and the Eliashberg equations alone, and thus it is as substantial as high T_c itself in high-T_c problem. Relying on the mesa results, we (except N. M.) consider that T_c is a temperature at which the amplitude of the superconducting order parameter becomes zero at least for not much reduced cuprates. Since a magnetic moment effectively reduces T_c, we presented a magnetic pair-break model in which a shorter magnetic correlation length than a superconducting coherence length was noted to be important for pair break. Another possibility for explaining the large ratio may be in a temperature dependent spin enhancement of the e-ph interaction at lower temperatures, and we proposed an experimental method for detecting the spin contribution to the pairing if it really exists.

Acknowledgments

The authors would like to express their sincere gratitude to U. Paltzer and F. W. de Wette for their cooperation in research and to J. C. Phillips for digging out our works and having widened our view. The study was greatly assisted by Y. Shiina who allowed us a usage of his computer program of the MSC-Eliashberg equations.

References

[1] Bednorz, J. D.; Müller, K. A. *Z. Phys*. 1986, B 64, 189-193.

[2] Scalapino, D. J.; Schrieffer, J. R.; Wilkins, J. W. *Phys. Rev*. 1966, 148, 263-279.

[3] McMillan, W. L.; Rowell, J. M.; *Phys. Rev. Lett*. 1965, 14, 108-112.

[4] Shimada, D.; Shiina, Y.; Mottate, A.; Ohyagi, Y.; Tsuda, N. *Phys. Rev*. 1995, B 51, 16495-16498.

[5] Tsuda, N. *Jap. J. Appl. Phys*. 1969, 8, 582-587.

[6] Miyakawa, N.; Shimada, D.; Kido, T.; Tsuda, N. *J. Phys. Soc. Jpn*. 1989, 58, L383-L386.

[7] Tsuda, N.; Saitoh, M.; Shimada, D. in *Proceedings of SPIE*; Bozovic, I.; Pavuna, D. Ed.; SPIE, Bellingham, WA, 2002, Vol. 4811, p. 206-215.

[8] Renker, B.; Gompf, F.; Evert, D.; Adelman, P.; Schmidt, H.; Gering, E.; Hinks, H. *Z. Phys*. 1989, B 77, 65-68.

[9] Shiina, Y.; Shimada, D.; Mottate, A.; Ohyagi, Y.; Tsuda, N. *J. Phys. Soc. Jpn*. 1995, 64, 2577-2584.

[10] Marsiglio, F.; Schossmann, M.; Carbotte, J. P. *Phys. Rev*. 1988, B 37, 4965-4969.

[11] Carbotte, J. P.; *Rev. Mod. Phys*. 1990, 62, 1027-1157.

[12] Schrieffer, J. R. *The Theory of Superconductivity*; Revised printing; Addison Wesley: Redwood City, Ca, 1971.

[13] Blonder, G. E.; Tinkham, M.; Klapwijk, T. M. *Phys. Rev*. 1982, B 25, 4515-4532.

[14] Shimada, D.; Tsuda, N.; Paltzer, U. ; de Wette, F. W. *Physica* 1998, C 298 , 195-202.

[15] Vedeneev, S. I.; Tsvetkov, A. A.; Jansen, A. G. M.; Wyder, P. *Physica* 1994, C 235-240, 1851-1852.

[16] Gonnelli, R. S.; Ummarino, G. A.; Stepanov, V. A. *Physica* 1997, C 275, 162-171.

[17] Zhou, X. J.; Shi, Junren; Yoshida, T.; Chuk, T.; Yang, W. L.; Brouet, V.; Nakamura, J.; Mannella, N.; Komiya, S.; Ando, Y.; Zhou, F.; Ti, W. X.; Xiong, J. W.; Zhao, Z. X.; Sasagawa, T.; Kakeshita, T.; Eisaki, H.; Uchida, S.; Fujimori, A.; Zhang, Z.; Plummer, E. W.; Laughlin, R. B.; Hussain, Z.; Shen, Z.-X. *Phys. Rev. Lett*. 2005, 95, 117001(1~4).

[18] Zeyher, R. *Z. Phys*. 1990, B 80, 187-192.

[19] McMillan, W. L. *Phys. Rev*. 1968, 167, 331-344.

[20] Shiina, Y.; Nakamura, Y. Oi, *Solid State Commun*. 1990, 76, 1189-1196.

[21] Chung, J.-H.; Egami, T.; McQueeney, R. J.; Yethiraj, M.; Arai, M.; Yokoo, T.; Petrov, Y.; Mook, H.A.; Endoh, Y.; Tajima, S.; Frost, C.; Dogan, F. *Phys. Rev.* 2003, B 67, 014517(1~9).

[22] Lee, J.; Fujita, K.; McElroy, K.; Slezak, J. A.; Wang, M.; Aiura, Y.; Bando, H.; Ishikado, M.; Masui, T.; Zhu, J.-X.; Balatsky, A. V.; Eisaki, H.; Uchida, S.; Davis, J. C. *Nature* 2006, 442, 546-550.

[23] Thomsen, C.; Cardona, M.; Friedl, B.; Rodriguez, C. O.; Mazin, I. I.; Andersen,O. K. *Solid State Commun.* 1990, 75, 219-223.

[24] Altendorf, E.; Chrzanowski, J.; Irwin, J. C.; O'Reilly, A.; Hardy, W. N. *Physica* 1991, C 175, 47-57.

[25] Kamimura, H.; Matsuno, S.; Suwa, Y.; Ushio, H. *Phys. Rev. Lett.* 1996, 77, 723-726.

[26] Cohen, R. E.; Pickett, W. E.; Krakauer, H. *Phys. Rev. Lett.* 1990, 64, 2575-2578.

[27] Krakauer, H.; Pickett, W. E.; Cohen, R. E. *Phys. Rev.* 1993, B 47, 1002-1015.

[28] Krakauer, H.; Pickett, W. E.; Cohen, R. E. In *Manifestations of the Electron-Phonon Interaction*; Baquero, R.; Ed.; World Scientific, Singapore, 1994, p. 38-45.

[29] Tsuda, N.; Shimada, D. *Physica* 2003, C 390, 95-100.

[30] Iyo, A.; Tanaka, Y.; Kodama, Y.; Kito, H.; Tokiwa, K.; Watanabe, T. *Physica* 2006, C 445-448, 17-22.

[31] Salchow, R.; Liebmann, R.; Appel, J. *J. Phys. Chem. Solids* 1983, 44, 245-259.

[32] Tachiki, M.; Machida, M.; Egami, T. *Phys. Rev.* 2003, B 67, 174506(1~11).

[33] Bishop, A. R.; Martin, R. L.; Müller, K. A.; Tesanovic, Z. *Z. Phys.* 1989, B 76, 17-24.

[34] Rodriguez, C. O.; Liechtenstein, A. I.; Mazin, I. I.; Jepsen, O.; Andersen, O. K.; Methfessel, M.; *Phys. Rev.* 1990, B 42, 2692-2695.

[35] Oda, M.; Hoya, K.; Kubota, R.; Manabe, C.; Momono, N.; Nakao, T.; Ido, M. *Physica* 1997, C 281,135-142.

[36] Miyakawa, N.; Guptasarma, P.; Zasadzinski, J. F.; Hinks, D. G.; Gray, K. E. *Phys. Rev. Lett.* 1998, 80, 157-160.

[37] Miyakawa, N.; Zasadzinski, J. F.; Ozyuzer, L.; Kaneko, T.; Gray, K. E. *Trends in Applied Spectroscopy* 2002, 4, 47-74.

[38] Marsiglio, F.; Akis, R.; Carbotte, J. P. *Phys. Rev.* 1987, B 36, 5245-5250.

[39] Shimada, D.; Miyakawa, N.; Tsuda, N. *Physica* 1997, C 282-287, 1517-1518.

[40] Won, H.; Maki, K. *Phys. Rev.* 1994, B 49, 1397-1402.

[41] Shiina, Y.; Matsuda, N.; Nakamura, Y. Oi. *Physica* 1993, C 212, 173-184.

[42] Tsuda, N.; Arao, T.; Hosokawa, T.; Shiina, Y.; Matsuda, N.; Shimada, D. *Physica* 1997, C 282-287, 1489-1490.

[43] Oki, T.; Tsuda, N.; Shimada, D. *Physica* 2001, C 353, 213-220.

[44] Rossat-Mignod, J.; Regnault, L. P.; Vettier, C.; Bouges, F.; Burlet, P.; Bossy, J.; Henty, J. Y.; Laperto, G. *Physica* 1991, C 185-189, 86-92.

[45] Emery, V. J.; Kivelson, S. A. *Nature* 1995, 374, 434-437.

[46] Deutche, G. *Nature* 1999, 397, 410-412.

[47] Renner, Ch.; Revaz, B.; Genoud, J.-Y.; Kadowaki, K.; Fisher, Φ. *Phys. Rev. Lett.* 1997, 80, 149-152.

[48] Kasiwaya, S.; Y. Tanaka. *Rep. Prog. Phys.* 2000, 63, 1641-1724.

[49] Phillips, J. C. *Phys. Rev.* 2005, B 71, 184505(1~10), and references therein.

[50] Pan, S. H.; O'Neil, J. P.; Badzey, R. L.; Chamon, C.; Ding, H.; Engelbrecht, J. R.; Wang, Z.; Eisaki, H.; Uchida, S.; Gupta, A. K.; Ng, K.-G.; Hudson, E. W.; Lang, K. M.; Davis, J. C. *Nature* 2001, 413, 282-285.

[51] Hanaguri, T.; Lupien, C.; Kohsaka, Y.; Lee, D.-H.; Azuma, A.; Takano, M.; Takagi, H.; Davis, J. C. *Nature* 2004, 430, 1001-1005.

[52] Shimada, D.; Tsuda, N. *Physica* 2002, C 371, 52-56.

[53] Bornemann, H. J.; Morris, D. *Phys. Rev.* 1991, B 44, 5322-5325.

[54] Zhao, G.; Hunt, M. B.; Keller, H.; Müller, K. A.; *Nature* 1997, 385, 236-239.

[55] Franck, J. P.; Harker, S.; Bwerer, J. H. *Phys. Rev. Lett.* 1993, 71, 283-286.

[56] Williams, G. V. M.; Pringle, D. J.; Tallon, J. L. *Phys. Rev.* 2000, B 61, R9257-R9260.

[57] Franck, J. P.; Jung, J.; Mohamed, M. A.-K.; Gygax, S.; Sproule, G. I. *Phys. Rev.* 1991, B 44, 5318-5321.

[58] Inada, Y.; Nasu, K. *J. Phys. Soc. Jpn.* 1992, 61, 4511-4520.

[59] Allen, P. B.; Rainer, D. *Nature* 1991, 349, 396-398.

[60] Shimada, D.; Shiina, Y.; Miyakawa, N.; Tsuda, N. *Physica* 1996, B 219, 192-194.

[61] Takigawa, M., Mitzi, D. B. *Phys. Rev. Lett.* 1994, 73, 1287-1290.

[62] Tralshawala, N.; Zasadzinski, J. F.; Coffey, L.; Gai, W.; Romalis, M.; Huang, Q.; Vaglio, R.; Gray, K. E. *Phys. Rev.* 1995, B 51, 3812-3819.

[63] Honma, T.; Yamaya, K.; Minami, F.; Takekawa, S. *Physica* 1991, C 176, 209-215.

[64] Bozovic, I. *Phys. Rev.* 1990, B 42, 1969-1984.

[65] Zeyher, R.; Kulić, M. L.; *Phys. Rev.* 1996, B 53, 2850-2862.

[66] Kulić, M. L. *Physics Reports* 2000, 338, 1-264.

[67] Ando, Y.; Boebinger, G. S.; Passner, A.; Wang, N. L.; Geibel, C.; Steglich, F. *Phys. Rev. Lett.* 1996, 77, 2065-2068.

[68] Watanabe, T.; Fujii, T.; Matsuda, A. *Phys. Rev. Lett.* 1997, 79, 2113-2116.

[69] Lanzara, A.; Bogdanov, P. V.; Zhou, X. J.; Keller, S. A.; Feng, D. L.; Lu, E. D.; Yoshida, T.; Eisaki, H.; Fujimori, A.; Kishio, K.; Shimiyama, J. -I.; Noda, T.; Uchida, S.; Hussain, Z.; Shen, Z. -X. *Nature* 2001, 412, 510-514.

[70] Gweon, G. -H.; Sasagawa, T.; Zhou, S. Y.; Graf, J.; Takagi, H.; Lee, D. -H.; Lanzara, A. *Nature* 2004, 430, 187-190.

[71] Zhou, X. J.; Yoshida, T.; Lanzara, A.; Bogdanov, P. V.; Keller, S. A.; Shen, K. M.; Yang, W. L.; Ronning, F.; Sasagawa, T.; Kakeshita, T.; Noda, T.; Eisaki, H.; Uchida, S.; Lin, C. T.; Zhou, F.; Xiong, J. W.; Ti, W. X.; Zhao, Z. X.; Fujimori, A.; Hussain, Z.; Shen, Z.-X. *Nature* 2003, 423, 398-398.

[72] Zhou, X. J.; Yoshida, T.; Lee, D. -H.; Yang, W. L.; Brouet, V.; Zhou, F.; Ti, W. X.; Xiong, J. W.; Zhao, Z. X.; Sasagawa, T.; Kakesghita, T.; Eisaki, H.; Uchida, S.; Fujimori, A.; Hussain, Z.; Shen, Z. -X. *Phys. Rev. Lett.* 2004, 92, 187001(1~4).

[73] Verga, S.; Knigavko, A.; Marsiglio, F. *Phys. Rev.* 2003, B 67, 054503(1~5).

[74] Engelsberg, S.; Schrieffer, J. R. *Phys. Rev.* 1963, 131, 993-1008.

[75] Shimada, D.; Umeyama, N.; Ishihara, T.; Tsuda, N. *Physica* 2006, C 439, 105-110.

[76] Tang, D. M.; Li, J.; Gong, C-D. *Phys. Rev.* 2003, B 67, 235421(1~4).

[77] Pietronero, L.; Strässler, S.; Grimaldi, C. *Phys. Rev.* 1995, B 52, 10516-10529.

[78] Grimaldi, C.; Pietronero, L.; Strässler, S. *Phys. Rev.* 1995, B 52, 10530-10535.

[79] Coffey, D.; Coffey, L. *Phys. Rev. Lett.* 1993, 70, 1529-1532.

[80] Sato, T.; Matsui, H.; Terashima, K.; Takahashi, T.; Ding, H.; Yang, H.-B.; Wang, S.-C.; Fujii, T.; Watanabe, T.; Matsuda, A.; Terashima, T.; Kadowaki, K. *Physica* 2004, C 412-414, 51-58.

[81] Shimada, D.; Miyakawa, N.; Kido, T.; Tsuda, N. *J. Phys. Soc. Jpn.* 1989, 58, L387-L389.

[82] For instance, Tsuda, N., Nasu, K.; Fujimori, A.; Siratori, K. *Electronic Conduction in Oxides*, 2nd ed.; Cardon, M.; Fulde, P.; von Klitzing, K.; Merlin, R.; Queisser, H.-J.; Störmer, H.; Ed.; Springer Series in Solid-State Sciences, Springer, Heidelberg, 1990, Vol. 94.

[83] Allen, P. B.; Cohen, M. L. *Phys. Rev.* 1970, B 1, 1329-1336.

[84] Allen, P. B.; Mitrović, B. in *Solid State Physics*; Ehrenreich, H.; Turnbull, D.; Ed.; Academic, New York, 1982, 37, 1-92.

[85] Shimada, D.; Tsuda, N. To appear in *Physica* 2007 C.

[86] Bogdanov, P. V.; Lanzara, A.; Keller, S. A.; Zhou, X. J.; Lu, E. D.; Zheng, W. J.; Gu, G.; Shimiyama, J.-I.; Kishio, K.; Ikeda, H.; Yoshizaki, R.; Hussain, Z.; Shen, Z. X. *Phys. Rev. Lett.* 2000, 85, 2581-2584.

[87] Puchkov, A. V.; Fournier, P.; Basov, D. N.; Timusk, T.; Kapitulnik, A.; Kolesnikov, N. N. *Phys. Rev. Lett.* 1996, 77, 3212-3215.

[88] Shulga, S. V.; Dolgov, O. V.; Maksimov, E. G. *Physica* 1991, C 178, 266-274.

[89] Allen, P. B. *Phys. Rev.* 1971, B 3, 305-320.

[90] Suzuki, M.; Watanabe, T.; Matsuda, A. *Phys. Rev. Lett.* 1999, 82, 5361-5364.

[91] Krasnov, V. M.; Yugens, A.; Winkler, D.; Delsing, P.; Claeson, T. *Phys. Rev. Lett.* 2000, 84, 5860-5863.

[92] Boekholt, M.; Bollmeier, Th.; Buschmann, L.; Fleuster, M.; Güntherodt, G. *Physica* 1992, C 198, 33-41.

[93] Wada, N.; Muro-oka, H.; Nakamura, Y.; Kumagai, K. *Physica* 1989, C 157, 453-459.

[94] Dipasupil, R. M.; Oda, M.; Momono, N.; Ido, M.; *J. Phys. Soc. Jpn.* 2002, 71, 1535-1540.

[95] Miyakawa, N. To be submitted.

In: New Research on Superconductivity
Editor: Barry P. Martins, pp. 143-157
ISBN 978-1-59454-197-1

Chapter 4

RELATIVISTIC BEC AND THE GENERALIZED BEC MODEL OF SUPERCONDUCTIVITY

***M. de Llano*[a] *and M. Grether*[b]**
[a]Texas Center for Superconductivity, University of Houston, Houston, TX 77204, USA and Instituto de Investigaciones en Materiales, Universidad Nacional Autónoma de México, 04510 México, DF, Mexico
[b]Facultad de Ciencias, UNAM, 04510 México, DF, Mexico

1. Introduction

We describe the behavior of the Bose-Einstein condensation (BEC) critical temperature in a relativistic ideal Bose gas with and without the antibosons expected to be pair-produced abundantly at sufficiently hot temperatures. Based on the respective similarities between bosons/antibosons and two-electron (2e)/two-hole (2h) Cooper pairs (CPs) in an interacting nonrelativistic many-fermion system as described by a generalized BEC theory of superconductivity, we hope to elucidate the empirically exceptionally efficient role of hole charge carriers in elevating the critical temperature of superconductors (including room-temperatures *with electron-phonon interactions*) as compared with electron charge carriers. The latter is done within a so-called *ideal* complete boson-fermion (BF) model of superconductivity which is a non-interacting ternary BF mixture consisting of electrons and *both* 2e- and 2h-CPs. It is the BE condensate of the 2h-CPs that emerges with the highest T_cs. This is consistent with empirical trends for either "exotic" or conventional superconductors.

2. Relativistic Bose-Einstein Condensation (RBEC)

The number equation for an ideal boson gas of boson masses m_B in d dimensions (*without* antibosons) is given by

$$N = \sum_{\mathbf{k}} [\exp \beta(E_k - \mu_B) - 1]^{-1} \tag{1}$$

where $\beta = 1/k_B T$, k_B is the Boltzmann constant and $\mu_B(T)$ is the boson chemical potential. The total boson energy is

$$E_k = \sqrt{c^2\hbar^2k^2 + m_B^2\,c^4} \rightarrow \begin{cases} \hbar^2k^2/2m_B + m_Bc^2 + O(k^4), & \text{if } \hbar k\,c << m\,c^2, \quad NR \\ \hbar k\,c\,[1 + \frac{1}{2}(m_B\,c/\hbar k)^2 + \cdots], & \text{if } \hbar k\,c >> m\,c^2, \quad UR \end{cases} \tag{2}$$

where c is the speed of light; the two limits refer to the nonrelativistic (NR) and ultrarelativistic (UR) extremes. For a cubic box of sides L in the continuous limit the sum in (1) over the d-dimensional wavevector **k** becomes an integral, namely

$$\sum_{\mathbf{k}} \rightarrow (L/2\pi)^d \int d^dk. \tag{3}$$

At the Bose-Einstein condensation (BEC) critical temperature $T = T_c$, $\mu_B(T_c) = m_B\,c^2$, and the number density can be expressed as

$$n_B \equiv \frac{N}{L^d} = \frac{1}{(2\pi)^d}\int d^d\,k\,\frac{1}{\exp\left[\beta_c\,(E_k - m_Bc^2)\right] - 1} \tag{4}$$

where $\beta_c \equiv 1/k_BT_c$.

The nonrelativistic limit (NR) follows by substituting the top limiting expression in (2) into (4), which gives

$$n_B = \frac{1}{(2\pi)^d}\int d^dk\,\frac{1}{\exp\left[\beta_c\left(\hbar^2k^2/2m_B\right)\right] - 1}. \tag{5}$$

Using $d^dk = [2\pi^{d/2}/\Gamma(d/2)]k^{d-1}dk$ when integrating over terms independent of angles, leaves

$$n_B = \frac{m_B^d\,c^d}{\hbar^d\,2^{d/2}\pi^{d/2}}\,g_{d/2}(1)\left(\frac{k_BT_c}{m_Bc^2}\right)^{d/2} \tag{6}$$

where

$$g_\sigma(z) \equiv \sum_{l=1}^{\infty}\left(z^l/l^\sigma\right) = \frac{1}{\Gamma(\sigma)}\int_0^\infty dx\frac{x^{\sigma-1}}{z^{-1}\exp x - 1} \tag{7}$$

is the Bose integral which for $z = 1$ diverges $g_\sigma(1) \rightarrow \infty$ when $\sigma \leq 1$, but becomes the Riemann Zeta function $\zeta(\sigma) < \infty$ when $\sigma > 1$. Here $\Gamma(\sigma)$ is the gamma function. Solving for the critical temperature one gets

$$k_BT_c^{NR} = \frac{2\pi\hbar^2}{m_B\left[\zeta(d/2)\right]^{2/d}}n_B^{2/d} \tag{8}$$

which in 3D reduces to the familiar textbook result

$$k_B\,T_c^{NR} = \frac{2\pi\hbar^2}{m_B\left[\zeta(3/2)\right]^{2/3}}n_B^{2/3} \simeq 3.31\frac{\hbar^2}{m_B}n_B^{2/3} \tag{9}$$

since $\zeta(3/2) \simeq 2.612$.

In the ultrarelativistic (UR) limit (1) and (2) lead to

$$n_B = \frac{1}{(2\pi)^d} \int d^d\, k \, \frac{1}{\exp(\beta_c \hbar c k) - 1} \tag{10}$$

which on integration yields

$$n_B = \frac{\Gamma(d)}{\Gamma(d/2) 2^{d-1} \pi^{d/2}} g_d(1) \left(\frac{k_B T_c}{\hbar c} \right)^d . \tag{11}$$

Solving for T_c gives

$$k_B T_c^{UR-B} = \left[\frac{\hbar^d c^d 2^{d-1} \pi^{d/2} \Gamma(d/2)}{\Gamma(d) \zeta(d)} \right]^{1/d} n_B^{1/d} \qquad UR, \quad \textit{no antibosons} \tag{12}$$

where the superscript UR-B stands for the ultrarelativistic limit with bosons (B) only and no antibosons (A). In 3D this becomes

$$k_B T_c^{UR-B} = \frac{\hbar c \pi^{2/3}}{[\zeta(3)]^{1/3}} n_B^{1/3} \simeq 2.017 \hbar c n_B^{1/3} \qquad 3D\, UR, \quad \textit{no antibosons} \tag{13}$$

as $\zeta(3) \simeq 1.20206$. In 2D $T_c^{UR-B} \neq 0$ unlike the common textbook case with quadratic dispersion where T_c^{UR-B} vanishes since $g_1(1)$ diverges. Specifically

$$k_B T_c^{UR-B} = \hbar c [2\pi / \zeta(2)]^{1/2} n_B^{1/2} \simeq 1.954 \hbar c n_B^{1/2} \qquad 2D\, UR, \quad \textit{no antibosons} \tag{14}$$

since $\zeta(2) = \pi^2/6$.

3. Relativistic BEC, Antibosons Included

At sufficiently high temperatures such that $k_B T >> m_B c^2$ boson-antiboson pair-production will occur abundantly; this has been stressed by Huang [1]. The complete number equation is then [2]

$$\begin{aligned} N - N^- &\equiv \sum_{\mathbf{k}} \left(n_{\mathbf{k}} - n_{\mathbf{k}}^- \right) \\ &= \sum_{\mathbf{k}} \left[\frac{1}{\exp\left[\beta(E_k - \mu_B)\right] - 1} - \frac{1}{\exp\left[\beta(E_k + \mu_B)\right] - 1} \right] \end{aligned} \tag{15}$$

where $n_{\mathbf{k}}$, $\left(n_{\mathbf{k}}^-\right)$ is the average number of bosons (anti-bosons) in the state of energy E_k at a given temperature T and N, N^- their total number at that temperature. Since $n_{\mathbf{k}}$, $n_{\mathbf{k}}^- > 0$ for all $\mathbf{k}$, the chemical potential must be bounded by

$$-m_B c^2 \leq \mu_B \leq m_B c^2. \tag{16}$$

Further, (15) simplifies to

$$N - N^- = \sum_{\mathbf{k}} \left[\frac{\sinh(\beta \mu_B)}{\cosh(\beta E_k) - \cosh(\beta \mu_B)} \right]. \tag{17}$$

Instead of N constant we must now impose the constancy of $N - N^-$ to extract the correct BEC critical temperature which we call T_c^{BA}. At $T = T_c^{BA}$, $\left|\mu_B(T_c^{BA})\right| = m_B c^2$ and the number density in this case can be expressed as

$$\begin{aligned} n_B &\equiv (N - N^-)/L^d \\ &= \frac{2\pi^{d/2}}{\Gamma(d/2)(2\pi)^d} \int_0^\infty dk k^{d-1} \frac{\sinh(\beta_c m_B c^2)}{\cosh\left(\beta_c \sqrt{c^2\hbar^2 k^2 + m_B^2 c^4}\right) - \cosh(\beta_c m_B c^2)} \end{aligned} \tag{18}$$

where (3) was employed in (17). Equation (18) is an exact expression for the ideal Bose gas at *any* temperature.

If $k_B T_c << m_B c^2$ antibosons can be neglected making (18) reduce to

$$\begin{aligned} n_B &\simeq \frac{1}{\Gamma(d/2) 2^{d-1} \pi^{d/2}} \int_0^\infty dk\, k^{d-1} \frac{1}{\exp\left(\beta_c \hbar^2 k^2 / 2m_B\right) - 1} \\ &= \frac{m_B^d c^d}{\hbar^d 2^{d/2} \pi^{d/2}} \zeta(d/2) \left(\frac{k_B T_c^{NR-BA}}{m_B c^2}\right)^{d/2}. \end{aligned} \tag{19}$$

This result coincides with (6) if $d/2 > 1$, when $g_{d/2}(1) \equiv \zeta(d/2)$.

If $k_B T_c >> m_B c^2$ then (18) becomes

$$\begin{aligned} n_B &\simeq \frac{2\beta_c m_B c^2 \pi^{d/2}}{\Gamma(d/2)(2\pi)^d} \int_0^\infty dk \frac{k^{d-1}}{\cosh(\beta_c \hbar c k) - 1} \\ &= \frac{4\beta_c m_B c^2 \pi^{d/2}}{\Gamma(d/2)(2\pi)^d} \int_0^\infty dk k^{d-1} \frac{e^{\beta_c \hbar c k}}{(\exp \beta_c c\hbar k - 1)^2} \end{aligned} \tag{20}$$

which on integrating gives

$$n_B \simeq \frac{4 m_B \pi^{d/2} \Gamma(d) \zeta(d-1)}{\hbar^d c^{d-2} \Gamma(d/2+1)(2\pi)^d} (k_B T_c)^{d-1}. \tag{21}$$

Then T_c is given by the limiting expression

$$k_B T_c^{UR-BA} = \left[\frac{\hbar^d c^{d-2} \Gamma(d/2)(2\pi)^d}{4 m_B \pi^{d/2} \Gamma(d) \zeta(d-1)}\right]^{1/(d-1)} n_B^{1/(d-1)} \quad UR,\ with\ antibosons \tag{22}$$

and sharply differs from (12). In 3D this becomes

$$k_B T_c^{UR-BA} = \left[\frac{3\hbar^3 c}{m_B}\right]^{1/2} n_B^{1/2} \simeq 1.732 \left[\frac{\hbar^3 c}{m_B}\right]^{1/2} n_B^{1/2} \quad UR,\ with\ antibosons. \tag{23}$$

In Fig. 1 the behavior of the exact T_c numerically extracted from (18) for $d = 3$ is compared with the nonrelativistic T_c^{NR-BA} from (19) and with the limit (23) for the ultrarelativistic T_c^{UR-BA}, all as functions of the dimensionless number density $\hbar^3 n_B / m_B^3 c^3$.

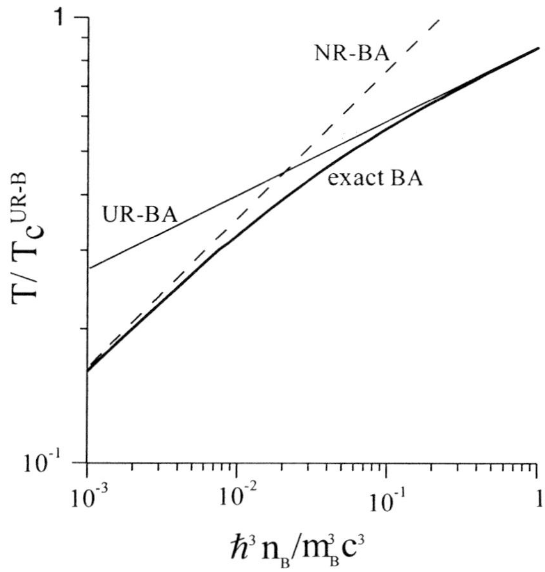

Figure 1. Critical temperature [in units of T_c^{UR-B} defined by (13)] in 3D as function of the dimensionless boson-number density variable $\hbar^3 n_B/m_B^3 c^3$. Thick curve is the exact numerical result extracted from (18) representing the BEC in a relativistic ideal Bose gas with antibosons. Thin full curve is the ultrarelativistic limit (23). Dashed curve is the nonrelativistic limit implied by (19).

Fig. 2 shows how, at sufficiently high densities the exact T_c^{BA} is the *first* BEC singularity encountered as the system is cooled, compared with the later BEC singularity in the system *without* antibosons at T_c^B extracted numerically from (4) and labeled in the figure "exact B." It is precisely this distinctly different behavior that puzzlingly also appears in the generalized nonrelativistic BEC theory of superconductors to which we now turn.

4. Generalized BEC for Charged Boson-Fermion Mixtures

Boson-fermion (BF) models of superconductivity (SC) as a BEC go back to the mid-1950's [3]-[6], pre-dating even the BCS-Bogoliubov theory [7]-[10]. Although BCS theory only contemplates the presence of "Cooper correlations" of single-particle states, *binary* BF models [3]-[6],[11]-[20] posit the existence of actual bosonic CPs. In spite of the central role played by CPs in both low- and high-T_c superconductivity, however, they are poorly understood. The fundamental drawback of early [3]-[6] BF models, which took 2e bosons as analogous to diatomic molecules in a classical atom-molecule gas mixture and thus with a quadratic dispersion, is the notorious absence of an electron energy gap $\Delta(T)$. "Gapless" models are useful in locating transition temperatures if approached from above, i.e., $T > T_c$. Even so, we are not aware of any calculations with the early BF models attempting to reproduce any empirical T_c values. The gap first began to appear in later BF models [11]-[17].

With two [14, 15] exceptions, however, all BF models neglect the effect of *hole* CPs

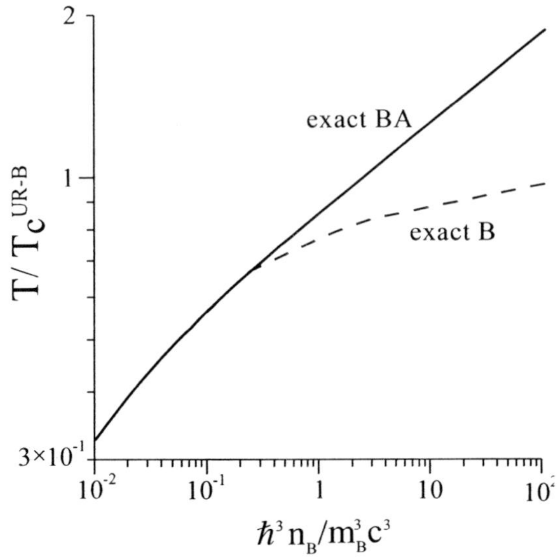

Figure 2. Critical temperature [in units of T_c^{UR-B} defined by (13)] in 3D as function of the dimensionless variable $\hbar^3 n_B/m_B^3 c^3$. Full curve is the exact numerical result of (18) representing the BEC T_c in the relativistic ideal Bose gas with antibosons included. Dashed curve represents the same quantity without antibosons as extracted numerically from (4).

included on an equal footing with electron CPs to give a generalized BEC (GBEC) theory consisting of *both* bosonic CP species coexisting with unpaired electrons in a *ternary* BF model. Unfortunately, no experiment has yet been done, to our knowledge, that distinguishes between electron and hole CPs. But for two distinct interfermion interactions (the δ-potential well [21][22] or the Cooper/BCS model [7][8]interactions) the "ordinary" CP problem [7] neglects the effect of 2h CPs treated on an equal footing with 2p CPs—as Green's functions [23], on the other hand, can naturally guarantee. A crucial observation [14, 15] is that the BCS condensate consists of *equal numbers*of 2p and 2h CPs. This was already suggested, if not evident, though widely unrecognized, from the perfect symmetry about $\epsilon = \mu$, the electron chemical potential, of the well-known Bogoliubov [24] $v^2(\epsilon)$ and $u^2(\epsilon)$ coefficients, where ϵ is the electron energy. The GBEC theory (also appropriately designated elsewhere as a "complete BF model," or CBFM) "unifies" [16] both BCS and ordinary BEC theories as special cases, and produces substantially higher T_c's than BCS theory without abandoning electron-phonon dynamics.

The GBEC theory [14, 15] is described in d dimensions by the Hamiltonian $H = H_0 + H_{int}$. The unperturbed Hamiltonian H_0 corresponds to a non-Fermi-liquid "normal" state which is an *ideal* (i.e., noninteracting) ternary gas mixture of unpaired fermions and both types of CPs, two-electron (2e) and two-hole (2h), and is

$$H_0 = \sum_{\mathbf{k}_1, s_1} \epsilon_{\mathbf{k}_1} a^+_{\mathbf{k}_1,s_1} a_{\mathbf{k}_1,s_1} + \sum_{\mathbf{K}} E_+(K) b^+_{\mathbf{K}} b_{\mathbf{K}} - \sum_{\mathbf{K}} E_-(K) c^+_{\mathbf{K}} c_{\mathbf{K}} \tag{24}$$

where $\mathbf{K} \equiv \mathbf{k_1} + \mathbf{k_2}$ is the CP CMM wavevector while $\epsilon_{\mathbf{k}_1} \equiv \hbar^2 k_1^2/2m$ are the single-electron, and $E_\pm(K)$ the 2e-/2h-CP *phenomenological*, energies. Here $a^+_{\mathbf{k}_1,s_1}$ $(a_{\mathbf{k}_1,s_1})$ are creation (annihilation) operators for fermions and similarly $b^+_{\mathbf{K}}$ $(b_{\mathbf{K}})$ and $c^+_{\mathbf{K}}$ $(c_{\mathbf{K}})$ for 2e- and 2h-CP bosons, respectively, although we do not attempt to construct them starting from the fermion operators. One postulates CPs in (24) based on magnetic-flux quantization experiments [25]-[27] that establish the presence of charged pairs—albeit without being able [28] to specify the *sign* of those charges. Unlike single, unpaired electrons and holes, two-hole CPs are further postulated to be *distinct* and *kinematically independent* from 2e-CPs, all of which provides a *ternary* BF gas mixture.

The interaction Hamiltonian H_{int} consists of four distinct BF interaction vertices each with two-fermion/one-boson creation or annihilation operators, depicting how unpaired electrons (subindex +) [or holes (subindex $-$)] combine to form the 2e- (and 2h-) CPs assumed in the d-dimensional system of size L, namely

$$H_{int} = L^{-d/2}\sum_{\mathbf{k},\mathbf{K}} f_+(k)\{a^+_{\mathbf{k}+\frac{1}{2}\mathbf{K},\uparrow}a^+_{-\mathbf{k}+\frac{1}{2}\mathbf{K},\downarrow}b_{\mathbf{K}} + a_{-\mathbf{k}+\frac{1}{2}\mathbf{K},\downarrow}a_{\mathbf{k}+\frac{1}{2}\mathbf{K},\uparrow}b^+_{\mathbf{K}}\}$$
$$+ L^{-d/2}\sum_{\mathbf{k},\mathbf{K}} f_-(k)\{a^+_{\mathbf{k}+\frac{1}{2}\mathbf{K},\uparrow}a^+_{-\mathbf{k}+\frac{1}{2}\mathbf{K},\downarrow}c^+_{\mathbf{K}} + a_{-\mathbf{k}+\frac{1}{2}\mathbf{K},\downarrow}a_{\mathbf{k}+\frac{1}{2}\mathbf{K},\uparrow}c_{\mathbf{K}}\}. \qquad (25)$$

where $\mathbf{k} \equiv \frac{1}{2}(\mathbf{k_1} - \mathbf{k_2})$ is the relative wavevector of a CP. The form factors $f_\pm(k)$ in (25) are essentially the Fourier transforms of the 2e- and 2h-CP intrinsic wavefunctions, respectively, in the relative coordinate of the two fermions. In Refs. [14][15] they are taken as

$$f_\pm(\epsilon) = \begin{cases} f & \text{if } \frac{1}{2}[E_\pm(0) - \delta\varepsilon] < \epsilon < \frac{1}{2}[E_\pm(0) + \delta\varepsilon] \\ 0 & \text{if} \qquad\qquad \text{otherwise} \end{cases} \qquad (26)$$

We now introduce the quantities E_f and $\delta\varepsilon$ as *new* phenomenological dynamical energy parameters (in addition to the positive BF vertex coupling parameter f) that replace the previous such $E_\pm(0)$, through the definitions

$$E_f \equiv \frac{1}{4}[E_+(0) + E_-(0)] \qquad \textit{and} \qquad \delta\varepsilon \equiv \frac{1}{2}[E_+(0) - E_-(0)] \qquad (27)$$

where $E_\pm(0)$ are the (empirically *un*known) zero-CMM energies of the 2e- and 2h-CPs, respectively. Clearly

$$E_\pm(0) = 2E_f \pm \delta\varepsilon. \qquad (28)$$

The quantity E_f serves as a convenient energy scale; it is not to be confused with the Fermi energy $E_F = \frac{1}{2}mv_F^2 \equiv k_BT_F$ where T_F is the Fermi temperature. The Fermi energy E_F equals $\pi\hbar^2n/m$ in 2D and $(\hbar^2/2m)(3\pi^2n)^{2/3}$ in 3D, with n the total number-density of charge-carrier electrons, while E_f is the same with n replaced by, say, n_f. The quantities E_f and E_F coincide *only* when perfect 2e/2h-CP symmetry holds, i.e. when $n = n_f$. The GBEC theory reduces to BCS theory in weak coupling provided we identify $\delta\varepsilon$ with $\hbar\omega_D$, as will be done henceforth.

The interaction Hamiltonian (25) can be further simplified by dropping all $\mathbf{K} \neq 0$ terms. This is done in the *full* Hamiltonian of BCS theory, but only in the H_{int} part of the H in the GBEC theory, *not* in its H_0 portion. The boson operators b_0^+, b_0 and c_0^+, c_0 remaining in (25) are then replaced, following Bogoliubov [29], by the "c-numbers" $\sqrt{N_0(T)}$ and $\sqrt{M_0(T)}$, respectively, where N_0 and M_0 are the number of $K = 0$ 2e- and 2h-CPs. The GBEC theory H thus resulting is *bilinear* in the a^+, a single-electron operators; this enables one to *exactly* diagonalize H via a Bogoliubov transformation. Hence, unlike BCS theory, the GBEC theory goes beyond mean-field. It can further be argued [30] that, unlike BCS theory, it is not a weak-coupling theory *a priori*. Let $\hat{N}$ be the electron number operator

$$\hat{N} = \sum_{\mathbf{k}_1, s_1} a^+_{\mathbf{k}_1, s_1} a_{\mathbf{k}_1, s_1} + 2\sum_{\mathbf{K}} b^+_{\mathbf{K}} b_{\mathbf{K}} - 2\sum_{\mathbf{K}} c^+_{\mathbf{K}} c_{\mathbf{K}}. \tag{29}$$

Then, along with (24) and (25), the grand potential for the GBEC theory is constructed as

$$\Omega(T, L^d, \mu, N_0, M_0) = -k_B T \ln \left[Tr\ \exp\{-\beta(H - \mu\hat{N})\} \right] \tag{30}$$

where “Tr” stands for “trace.” An *equilibrium* thermodynamic state of the system with volume L^d and temperature T is stationary with respect to N_0 and M_0. Thus, one imposes that

$$\frac{\partial\Omega}{\partial N_0} = 0, \qquad \frac{\partial\Omega}{\partial M_0} = 0 \qquad and \qquad \frac{\partial\Omega}{\partial\mu} = -N. \tag{31}$$

where the last relation fixes the total number N of electrons, both paired and unpaired, by introducing the electron chemical potential μ. Some algebra then leads to the three coupled integral Eqs. (7)-(9) of Ref. [14]. The relation between the fermion spectrum $E(\epsilon)$ and fermion energy gap $\Delta(\epsilon)$ turns out to be BCS-like

$$E(\epsilon) = \sqrt{(\epsilon - \mu)^2 + \Delta^2(\epsilon)} \quad where \quad \Delta(\epsilon) \equiv \sqrt{n_0} f_+(\epsilon) + \sqrt{m_0} f_-(\epsilon). \tag{32}$$

This last expression for the gap $\Delta(\epsilon)$ implies a simple T-dependence rooted in the 2e-CP $n_0(T) \equiv N_0(T)/L^d$ and 2h-CP $m_0(T) \equiv M_0(T)/L^d$ number densities of BE-condensed bosons, i.e.,

$$\Delta(T) = \sqrt{n_0(T)} f_+(\epsilon) + \sqrt{m_0(T)} f_-(\epsilon). \tag{33}$$

Self-consistent (at worst, numerical) solution of the aforementioned *three coupled equations* then yields the three thermodynamic variables of the GBEC theory

$$n_0(T, n, \mu), \quad m_0(T, n, \mu), \quad and \quad \mu(T, n). \tag{34}$$

Most significantly, the three GBEC theory equations contain the key equations of *five* different statistical theories as special cases. Perfect 2e/2h CP symmetry signifies equal number of 2e- and 2h-CPs, i.e., $n_B(T) = m_B(T)$ *as well as* $n_0(T) = m_0(T)$. With (28) this implies that E_f coincides with μ, and the GBEC theory then reduces to:

i) the gap and number equations of the *BCS-Bose crossover picture* for the BCS model interaction—if the BCS parameters V and Debye energy $\hbar\omega_D$ are identified with the BF interaction Hamiltonian H_{int} parameters $f^2/2\delta\varepsilon$ and $\delta\varepsilon$, respectively. The crossover picture for unknowns $\Delta(T)$ and $\mu(T)$ is now supplemented by the key relation

$$\Delta(T) = f\sqrt{n_0(T)} = f\sqrt{m_0(T)}. \tag{35}$$

The crossover picture is associated with many writers beginning with Friedel and coauthors [31] in 1967; for a review see Ref. [32]. However, it is widely unrecognized to be a very modest improvement, at least for the Cooper/BCS model interaction, over BCS theory *per se* since an unphysically large λ of about 8 is required to bring $\mu(T_c)/E_F$ down from 1.00 to 0.998. As usual, $\lambda \equiv VN(E_F)$ where $N(E_F)$ is the electronic density of states (for one spin) at the Fermi surface. In fact, T_c-values differ very slightly [30] all the way up to $\lambda \sim 50$ when the Fermi surface pinned at μ disappears so that the model interaction breaks down. Indeed, room-temperature superconductivity (RTSC) is formally possible within the BCS-Bose crossover picture but only for unphysically large values of $\lambda \gtrsim 20$. If one imposes that $\mu(T_c) = E_F$ exactly, as follows from the number equation for weak BF coupling f, the crossover picture is well-known to reduce to:

ii) *ordinary BCS theory* which is characterized by a *single* equation, the gap equation for any T. Thus, *the BCS condensate is precisely a BE condensate* whenever both $n_B(T) = m_B(T)$ and $n_0(T) = m_0(T)$ *and* the BF coupling f is small. For small coupling λ the GBEC theory $T = 0$ superconducting state has the *same* condensation energy in lowest order in λ, or equivalently Δ, as the BCS state, and is in fact *lower* in next-to-lowest order [30]. On the other hand, for no 2h-CPs present the GBEC theory reduces [14] also to:

iii) the *BEC BF model* in 3D of Friedberg and T.D. Lee [12, 13] characterized by the relation $\Delta(T) = f\sqrt{n_0(T)}$ first reported in Ref. [11]. But lacking 2h-CPs this model cannot strictly be related to BCS theory. When $f = 0$ it reduces to:

iv) the *ideal BF model* (IBFM) of Refs. [19, 20] that predicts nonzero 2e-CP BEC T_c's even in 2D. The "gapless" IBFM cannot describe the superconducting phase. But considered as a model for the *normal state* it should provide feasible T_c's as singularities within a BE scenario that are approached from *above* T_c, and this is indeed [20] the case. Finally, in 3D this model in the limit of no unpaired electrons reduces to:

v) the familiar T_c-formula of ordinary BEC in 3D.

The very general GBEC theory has been applied in both 2D and 3D and gives sizeable enhancements in T_c's over BCS theory that emerge for moderate departures from perfect 2e/2h-pair symmetry. This is attained for the *same* Cooper/BCS interaction model (coupling strength λ and cutoff $\hbar\omega_D$) parameters often used in conventional superconductors. The three coupled equations of the GBEC theory that determine the d-dimensional BE-condensate number-densities $n_0(T)$ and $m_0(T)$ of 2e- and 2h-CPs, respectively, as well as the electron chemical potential $\mu(T)$, were solved numerically. At $n/n_f = 1$ one has perfect 2e/2h-CP symmetry; n_f can be seen [15] to be the number $n_f(T)$ of unpaired electrons per unit volume when $\Delta = 0$ and $T = 0$. The third, or "complete" number, equation then explicitly reads

$$2n_B(T) + 2m_B(T) + n_f(T) = n \tag{36}$$

where $n_B(T) \equiv n_0(T) + n_{B+}(T)$ and $m_B(T) \equiv m_0(T) + m_{B+}(T)$, with $n_{B+}(T)$ and $m_{B+}(T)$ being precisely the *total* number of "pre-formed" $K > 0$ 2e- and 2h-CPs, respectively. Besides the *normal* phase consisting of the ideal BF ternary gas described by H_0, three different stable BEC phases emerge, two pure and one mixed phase.

We shall here focus on *linearly*-dispersive bosons. This occurs in the leading term or "ordinary" CPs in a Fermi sea as well as for "generalized" CPs in a BCS-correlated Fermi sea. For the latter, the boson excitation energy ε to be used has a leading term in the many-body Bethe-Salpeter (BS) CP dispersion relation given by $\varepsilon \simeq (\lambda/4)\hbar v_F K$ in 3D [33]. As usual, $\lambda \equiv VN(E_F)$ where $N(E_F)$ is the electron DOS (for one spin) at the Fermi surface. This ε contrasts sharply with the quadratic $\hbar^2K^2/2(2m)$ appropriate for a composite boson of mass $2m$ moving not in the Fermi sea but in vacuum, and which was assumed in Refs.[3]-[6], [12]-[15].

5. Ideal Complete Boson-Fermion Model (ICBFM)

The ICBFM is described in d dimensions by the Hamiltonian H_0 defined by (24). One can construct its associated grand potential

$$\Omega_0(T, L^d, \mu, N_0, M_0) = -k_B T \ln\left[Tr \exp\{-\beta(H_0 - \mu\hat{N})\}\right] \tag{37}$$

with N_0 and M_0 as before the number of zero-CMM 2e- and 2h-CPs, respectively. One gets

$$\begin{aligned} \Omega_0(T,\, L^d, \mu,\, N_0,\, M_0) &= [E_+(0) - 2\mu]N_0 + [2\mu - E_-(0)]M_0 \\ &- 2k_B T B_d L^d \int_0^\infty k^{d-1} dk \ln[1 + \exp\{-\beta(\epsilon_k - \mu)\}] \\ &+ k_B T B_d L^d \int_{0+}^\infty K^{d-1} dK \ln(1 - \exp\{-\beta[2\mu - E_-(K)]\}) \end{aligned} \tag{38}$$

where $B_d = 1/\pi$, $1/2\pi$ and $1/2\pi^2$ for $d = 1$, 2 and 3, respectively. An equilibrium thermodynamic state makes $\Omega(T, L^d, \mu, N_0, M_0)$ stationary with respect to N_0 and to M_0 and requires that the number density of electrons $n = N/L^d$ remain constant. Namely, one imposes that

$$\frac{\partial\Omega_0}{\partial N_0} = 0, \qquad \frac{\partial\Omega_0}{\partial M_0} = 0 \qquad and \qquad -\frac{\partial\Omega_0}{\partial\mu} = N. \tag{39}$$

This leads to the three equations

$$[E_+(0) - 2\mu] = 0, \quad [2\mu - E_-(0)] = 0 \quad and \quad 2n_B(T) - 2m_B(T) + n_f(T) = n \tag{40}$$

with the latter being the *"number equation"* that ensures charge conservation in the ternary mixture. In the last relation

$$n_B(T) \equiv n_0(T) + B_d \int_{0+}^\infty dK K^{d-1}[\exp\beta[E_+(0) - 2\mu + \varepsilon_K] - 1]^{-1} \tag{41}$$

$$m_B(T) \equiv m_0(T) + B_d \int_{0+}^\infty dK K^{d-1}[\exp\beta[2\mu - E_-(0) + \varepsilon_K] - 1]^{-1} \tag{42}$$

while

$$n_f(T) \equiv 2B_d \int_0^\infty dk k^{d-1}[\exp\{\beta(\epsilon_k - \mu)\} + 1]^{-1} \tag{43}$$

is the number density of *unpaired* electrons at any T and $\epsilon_k = \hbar^2 k^2/2m$ is the electron energy.

Consider the three equations (40) assuming only that $E_-(0) < E_+(0)$. The third equation of (40) explicitly becomes

$$
\begin{aligned}
n \equiv k_F^3/3\pi^2 &= 2B_d \int_0^\infty k^{d-1}dk\,[\exp\{\beta(\epsilon_k-\mu)\}+1]^{-1} \\
&+ 2n_0(T) + 2B_d \int_0^\infty K^{d-1}dK(\exp\{\beta[E_+(0)-2\mu+\varepsilon_K]\}-1)^{-1} \\
&- 2m_0(T) - 2B_d \int_0^\infty K^{d-1}dK(\exp\{\beta[2\mu-E_-(0)+\varepsilon_K]\}-1)^{-1}.
\end{aligned} \tag{44}
$$

Take first the limiting case $2\mu = E_+(0)$ when the first equation of (40) is satisfied but not the second. From (39) this implies that $M_0(T) = 0$ for all T but that $N_0(T) \neq 0$ so that a pure 2e-CP BEC phase may occur below a critical temperature T_c determined by $n_0(T_c, n) = 0$. After substituting (28) in (44) with $\mu = E_+(0)/2$, thus eliminating $E_\pm(0)$ in favor of $E_f \equiv (\hbar^2/2m)(3\pi^2 n_f)^{2/3}$and $\delta\varepsilon \equiv \hbar\omega_D$, we obtain a dimensionless "working number equation" for the *pure 2e-CP BEC phase*

$$
\begin{aligned}
1/3 &= \int_0^\infty \tilde{k}^2 d\tilde{k}[\exp\{(\tilde{k}^2 - \tilde{n}^{-2/3} - \nu/2)/\tilde{T}_c\}+1]^{-1} + \int_0^\infty \tilde{K}^2 d\tilde{K}[\exp\{\lambda\tilde{K}/2\tilde{T}_c\}-1]^{-1} \\
&- \int_0^\infty \tilde{K}^2 d\tilde{K}[\exp\{(\lambda\tilde{K}+4\nu)/2\tilde{T}_c\}-1]^{-1}
\end{aligned} \tag{45}
$$

with $\tilde{k} \equiv k/k_F$, $\tilde{K} \equiv K/k_F$, $\tilde{n} \equiv n/n_f$, $\tilde{T}_c \equiv T_c/T_F$, $\nu \equiv \hbar\omega_D/E_F$, and we took $\varepsilon_K \simeq \lambda\hbar v_F K/4$, Ref. [33], Eq. (12), for $d = 3$. The integrals are exact, the first and last being expressible as polylogarithm functions $Li_\sigma(z)$ (designated by PolyLog[σ,z] in Ref. [34]) where

$$
-aLi_\sigma(-az) \equiv \frac{1}{\Gamma(\sigma)}\int_0^\infty dx \frac{x^{\sigma-1}}{z^{-1}\exp x + a} = -\frac{1}{a}\sum_{l=1}^\infty \frac{(-az)^l}{l^\sigma}. \tag{46}
$$

with z a true (or effective) fugacity. For $a = -1$ (46) reduces to the Bose integral $g_\sigma(z)$ defined in (7), which for $z = 1$ and $\sigma \geq 1$ becomes the Riemann Zeta function $\zeta(\sigma)$ of order σ; for $a = 1$ it becomes the Fermi integral $f_\sigma(z)$. Both integrals are as defined in Appendices D and E of Ref. [35]. The second integral in (45) gives just $\zeta(3)$, and the working number equation can be reduced to

$$
\begin{aligned}
1/3 &= -\frac{\sqrt{\pi}\tilde{T}_c^{\,3/2}}{4} PolyLog[3/2, -\exp\{(\nu/2+\tilde{n}^{-2/3})/\tilde{T}_c\}] \\
&+ \frac{16\tilde{T}_c^{\,3}}{\lambda^3}\{\zeta(3) - \{PolyLog[3, \exp(-2\nu/\tilde{T}_c)]\}.
\end{aligned} \tag{47}
$$

This can be solved for $\tilde{T}_c$ as a function of $\tilde{n}$ which is plotted as the dashed curve in Fig. 3 for $\lambda = \frac{1}{2}$ and $\hbar\omega_D = 0.0002$.

The *pure 2h-CP BEC phase* comes from the limiting case $2\mu = E_-(0)$ as then second, but not the first, relation in (40) is satisfied, so that, again from (39), $N_0 = 0$ and $M_0 \neq 0$. A working equation different from (45) ensues and the critical temperature is now determined by $m_0(T_c, n) = 0$ which eventually gives

$$\begin{aligned} 1/3 = &-\frac{\sqrt{\pi}\tilde{T_c}^{3/2}}{4} PolyLog[3/2, -\exp\{(\tilde{n}^{-2/3} - \nu/2)/\tilde{T_c}\}] \\ &+ \frac{16\tilde{T_c}^3}{\lambda^3}\{PolyLog[3, \exp(-2\nu/\tilde{T_c})] - \zeta(3)\}. \end{aligned} \tag{48}$$

Solving this for $\tilde{T}_c$ leads to the full curve in Fig. 3. The two uppermost curves, dashed and full, are the phase-boundary curves for the two pure, stable BEC phases for $\lambda = 1/2$, for $\hbar\omega_D/E_F = 0.0002$. The BCS value from their formula $T_c/T_F \simeq 1.134(\hbar\omega_D/E_F)\exp(-1/\lambda)$ gives $\simeq 3.1 \times 10^{-5}$ for $\lambda = 1/2$, for $\hbar\omega_D/E_F = 0.0002$ (large black dot); it lies well within the range $T_c/T_F \lesssim 10^{-3}$ empirically found for conventional, elemental superconductors. Empirical data [36] for cuprate superconductors fall within the range $T_c/T_F \simeq 0.01 - 0.05$. Thus, moderate departures from perfect 2e/2h-CP symmetry enable the ICBFM to predict, unlike BCS theory, empirical T_c values for cuprates *without abandoning electron-phonon dynamics*. The figure also shows that *room-temperature superconductivity* (RTSC) is possible in principle within the ICBFM, but only via 2h-CP BE condensates for $\hbar\omega_D/E_F = 0.0002$ as the lower-T_c 2e-CP BE condensate arguably [15] having higher Helmholtz free-energy is metastable with respect to the stable 2h-CP BEC phase. A more realistic value of $\hbar\omega_D/E_F$ is 0.005 and yields no pure 2h-CP BEC phase, only the pure 2e-CP BEC phase shown as the lowest curve (dot-dashed). In this case, the BCS value 7.7×10^{-4} is marked by an open circle.

Finally, for intermediate values of μ, namely for $E_-(0) < 2\mu < E_+(0)$, neither the first nor second equations of (39) are satisfied so that $n_0(T) = 0 = m_0(T)$ for all T and this implies no condensed phases. Thus, the ICBFM (characterized by zero BF coupling f) *contains no mixed phase* in contrast to the CBFM where $f \neq 0$ which will be treated elsewhere.

The GBEC theory rests on four essential ingredients: 1) 2h-CPs cannot and must not be neglected in a fully self-consistent treatment in any many-*fermion* system, otherwise a spurious value of T_c may result; 2) CPs are *bosons*, even though BCS pairs not; 3) CPs are *linearly-dispersive* for small K; 4) to achieve higher T_c's one must depart from the perfect 2e-/2h-CP symmetry of the BE condensate of BCS. Neglected here thus far, however, are: a) $K > 0$ terms in the boson-fermion vertex interactions; b) boson-boson interactions (as is BCS); c) a $T > 0$ Bethe-Salpeter CP treatment; d) different hole and electron effective masses; and e) ionic-lattice crystallinity effects which might initially be dealt with via Van Hove singularities introduced in the electronic density of states or via "bipolarons" instead of CPs.

6. Conclusions

To summarize, the new generalized BEC (GBEC) theory includes as limiting cases the following theories: i) BCS and ii) BCS-Bose "crossover" when the BE condensate consists of

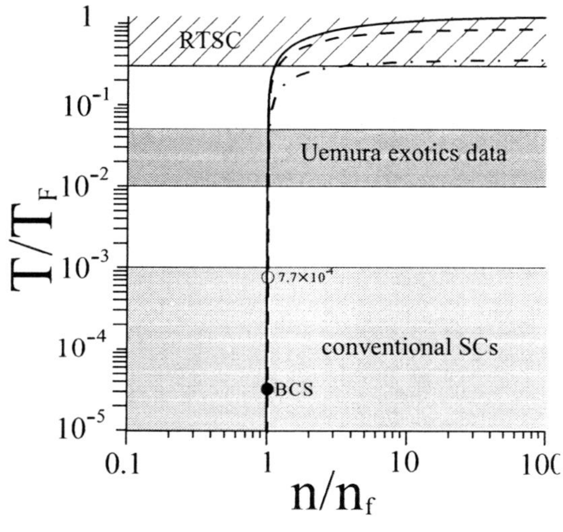

Figure 3. Temperature T (in units of T_F) vs electron density n [in units of n_f as defined just below (28)] ICBFM phase diagram in 3D showing the T_c phase boundaries for the pure 2e-CP BEC phase (dashed curve) and the pure 2h-CP BEC phase (full curve) for $\lambda = 1/2$ with $\nu \equiv \hbar\omega_D/E_F = 0.0002$. The BCS T_c's are determined as explained in text are marked by a black dot (for $\hbar\omega_D/E_F = 0.0002$) and by an open circle (for $\hbar\omega_D/E_F = 0.005$). For this latter value the only pure phase found is the 2e-CP BEC phase (dot-dashed). RTSC refers to a material with $T_F = 10^3\ K$.

equal numbers of electron- and hole-pairs, as well as iii) the Friedberg-Lee BEC model, iv) the "ideal boson-fermion model" (IBFM), and v) BEC theory, when there are no unpaired electrons. The BCS condensate is precisely a BE condensate of equal numbers of 2e/2h-pairs and weak coupling. Without abandoning electron-phonon dynamics the GBEC theory leads to up to four orders-of-magnitude higher T_c's—including room-temperature superconductivity. It also suggests an "explanation" for the puzzling empirical skewed role of normal-state hole charge-carriers in producing higher T_c's, in full analogy with the higher-T_c beyond a certain total boson number density in the RBEC case *with* antiboson as compared to the case without them.

Acknowledgments

We acknowledge UNAM-DGAPA-PAPIIT (Mexico) grant # IN106401 and CONACyT (Mexico) grant # 41302F for partial support. MdeLl appreciates the hospitality extended him at the Texas Center for Superconductivity, University of Houston during a sabbatical leave, and is grateful for travel support to attend the CMT29 Workshop through a grant to Southern Illinois University at Carbondale from the U.S. Army Research Office.

References

[1] K. Huang, *Statistical Mechanics* (Wiley, NY, 1987, 2nd Ed.) pp. 154-7 and esp. p. 294.

[2] H.E. Haber and H.A. Weldon, *Phys. Rev. Lett.* **46**, 1497 (1981).

[3] M.R. Schafroth, *Phys. Rev.* **96**, 1442 (1954).

[4] M.R. Schafroth, et al., *Helv. Phys. Acta* **30**, 93 (1957).

[5] M.R. Schafroth, *Sol. State Phys.* **10**, 293 (1960).

[6] J.M. Blatt, *Theory of Superconductivity* (Academic, New York, 1964).

[7] L.N. Cooper, *Phys. Rev.***104**, 1189 (1956).

[8] J. Bardeen, L.N. Cooper and J.R. Schrieffer, *Phys. Rev.* **108**, 1175 (1957).

[9] N.N. Bogoliubov, *JETP* **34**, 41 (1958).

[10] N.N. Bogoliubov, V.V. Tolmachev and D.V. Shirkov, *Fortschr. Phys*. **6**, 605 (1958); and also in *A New Method in the Theory of Superconductivity* (Consultants Bureau, NY, 1959).

[11] J. Ranninger, R. Micnas and S. Robaszkiewicz, *Ann. Phys. Fr.* **13**, 455 (1988).

[12] R. Friedberg and T.D. Lee, *Phys. Rev. B* **40**, 6745 (1989).

[13] R. Friedberg, et al., *Phys. Lett. A* **152**, 417 and 423 (1991).

[14] V.V. Tolmachev, *Phys. Lett. A* **266**, 400 (2000).

[15] M. de Llano and V.V. Tolmachev, *Physica A* **317**, 546 (2003).

[16] J. Batle, et al., *Cond. Matter Theories* **18**, 111 (2003). Cond-mat/0211456.

[17] M. de Llano, in *Frontiers in Superconductivity Research* (ed. B.P. Martins) (Nova Science Publishers, NY, 2004). Available in cond-mat/0405071.

[18] M. Casas, et al., *Phys. Lett. A* **245**, 5 (1998).

[19] M. Casas, et al., *Physica A* **295**, 146 (2001).

[20] M. Casas, et al., *Sol. State Comm.* **123**, 101 (2002).

[21] S.K. Adhikari, et al., *Phys. Rev. B* **62**, 8671 (2000).

[22] S.K. Adhikari, et al., *Physica C* **351**, 341 (2001).

[23] A.L. Fetter and J.D. Walecka, *Quantum Theory of Many-Particle Systems* (McGraw-Hill, New York, 1971).

[24] N.N. Bogoliubov, *N. Cim.* **7**, 794 (1958).

[25] B.S. Deaver, Jr. and W.M. Fairbank, *Phys. Rev. Lett.* **7**, 43 (1961).

[26] R. Doll and M. Näbauer, *Phys. Rev. Lett.* **7** , 51 (1961).

[27] C.E. Gough, M.S Colclough, E.M. Forgan, R.G. Jordan, M. Keene, C.M. Muirhead, I.M. Rae, N. Thomas, J.S. Abell, and S. Sutton, *Nature* **326,** 855 (1987).

[28] C.E. Gough, *Priv. Comm.*

[29] N.N. Bogoliubov, *J. Phys. (USSR)* **11**, 23 (1947).

[30] M. de Llano et al., *Cond. Mat. Phys.* **20**, xxx (2005). Also in cond-mat/0509118.

[31] J. Labbé, S. Barisic, and J. Friedel, *Phys. Rev. Lett.* **19**, 1039 (1967).

[32] M. Randeria, in: *Bose-Einstein Condensation*, ed. A. Griffin *et al.* (Cambridge University, Cambridge, 1995) p. 355.

[33] M. Fortes, et al., *Physica C* **364**, 95 (2001).

[34] S. Wolfram, *The MATHEMATICA Book*, 3rd. Ed. (Wolfram Media, IL, 1996) p. 743.

[35] R.K. Pathria, *Statistical Mechanics*, 2nd Ed. (Pergamon, Oxford, 1996).

[36] Y.J. Uemura, *J. Phys.: Cond. Matter* **16**, S4515 (2004).

In: New Research on Superconductivity
Editor: Barry P. Martins, pp. 159-171

ISBN: 978-1-59454-197-1

Chapter 5

SUPERCONDUCTING FLUCTUATIONS ABOVE T_C IN OVERDOPED $BI_2SR_2CACU_2O_{8+\delta}$

Z.A. Xu[1], J.Q. Shen[1], Z.W. Zhu[1], S. Ooi[2], T. Mochiku[2] and K. Hirata[2]

[1] Department of Physics, Zhejiang University, Hangzhou 310027, China,

[2] National Institute for Materials Science, 1-2-1 Sengen, Tsukuba, Ibaraki 305-0047, Japan

Abstract

The resistivity, Hall effect and Nernst effect of overdoped $Bi_2Sr_2CaCu_2O_{8+\delta}$ single crystals were studied to investigate superconducting fluctuations above T_c. The superconducting transition temperature (T_{c0}) for the three measured samples with various oxygen content is 86.5, 83 and 77 K, and their doping level (hole concentration, p) is estimated to be 0.180, 0.193 and 0.204 respectively. The pseudogap opening temperature (T^*) determined from the temperature dependence of resistivity decreases slightly from 126 K to 100 K with increasing doping level p. The temperature scale (T_1) determined from the cotangent of Hall angle ($\cot\theta$) is consistent with the onset temperature (T_v) of vortex-like excitation detected by Nernst effect, and both decrease with increasing p. Moreover, none of the three temperature scales (T^*, T_1 and T_v) drops to zero as p increases across the proposed quantum critical point (QCP) of p ~0.19. This result indicates that both pseudogap opening and superconducting fluctuations exist above T_c in the overdoped region of electronic phase diagram of cuprates and there is no evidence for QCP at p ~ 0.19. We suggest that the temperature range between T_c and T_v (or T_1) is characterized by finite pairing amplitude with strong phase fluctuations. As T is above T_v, the phase correlation disappears completely and only the pairing amplitude fluctuations remain. The strong superconducting phase fluctuation in the pseudogap regime implies that the pseuodgap is a precursor to superconductivity.

PACS: 74.72.-h, 74.40.+k, 74.25.Fy, 74.72.Hs

Keywords: High-T_c compounds, superconducting fluctuations, pseudogap, Nernst effect

1. Introduction

High-T_c superconductors (HTS) have been a focus of interests for about 20 years since their discovery in 1986. Not only the superconducting properties of HTS are contrast to those of conventional BCS-type superconductors, but also their normal-state properties deviate severely from the predictions of Fermi Liquid theory. Up to now, the pairing mechanism is still an open issue though great progress has been made in probing the nature of HTS. One of the puzzles is that there exists a pseuodogap which opens at a certain characteristic temperature T^* above T_c in the underdoped regime [1]. The energy gap (called superconducting gap) is one of the defining properties of a superconductor. Experimental probes like the frequency-dependent conductivity [2], the angle-resolved photoemission spectra (ARPES) [3], the Raman scattering spectra [4], and the tunneling conductance [5] have found that the energy gap in the cuprates does not show the similar behavior as in a conventional superconductor. The depression of excitations is incomplete and often starts well above T_c in the normal state. Now this behavior is ascribed to the d-wave nature of the superconducting order parameter and the persistence of a pseudogap in the normal state.

The pseudogap opening has significant effect on both the in-plane and c-axis transport properties [6]. It is well known that the variation of in-plane resistivity is linear with temperature for a large temperature range extending from 10 –1000 K in optimally doped HTS. In the underdoped HTS, the presence of the pseudogap causes striking deviations from the linear behavior. Bucher et al. observe a reduced in-plane resistivity in $YBa_2Cu_4O_8$ (Y124) below T^* where the pseudogap opening is seen by NMR [7]. Complete studies of the doping dependence of the characteristic temperature T^* determined by the transport measurements have been done for $La_{2-x}Sr_xCuO_4$ (LSCO) [8], $YBa_2Cu_3O_{6+x}$ (YBCO) [9], and $Bi_2Sr_2CaCu_2O_{8+\delta}$ (Bi-2212) [10]. All the studies reach a consensus that T^* decreases monotonously with increasing doping level (p). The inverse of Hall coefficient (R_H^{-1}) is also linear with temperature above the pseudogap temperature T^*, whereas the cotangent of Hall angle cot(θ) displays a T^2 temperature dependence [11]. However, the characteristic temperature where cot(θ) deviates from the T^2 temperature dependence, found to lie between T_c and T^*, defines another crossover line in the temperature versus doping phase diagram [12]. Actually the Knight shift and the spin-lattice relaxation time in the Cu^{63} NMR measurement [7] also define two different characteristic temperatures.

An explanation of the pseudogap phenomenon is regarded as one of the most important unresolved questions in the theory of superconductivity. Due to the extremely complex structural and electronic nature of the cuprates, and the somewhat controversial nature of much of the current experimental data, there are many theoretical attempts to explain pseudogap behavior. Basically the theoretic models can be divided into two categories according to the relationship between the pseudogap state and superconducting state in these models. In one category of the theoretic models, the pseudogap state is regarded as a precursor to superconductivity. In the Anderson's spin-charge separation scenario [13-14], the spinons pair to form a gap at T^* in the spin excitations, identified as the pseudogap. Then the holons Bose-condense at T_c to form the superconducting state. Emery and Kivelson (EK model) [15] have developed a performed pair model of the pseudogap in which incoherent Cooper pairs form below T^* and phase order or Bose condensation of Cooper pairs establish superconductivity at T_c. These models predict that T^* line should merge with T_c line gradually in the overdoped regime. In the other category of theoretic models, the pseudogap is induced by a kind of fluctuation of other orders (such as antiferromagnetic fluctuations, spin waves, charge –density waves, or stripes) which competes or coexists with superconductivity. A quantum critical point (QCP) in the overdoped regime is predicted by some theoretic models [16-17] in this category and the T^* line should drop to zero temperature around QCP.

Recently Xu et al. and his colleagues [18] have surprisingly found that the Nernst signals due to vortex-like excitations still persist in the pseudogap region above T_c, which is a strong evidence supporting the theoretic model proposed by Emery and Kivelson [15]. The Nernst effect is the appearance of a transverse electric field E_y in response to a temperature gradient ∇T in x-axis direction, in the presence of a perpendicular magnetic field H in z-axis direction, and under open circuit conditions. Because of a so-called "Sondheimer cancellation"[19] the Nernst effect is usually small in the normal state of normal metals where transport by quasi-particles is dominant. However, for a type-II superconductor (in the vortex-liquid state), a new set of excitations - vortices- are driven by temperature gradient $\nabla T \parallel$ x. Vortices diffuse down the gradient with velocity v. As each vortex core cross the line between a pair of transverse voltage electrodes, the 2π phase slip of the condensate phase leads to a Josephson electric field given by $\mathbf{E}=\mathbf{B}\times\mathbf{v}$, which is called vortex Nernst effect. Nernst effect measurement is regarded as a very sensitive probe of vortex excitations.

A large anomalous Nernst signal in the peudogap state has been observed in the hole-doped cuprates such as LSCO[18], YBCO[20-21], and Bi-2212 [20, 22], and it is regarded as evidence for vortex-like excitations above T_c, which implies that there is a relationship or interaction between the pseudogap state and the superconducting state. The onset temperature T_v of vortex-like excitation defines a new crossover line between T_c and T^* in the temperature versus doping phase diagram. This discovery has inspired a revisit to the theory of superconducting fluctuations in the cuprates. The conjecture that significant superconducting fluctuations in the pseudogap region give rise to the large Nernst signal is accord with the idea on the pseudogap proposed by Emery and Kivelson [15]. Ussishkin et al.[23] suggest that Gaussian superconducting fluctuations above T_c are able to explain the Nernst effect for the optimally doped and overdoped regimes. The work of Carlson et al. [24] predicts that in the hole-doped cuprates the fluctuations in the phase of the order parameter would dominate the Nernst signal up to a certain temperature above T_c, and at still higher temperatures there should be contributions to the Nernst effect from fluctuations both in phase and the amplitude of the order parameter. Curty and Beck [25] propose that there are two different regimes in the pseudogap state: one is near the superconducting state, phase correlations of Cooper pairs are important up to the temperature T_ϕ; the other is above T_ϕ, the pseudogap region is only determined by the amplitudes, and phases are uncorrelated. In their model, the Nernst signal of vortex-like excitation should appear in the first regime ($T_c<T<T_\phi$).

Up to now, most investigations of Nernst effect have been performed on underdoped cuprates and few studies have been reported for overdoped cuprates. How the T^* line continues in the overdoped region, i.e., what is the "real phase diagram", is crucial to the theories. Moreover, the relationship between these characteristic temperatures T^*, T_v and $T\phi$, is worth more experimental investigations.

This paper reports the measurements of resistivity, Hall effect, and Nernst effect in the overdoped high-T_c superconducting $Bi_2Sr_2CaCu_2O_{8+\delta}$ (Bi-2212) single crystals with T_{c0} of 86.5, 83 and 77 K. It was found that the onset temperature T_v of anomalous Nernst signals due to vortex-like excitation is consistent with the characteristic temperature T_1 determined by the T^2 dependence of the cotangent of Hall angle. Our results show that T_v line is consistent with the T_ϕ proposed by Curty and Beck [25], implying that the regime between T_c and T_v is characterized by superconducting phase fluctuation. Moreover, none of the three temperature scales (T^*, T_1 and T_v) drops to zero as p increases across the proposed quantum critical point (QCP) of p ~0.19, which means that there is no evidence for QCP at p ~ 0.19. Our results imply that the pseudogap is a precursor to superconductivity.

2. Experimental

The Bi-2212 single-crystals were grown by traveling solvent floating-zone technique. The details of single crystal growth were reported elsewhere [26-27]. Usually as grown Bi-2212 crystals are overdoped and the oxygen distribution is not homogenous. The oxygen content of Bi-2212 can be changed by annealing under different partial pressure of oxygen atmosphere and temperatures. For example, Watanabe et al [10] reported an equilibrium phase diagram on a Bi-2212 single crystal for various oxygen contents, which can be used for the precise control of the carrier doping level. The crystals were cut into about 3 mm in length and 1 mm in width (the thickness is about 10 μm) by a dicing saw before annealing in oxygen. The doping concentration (p) can be estimated using the empirical relation T_c/T_{cmax} = 1- 82.6(p-0.16)^2 [28]. We take T_{cmax} = 95 K and take the mid-point temperature of resistive transition for T_c. The slightly overdoped sample (Sample A) was obtained by annealing the as grown crystals under a suitable oxygen partial pressure at 700 for about 30 hours. The hole-doping concentration p of Sample A is estimated to be 0.180. After performing the measurements of transport properties and Nernst effect, Sample A was annealed again in 1 atm flowing oxygen at 500 for 12 hours and the outcome sample is denoted as Sample B. The hole-doping concentration p of Sample B is about 0.193. After performing the similar measurements, Sample B was then annealed in 120 atm high pressure oxygen and the outcome sample is denoted as Sample C. The hole-doping concentration p of Sample C is about 0.204.

The in-plane resistivity and magnetoresistance were measured by standard four-probe method. The electric contacts were made by using golden paste and the contact resistance was less than 1 Ω. A LR 700 AC Resistance Bridge was used to measure both the resistivity and Hall effect. In the set-up of Nernst effect measurement, the temperature gradient was applied along longitudinal direction (x-direction), and the magnetic field was applied along z-direction (c-axis of the sample). Thus the Nernst electric field is along y-direction, which was measured by Keithley Model 2182 Nano-voltmeter. The temperature gradient was measured by two small Cernox thermometers (CX-1050-BR from Lake Shore Cryotronics) thermally anchored to the two ends of the sample and GE varnish (VGE-7031) was used for good thermal contact. A small heater with a size of 0.5×1 mm^2 in area was supported by a thin nylon pillar and closely attached to one end of the sample. The heater current was provided by a Keithley 2400 Digital Source-meter and the applied temperature gradient was about 5K/cm. The Nernst coefficient ν is defined as:

$$\nu = \frac{E_y}{|\partial_x T| B_z} \tag{1}$$

At fixed temperatures the Nernst signal was recorded while the magnetic field was sweeping slowly. The temperature was regulated by a Lakeshore temperature controller and the temperature drift was less than 5 mK during the measurement.

3. Results and Discussion

Figure 1 shows the temperature dependence of in-plane resistivity for Sample A, B and C. All three samples show quite sharp superconducting transitions, which implies high sample quality. The zero-point and mid-point transition temperatures are 86.5 K and 91.8 K for Sample A, 83 K and 86.5 K for Sample B, and 77 K and 79.7 K for Sample C, respectively. Because of the extremely low melting field of vortex lattice of Bi-2212, a very low magnetic field (like remanent flux trapped in superconducting magnet) can cause a flux-flowing resistivty and T_{c0} value determined by resistance measurement is often shifted to a lower value. Therefore, in order to estimate the doping level p, it is more reliable to use the mid-point transition temperature than T_{c0}. The doping concentration (p) is estimated to be 0.180, 0.193 and 0.204 respectively using the empirical relation $T_c/T_{cmax} = 1 - 82.6(p-0.16)^2$ [28], where $T_{cmax} = 95$ K. The high temperature resistivity changes linearly with temperature and a very slightly upward curvature is seen in Sample C. With increasing doping concentration (p), the normal state resistivity decreases and the slope of $\rho(T)$ decreases slightly. Although the

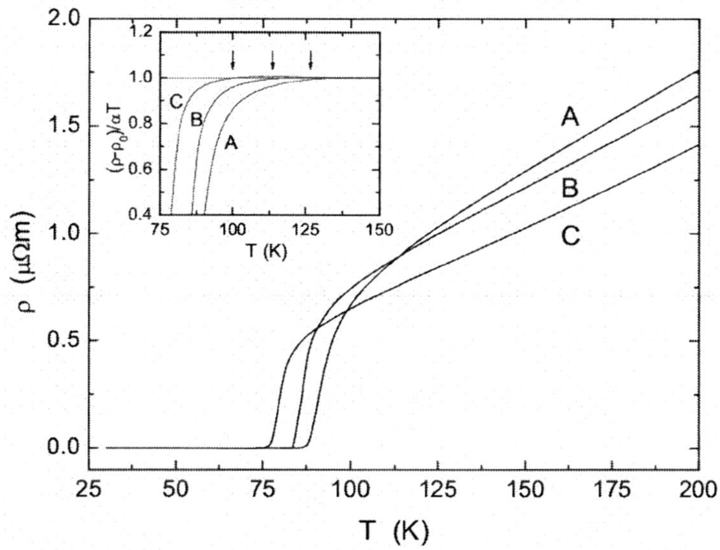

Figure 1. The temperature dependence of in-plane resistivity for Sample A, B and C. Inset: the plot of $(\rho-\rho_0)/\alpha T$ versus T, where ρ_0 is the T=0 intercept of the extrapolated T-linear high temperature curve and α the slope of the linear part of the resistivity. The arrows show the characteristic temperatures T^* at which $\rho(T)$ starts to decrease rapidly with decreasing temperature.

slightly upward curvature in $\rho(T)$ means the resistivity of Sample C starts to deviate from the T- linear dependence around optimally doped regime and approaches T^2 dependence in the heavily overdoped regime, the high temperature resistivity can still be fitted linearly. The inset of Figure 1 shows the plot of $(\rho-\rho_0)/\alpha T$ versus T, where ρ_0 is the T=0 intercept of the extrapolated T-linear high temperature curve and α the slope of the linear part of the resistivity. The arrows in the inset of Fig.1 show the characteristic temperatures T^* at which $\rho(T)$ starts to decrease rapidly with decreasing temperature. This characteristic temperature T^* should correspond to the pseudogap opening [7-9]. From Fig.1, T^* is determined to be about 126, 114, and 100 K for Sample A, B and C. The results are consistent with the previous reports [10].

It is easy to confuse the rapid decrease in $\rho(T)$ below T^* with the superconducting fluctuation induced conductivity enhancement $\Delta\sigma_{FL}$. One good way to distinguish the two effects is to perform magnetoresistance measurement. It has been found that the T^* and T_c behave differently as function of magnetic field. While T_c decreases rapidly with increasing magnetic field H, T^* is weakly dependent on H [29-30]. Such a weak H dependence of T^* can be explained within a precursor superconductivity scenario [31]. We have performed magnetoresistance measurements on Sample A, B and C. Figure 2 shows the temperature dependence of in-plane resistivity for Sample A under magnetic field H of 0, 10, 50 kOe. H is along the c-axis direction. Apparently there is no significant magnetoresistance (MR) around T^* and MR appears at a lower temperature (about 100 K). Similar results are also observed for Sample B and C (not shown here). The results confirm further that T^* determined in Fig.1 does correspond to the pseudogap opening temperature instead of conventional fluctuation effect.

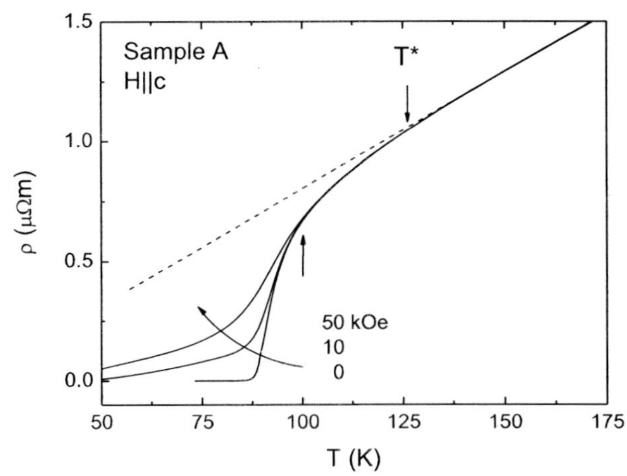

Figure 2. The temperature dependence of in-plane resistivity for Sample A under magnetic field H of 0, 10, 50 kOe. H is along the c-axis direction.

The temperature dependence of Hall coefficient R_H was measured under magnetic field H of 50 kOe for Sample A, B and C. Figure 3 shows the temperature dependence of Hall coefficient R_H for these 3 samples. Overall R_H in the normal state increases with doping level p. In conventional metals, the Hall coefficient R_H is temperature independent as T is above a fraction of the Debye temperature (usually $0.2\theta_D \sim 0.4\theta_D$) [32], and the charge carrier density n_H can be estimated from the Hall coefficient, i.e., $n_H = 1/(eR_H)$, where e is the charge of electron. Since R_H depends strongly on the temperature for HTS, the relation between the charge carrier concentration and R_H is not straightforward. But the Hall density n_H is still often used as a good parameter to describe the hole concentration for HTS even though the absolute value of n_H does not mean the real charge carrier density [33]. The n_H values are calculated to be 3.37×10^{21} cm^{-3}, 4.51×10^{21} cm^{-3} and 6.30×10^{21} cm^{-3} for Sample A, B and C by taking their 200 K R_H data. The increase of n_H is consistent with the increase of the doping level from Sample A, B to C.

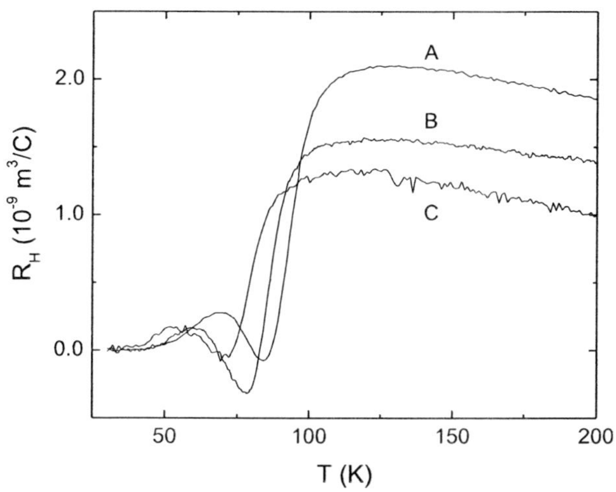

Figure 3. The temperature dependence of Hall coefficient R_H for Sample A, B and C.

Just below T_{c0}, the double sign revisals of Hall coefficient are observed for all the three samples, which have been observed in high-quality Bi-2212 crystals previously [9]. Several theoretical approaches based on the mechanism of vortex motion, superconducting fluctuations, and/or hydrodynamics interaction between vortices and the charge carriers have been attempted to explain the complex features of Hall effect below T_{c0} [10], but we do not attempt to discuss further the anomalous Hall effect in mixed state in this paper.

Many studies have suggested that the maximum in R_H around 100 K is also associated with the opening of the pseudogap at the temperature T^* [7, 9, 12, 33-34]. It has been suggested that the Hall angle is the more meaningful parameter than R_H [11]. The Hall angle θ is defined as $\tan\theta \equiv \rho_H(H)/\rho(H)$, where $\rho_H(H)$ is Hall resistivity (measured at $\mu_0 H = 5$ T in our experiments). Generally the Hall angle in the normal state of HTS satisfies the relation [11]:

$$\cot\theta = a + bT^2 \tag{2}$$

where the coefficient a is related to the impurity scattering, and b scales with the hole concentration. Eq.(2) can be interpreted in the Anderson's spin-charge separation model [35]. Figure 4 shows the plot of cotangent of Hall angle, $\cot\theta$, versus the square of temperature, T^2, for Sample A, B and C. By fitting the $\cot\theta$ data with Eq.(2), the parameter a and b are obtained. a =202.8, b = 0.019 for Sample A; a =291.7, b = 0.022 for Sample B; a =254.9, b = 0.035 for Sample C. The increase of b with the doping level is consistent with the change of the Hall density n_H calculated from Hall effect. Although the Hall coefficient R_H deviates downwards from $1/T$ dependence around T^*, there is no observable change in the $\cot\theta$ - T^2 curve as T decreases across T^*. As T further approaches to superconducting transition temperature T_{c0}, $\cot\theta$ deviates from the T^2 dependence and increases rapidly due to the strong superconducting fluctuation effect. We define this characteristic temperature as T_1 at which $\cot\theta$ starts to increase rapidly. From Fig. 4, we found that T_1 is 103, 95 and 88 K for Sample A, B and C. Apparently these T_1 values are smaller than the corresponding T^* values for the same doping level, defining a new crossover line between T_{c0} and T^*. Whereafter we will discuss the meaning of this T_1 line further.

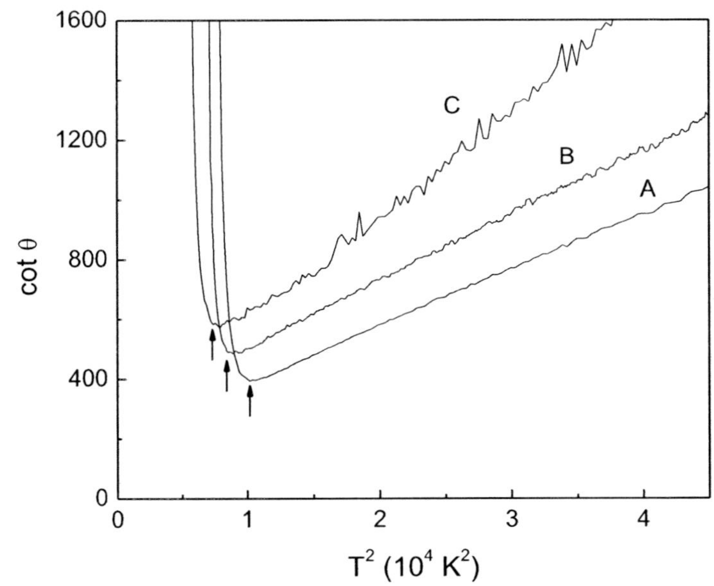

Figure 4. The plot of cotangent of Hall angle, $\cot\theta$, versus the square of temperature, T^2, for Sample A, B and C. The arrows indicate the points where $\cot\theta$ starts to deviate from the T^2 dependence.

Systematic Nernst effect measurements have been performed for Sample A and B. Figure 5 shows the Nernst signal $e_y \equiv E_y / \partial_x T$ versus magnetic field H at different temperatures for Sample A. At temperatures well below T_c, all the curves have the characteristic features of the vortex Nernst effect. Due to the extremely low melting field (H_m) of vortex lattice in Bi-2212, the vortex solid to liquid phase transition is not observed in the Nernst signal e_y. Even for the temperatures as low as 40 K, the vortices can still be driven to move by a temperature gradient, and e_y increases slowly with increasing magnetic field. As the temperature rises, the velocity of vortex also increases and the initial slope of $e_y(H)$ increases fast. At higher magnetic field, the entropy of vortex is suppressed and the thermal driving force decreases, therefore $e_y(H)$ exhibits a downward curvature.. As T is above T_c, the Nernst signal is still a sizeable fraction of low-T values. As the temperature increases further, the Nernst signal decreases gradually, approaching a straight line of negative slope, which we identify with the background signal v_n from the normal charge carriers (holes). The Nernst signals of Sample B exhibit similar traces of e_y versus H except that the maximum of Nernst signal moves to a lower temperature.

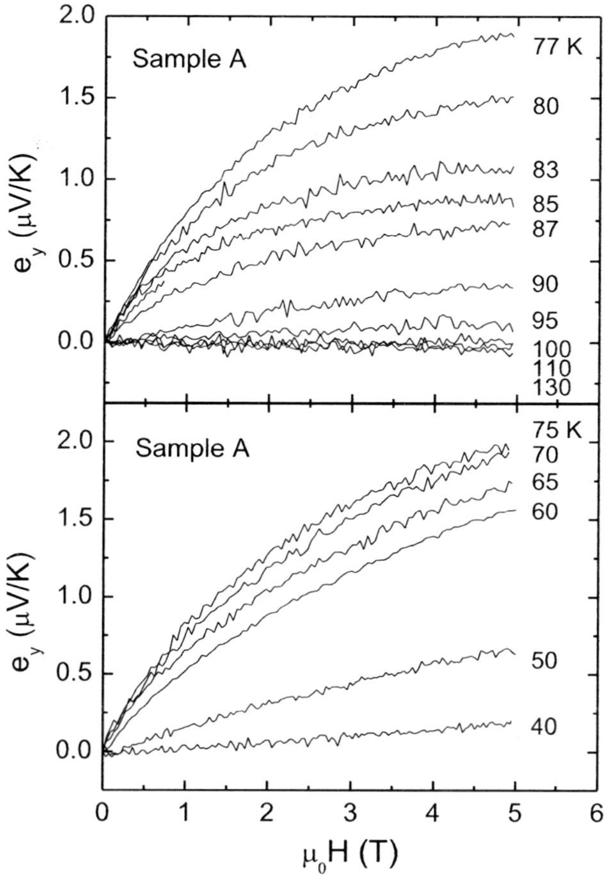

Figure 5. Lower panel: the Nernst signal e_y versus magnetic field H at different temperatures below T_c for Sample A. Upper panel: the Nernst signal e_y versus magnetic field H at different temperatures around and above T_c for Sample A. The numbers beside the curves indicate the temperatures.

From the traces of e_y versus H the Nernst coefficient ν is calculated, i.e., $\nu = \dfrac{E_y}{|\partial_x T| B_z}$.

Since the $e_y(H)$ curves are no longer linear for $T < T_c$, we use the initial slope of $e_y(H)$ for the Nernst coefficient ν. Figure 6 shows the temperature dependence of Nernst coefficient ν for Sample A and B. For both samples, the Nernst coefficient ν in high-temperature normal state is small and almost temperature independent, similar to the underdoped HTS [18]. With decreasing temperatures, ν deviates from the high-temperature background and increases sharply at a certain temperature T_ν, then reaches a maximum below T_{c0}, and finally decreases to zero as T<40 K. In the temperature range between T_{c0} and T_ν there exists anomalously large Nernst signal, indicating that there are vortex-like excitations in this region. T_ν is the onset temperature at which the Nernst signal starts to deviate from the high temperature background. T_ν is about 105 K and 95 K for Sample A and B respectively.

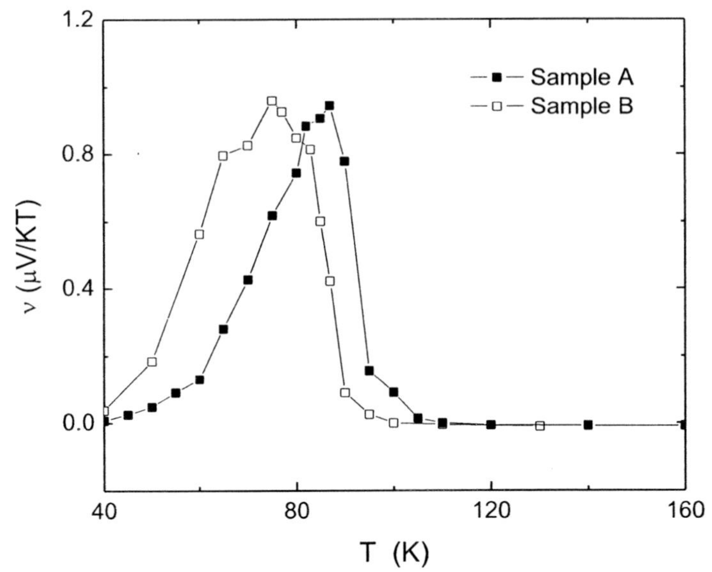

Figure 6. The temperature dependence of Nernst coefficient ν for Sample A and B. Since the $e_y(H)$ is nonlinear for $T < T_c$, we use the initial slope of $e_y(H)$ for the Nernst coefficient.

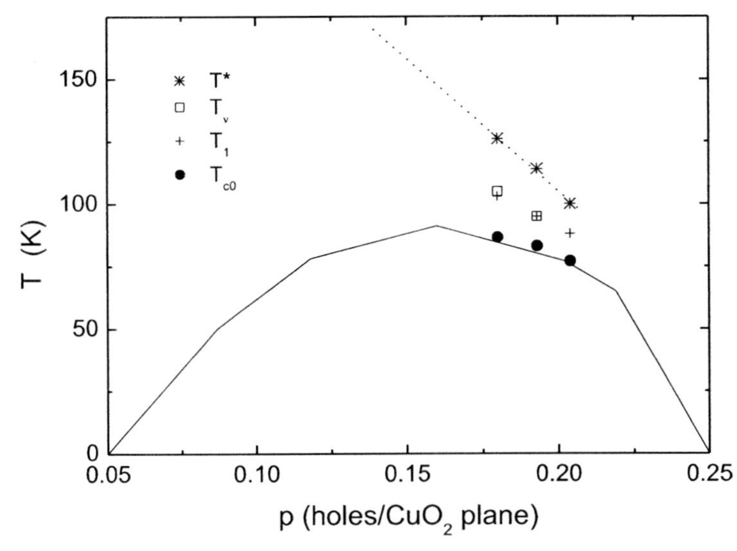

Figure 7. The variation of T_{c0}, T_1, T_v, and T^* with doping level p. The line shows T_{c0} versus p cited from other group [36]. The dash line is only for guide of eyes.

Anomalous Nernst signals due to vortex-like excitations have been observed at temperatures well above T_{c0} in underdoped HTS, such as $La_{2-x}Sr_xCuO_4$ ($0.05 \leq x \leq 0.17$) [18], $Bi_2Sr_2CaCu_2O_x$ ($8.05 \leq x \leq 8.1$) [13, 14], and $YBa_2Cu_3O_x$ ($6.4 \leq x \leq 6.9$) [14]. According to the general phase diagram of HTS, as the doping concentration decreases from optimally-doping, T_c decreases, but T^* increases on the contrary in a certain doping range. The varying trend of T_v line is similar to that of T^* line, but opposite to that of T_{c0} line in the underdoped region. Theses results indicate that the anomalous Nernst signals in the pseudogap region of underdopoed HTS can not be understood in the frame of conventional superconducting fluctuations, which implies that the mechanism of HTS is substantially different from the that of conventional superconductors. According to the scenario of precursor pairing proposed in EK model [15], the pairing amplitude $|\psi|$ is no longer zero when the pseudogap opens as $T<T^*$, and the loss of the Meissner effect (flux-expulsion) in the temperature range $T_c<T<T^*$ corresponds to a loss of long-range phase coherence of Cooper pairs caused by the vortex excitations. The broad range of temperatures above T_{c0} in which the vortex signal is observed by the highly sensitive Nernst effect experiment presents the most compelling evidence for this scenario. Curty and Beck [25] further propose that the phase correlations of Cooper pairs are important up to the temperature T_ϕ above T_c; as T is above T_ϕ, the pseudogap region is only determined by the amplitude, and the phase correlation disappears completely. The Nernst signal of vortex-like excitation should appear in the regime between T_c and T_ϕ.

Corson et al. [35] detected dynamic vortices by high-frequency ac conductivity and found that there exist dynamic vortices in a temperature range above T_{c0} in underdoped Bi-2212. The temperature regime where there are dynamic vortices is consistent with the regime where the Nernst signals due to vortex-like excitations are observed. From the fact that the temperature scale T_1 (determined from the $\cot\theta$) is same as T_v, we further suggest that the strong phase fluctuations have significant effect on both the Hall angle and Nernst effect. Therefore both the characteristic temperatures T_v and T_1 correspond to the phase correlation temperature T_ϕ proposed in Curty and Beck's theoretic work. We summarize the characteristic temperatures T_{c0}, T_1, T_v, and T^* in the (T, p) phase diagram plot as in Figure 7. The solid line shows the variation of T_{c0} with p for Bi-2212 which is taken from the work of the Princeton group [36]. The temperature scale T_v (or T_1) defines a new crossover line which divides the pseudogap regime into two domains. One domain is near the T_{c0} ($T_{c0}<T< T_v$) where the strong phase fluctuations cause the deviation from T^2 dependence of $\cot\theta$ and significant vortex-like excitations which can be probed by the Nernst effect measurement. The other domain is above T_v ($T_v < T < T^*$) where the phase correlation disappears and only pairing amplitude fluctuations remain. The T_v line is consistent with the temperature scale T_ϕ proposed in Curty and Beck's theoretic work.

4. Conclusion

From the studies of the resistivity, Hall effect and Nernst effect of 3 overdoped $Bi_2Sr_2CaCu_2O_{8+\delta}$, it has been found that the pseudogap opening temperature (T^*) determined from the temperature dependence of resistivity decreases slightly from 126 K to 100 K with increasing doping level p. The temperature scale (T_1) determined from the cotangent of Hall angle ($\cot\theta$) is consistent with the onset temperature (T_v) of vortex-like excitation detected by Nernst effect, and both decrease with increasing p. Moreover, none of the three temperature scales (T^*, T_1 and T_v) drops to zero as p increases across the proposed quantum critical point (QCP) of $p \sim 0.19$. This result indicates that both pseudogap opening and superconducting fluctuation exist above T_c in the overdoped region of electronic phase diagram of cuprates and there is no evidence for QCP at $p \sim 0.19$. We suggest that the temperature range between T_c and T_v (or T_1) is characterized by finite pairing amplitude with strong phase fluctuations. As T

is above T_v, the phase correlation disappears completely and only the pairing amplitude fluctuations remain. This picture is consistent with Curty and Beck's theoretic result [25]. The results imply that the pseudogap is a precursor to superconductivity.

Acknowledgements

This work was sponsored by the National Science Foundation of China (NSFC 10225417) and the National Basic Research Program of China (Grant No. 2006CB601003).

References

[1] For a reviw, see Timusk T.; Statt B., *Rep Prog Phys* 1999, 62, 61 - 122.

[2] Puchkov A.V., Fournier P., and Timusk T., *J. Phys:Condens.Matter* 1996, 8, 10049 - 10082.

[3] Damascelli A., Hussain Z., and Shen Z. X., *Rev. Mod. Phys.* 2003, 75, 473 –542.

[4] Hewitt K. C., and Irwin J. C., *Phys. Rev. B* 2002, 66, 054516-1- 9

[5] Matsuda A., Sugita S., and Watanabe T., *Phys. Rev. B* 1999, 60, 1377–1381.

[6] For a earlier review, see Batlogg B., Hwang H. Y., Takagi H., Cava R. J., Kao H. L., and Kwo J., *Physica C* 1994, 235-240, 130 -133.

[7] Bucher B., Steiner P., Karpinski J., Kaldis E., and Wachter P., *Phys. Rev. Lett* 1993, 70, 2012-2015.

[8] Takagi H., Batlogg B., Kao H. L., Kwo J., Cava. R. J., Krajewski J. J., and Peck. W. F., Jr., *Phys. Rev. Lett* 1992, 69, 2975-2978.

[9] Ito T., Takenaka K., and Uchida S., *Phys. Rev. Lett* 1993, 70, 3995-3998.

[10] Watanab. T., Fujii T., and Matsuda A., *Phys. Rev. Lett* 1997, 79, 2113 –2116.

[11] Chien T. R., Wang Z. Z., and Ong N. P., *Phys. Rev. Lett* 1991, 67, 2088 -2091.

[12] Matthey D., Gariglio S., Giovannini B., and Triscone J.-M., *Phys. Rev. B* 2001,64, 024513-1-5.

[13] Anderson P. W., *Science* 1987, 235, 1196-1199.

[14] Nagaosa N., Lee P. A., *Phys. Rev. B* 1992,45, 966 -970.

[15] Emery V. J. and Kivelson S. A., *Nature* (London),1995, 374, 434-437.

[16] Tallon J. L., Williams G.. V. M.,. Staines M. P., and Bernhard C., *Physica C* 1994, 235-240, 1821-1822.

[17] Tallon J. L., and Loram J. W., *Physica C* 2001, 349, 53-68.

[18] Xu Z. A.; Ong N. P.; WangY.; Kakeshita T.; Uchida S., *Nature* (London) 2000, 406, 486-488.

[19] Sondheimer E H., *Proc. R. Soc. London, Ser. A*, 1948, 193, 484-512.

[20] Ong N. P. Wang Yayu, Ono S., AndoY., and Uchida S., *Ann. Phys.* (Leipzig) 2004, 13, 9 – 14.

[21] Xu Z. A., Shen J. Q., Zhao S. R., Zhang Y. J., and Ong C. K., *Phys. Rev. B*, 2005, 72, 144527-1-5.

[22] Wang. Y., Ono. S., Onose. Y., GuG., Ando. Y., Tokura Y., Uchida S., and Ong N. P., *Science* 2003, 299, 86-89.

[23] Ussishkin I., Sondhi S. L., and Huse D. A., *Phys. Rev. Lett.* 2002, 89, 287001-1-4.

[24] Carlson E. W., Emery. V. J., Kivelson S. A., and Orgad D. (2002). Concepts in High Temperature Superconductivity. *cond-mat / 0206217*, http://xxx.arXiv.org.

[25] Curty P., and Beck H., *Phys. Rev. Lett*, 2003, 91, 257002-1-4.

[26] Mochiku T., and Kadowaki K., *Physica C*, 1994, 235-240, 523-524;

[27] Mochiku T., Hirata K., and Kadowaki K., *Physica C*, 1997, 282-287, 475-476

[28] Tallon J. L., Cooper J. R., de Silva P. S. I. P. N., Williams G. V. M., and Loram J. W., *Phys. Rev. Lett.* 1995, 75, 4114 -4117

[29] Zheng G., Clark W. G.,. Kitaoka Y, Asayama K., Kodama Y., Kuhns P., and Moulton W. G., *Phys. Rev. B* 1999, 60, R9947-1-4.

[30] Shibauchi T., Krusin-Elbaum L., Li Ming, Maley M. P., and Kes P. H., *Phys. Rev. Lett.* 2001, 86, 5763 –5766.

[31] Pieri P., Strinati G. C., and Moroni D., *Phys. Rev. Lett.* 2002, 89, 127003-1-4.

[32] Hurd C., in *The Hall effect and its applications*; Chien C. L., and Westgate C. R. ; Ed.; Plenum: New York, NY, 1980, pp.1 - 550.

[33] Hwang H. Y., Batlogg B., Takagi H., L. Kao H., Kwo J., Cava R. J., Krajewski J. J., and Peck W. F., Jr., *Phys. Rev. Lett.* 1994, 72, 2636-2639.

[34] Jin R., and Ott H. R, *Phys. Rev. B* 1998, 57, 13872-13877.

[35] Corson J., Mallozzi R., Orenstein J., Eckstein J.N., Bozovic I., *Nature* (London) 1999, 398, 221-223.

[36] Wang Y., *Ph. D thesis*, Princeton University, 2003 (unpublished).

In: New Research on Superconductivity
Editor: Barry P. Martins, pp. 173-186

ISBN: 978-1-59454-197-1

Chapter 6

PHYSICAL PROPERTIES OF LI_2PD_3B AND LI_2PT_3B SUPERCONDUCTORS

Hiroyuki Takeya
National Institute for Materials Science, Sengen 1-2-1,
Tsukuba, Ibaraki 305-0047, Japan

Abstract

Superconductivity has been recently found in two Li containing compounds, Li_2Pd_3B and Li_2Pt_3B. They show superconducting transition at the temperatures, 7.2 K and 2.6 K respectively. Those materials have the same cubic structure ($P4_332$) composed of distorted octahedrons, whose corners are occupied by Pd/Pt. This structure does not have mirror or inversion symmetry along any directions. It is a very interesting feature of those materials in relation to the symmetry of superconductivity. Their superconductivity is induced by Li addition. Controlling the Li stoichiometry is a key to synthesize high quality samples in the preparation process of arc-melting because Li is volatile at high temperatures. Specific heat of these compounds was measured for confirming their bulk superconductivity and investigating superconducting properties. The values of dC/T_C were evaluated to be around 17 mJ/mol K^2 for Li_2Pd_3B and 7.2 mJ/mol K^2 for Li_2Pt_3B. Electronic specific heat coefficient (γ) and Debye temperature (θ_D) were derived from the normal state data in a magnetic field over H_{C2} and lead to 8.3 mJ/mol K^2 and 202 K for Li_2Pd_3B, 9.6 mJ/mol K^2 and 240 K for Li_2Pt_3B, respectively. Pressure effect was also measured with a SQUID magnetometer using a low temperature hydrostatic micro pressure cell at pressures up to around 1 GPa. The superconducting transition temperatures of both Li_2Pd_3B and Li_2Pt_3B decreased linearly with pressure. The pressure dependence, dT_C/dP, were −0.091 and −0.090 K/GPa, respectively. These values are comparable to those of conventional metallic superconductors, but much smaller than the value, −1.03 K/GPa, of MgB_2.

1. Introduction

The discovery and research of new metal boride superconductors have been encouraged since the discovery of MgB_2. Some electronic band structure calculations suggest that higher superconducting transitions are possible in materials composed of light elements. For example, LiB_2, BeB_2[1] and LiBC[2] are the first natural candidates, since lighter atoms may help in providing higher phonon frequencies. They have been investigated theoretically and experimentally with great efforts, but extensive research into these new materials with light elements has not produced any positive results.

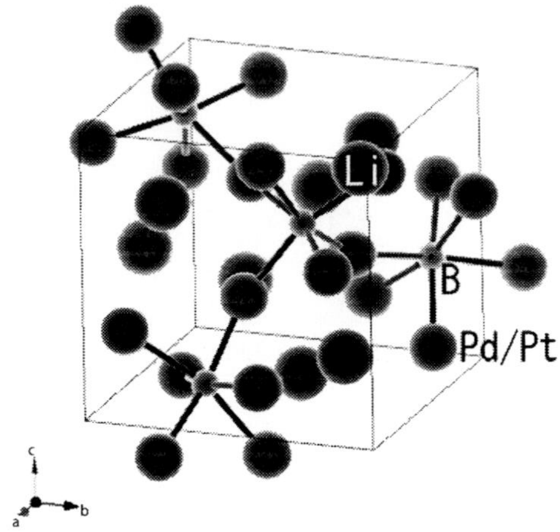

Figure 1. Structure of $Li_2(Pd/Pt)_3B$ illustrated using VICS [14].

We have been for a long time interested in multinary borides. The diborides of MB_2 (M=Nb, Mo, Ta, Ti, Hf, V, and Cr) were investigated in the 1970s, and superconductivity was found in some metal-deficient diborides of NbB_2 and MoB_2[3]. There have been some recent reports on the superconductivity of TaB_2 [4] and ZrB_2 [5], however, they have not been confirmed as superconductors. The reports have not excluded the high probability of the superconductivity being the result of the inclusion phase. We have found and reported a metastable superconducting phase in stoichiometric NbB_2 prepared by a combustion synthesis [6] and a rapid cooling process [7]. RRh_4B_4, RM_3B and RNi_2B_2C (R=Y, Lanthanoid Elements, M=Rh, Pd, Pt) show rich and interesting phenomena derived from the competition between superconductivity and magnetism at low temperatures. We have investigated shock loading effects on a magnetic superconductor, $ErRh_4B_4$ [8,9]. Some new superconductors have been found in borides, carbides and germanides late years. For example, $MgCNi_3$ (T_C=7.6 K) [10],Y_2C_3 (T_C=18 K)[11], $W_7Re_{13}X$ (X=B, T_C=7.1 K and X=C, T_C=7.3K)[12], $La_3Pd_4Ge_4$ (T_C=2.75K) [13] and B-doped diamond (T_C=4 K) [14]. We also have been searching new superconducting materials containing lithium and boron since they are recognized as interesting elements which react with many other metals. Superconductivity was confirmed in Li_2Pd_3B and Li_2Pt_3B around at 7.2 K and 2.6 K, respectively [15]. It is obvious that their superconductivity is induced by Li addition because Pd_3B and Pt_3B show no superconductivity at all. The two compounds consist of distorted octahedral coordinations of (Pd/Pt)-B in the cubic lattice structure with the space group of $P4_332$ as shown in Fig. 1 [16]. The distorted octahedral coordination of (Pd/Pt)-B is in contrast to the structure of the ideal octahedral coordination in RM_3B. Especially, since materials lacking mirror and inversion symmetries are uncommon among superconductors, we have an interest in their superconducting behaviors. These two compounds, Li_2Pd_3B and Li_2Pt_3B, will be compared to a novel heavy fermion superconductor of $CePt_3Si$ [17] lacking a mirror symmetry along c-axis. In this report, we have performed measurements of resistivity, magnetization, pressure effect and specific heat to investigate the different superconducting behaviors between them.

2. Experimental

$Li_2(Pd/Pt)_3B$ materials were prepared from Li lump (Kanto Kagaku Co., above 99%) , Pd plate (Rare Metallic Co., 99.95%) Pt wire (Furuuchi Metal Co., 99.99%) and B flake (Rare Metallic Co., 99.8%). First, stoichiometric Pd/Pt and B were molten together using an arc-furnace in a pure argon atmosphere. Then, molten materials were pressed using a copper hummer installed in the arc-furnace and cast into a plate. The plate of $(Pd/Pt)_3B$ and a lump of Li with some excess amount were weighed to produce stoichiometric $Li_2(Pd/Pt)_3B$. The mixed materials were heated and reacted in three different ways using the arc-melting furnace, a radio frequency (RF) furnace with a cold crucible and a conventional muffle furnace. (1) Arc-melting furnace: when the arc-flame slowly reached the plate of $(Pd/Pt)_3B$ and the lump of Li, both reacted quickly, and the heating continued for fewer than 10 seconds. (2) RF furnace: the starting materials of $(Pd/Pt)_3B$ and Li were placed in a cold crucible and heated by radio frequency in a pure argon atmosphere. When the temperature of $(Pd/Pt)_3B$ rose up to around 800 ^{o}C, the lump of Li was absorbed into the $(Pd/Pt)_3B$. The further heating was necessary to complete the reaction of Li and $(Pd/Pt)_3B$. (3) Muffle furnace: the lump of Li and $(Pd/Pt)_3B$ were put in a Ta or Mo tube with one end closed. Another end was closed by the arc-furncace in an argon atmosphere. The tube was sealed in a vacuum quartz tube followed by the heating process at around 700 ^{o}C for 48 hours in the muffle furnace.

The internal textures of reacted materials were observed using a scanning electron microscope (SEM, JEOL JSM-6301F). The microstructural observation was carried out for powder samples using a transmission electron microscopy (TEM, JEOL FX-2000). X-ray powder diffraction (XRD) analyses were performed using a diffractometer equipped with a rotating anode and a graphite monoclometer (JEOL, JDX-3500). Low-temperature X-ray powder diffraction was achieved down to 7.6 K for the investigation of a crystal structure transformation and lattice shrinking using RIGAKU RU-300 (50kW 300mA). Lattice parameters were calculated by the Rietveld method. The magnetic properties were measured using a SQUID magnetometer (MPMS-XL, Quantum Design Co., Ltd.). Electrical resistivity was measured by a conventional four-probe method using a ^{3}He refrigerator of Heliox (Oxford Instruments) at temperatures down to 0.4 K and a modified system for the resistivity measurement in MPMS-XL at temperatures down to 1.8 K. The current was set at 1mA and the magnetic field was applied in perpendicular to it. The composition of Li:(Pd/Pt):B was determined by the chemical analyses using an ICP (induced coupled plasma) method. The samples were kept in vacuum or argon gas atmosphere during conserving or handling.

Polycrystalline samples were cut into rectangular shapes with dimensions of approximately 1.7x3.0x0.4 mm^{3}. The samples of Li_2Pd_3B and Li_2Pt_3B with a weight of 11.1mg and 9.1mg were used for the specific heat measurements carried out on a relaxation type calorimeter (Heliox, Oxford Instruments). The calorimeter equipped with ^{3}He refrigerator was operated within the temperature range of 0.4 K and 15 K.

Pressure dependence of superconductivity was measured using a micro pressure cell in the SQUID magnetometer. The original idea of the cell has been described by Diederichs et al.[18]. The pressure cell was made of hardened beryllium-copper (Be-Cu) alloy. A sample was placed in the cell filled by Daphne oil as a pressure-transmitting fluid.. At room temperature, force was transmitted to high pressure region up to 1GPa by a Be-Cu pushing piston, which was clamped with a Be-Cu bolt. Sn as a manometer was also placed in the cell at a distance of 40mm from the sample. The pressure at a low temperature was determined from the superconducting transition temperature T_C of Sn whose pressure dependence is well-known [19].

3. Results and Discussion

The Li_2Pd_3B and Li_2Pt_3B materials were synthesized in three ways in this research. Arc-melting and RF heating methods are able to cause a reaction quickly. However, according to ICP analysis, a considerable amount of about 30% of Li was lost during heating due to the high volatility of the Li element. Therefore, a mixture of Li and an alloy of $(Pd/Pt)_3B$ with a ratio of Li:(Pd/Pt):B=2.6:3:1 was reacted by arc- and RF-heating for 10 seconds. The Li lump was absorbed in $(Pd/Pt)_3B$ with the reaction during heating. Another is a conventional heating process with materials in an argon-sealed Ta or Mo tube. The materials in the tube were heated up to between 900 °C and 600 °C for 48 h. The XRD patterns of those materials after the heating treatment showed considerable reactions with Ta over 800 °C. The 600 °C heating treatment showed a lack of reaction of the Li with $(Pd/Pt)_3B$. Therefore, we usually chose the reaction temperature at around 700 °C. According to ICP analysis, the loss of Li after the heating process compared to the starting composition of the mixed materials of Li and $(Pd/Pt)_3B$ was estimated to be 3%. The samples were kept in a vacuum or argon gas atmosphere during conserving or handling.

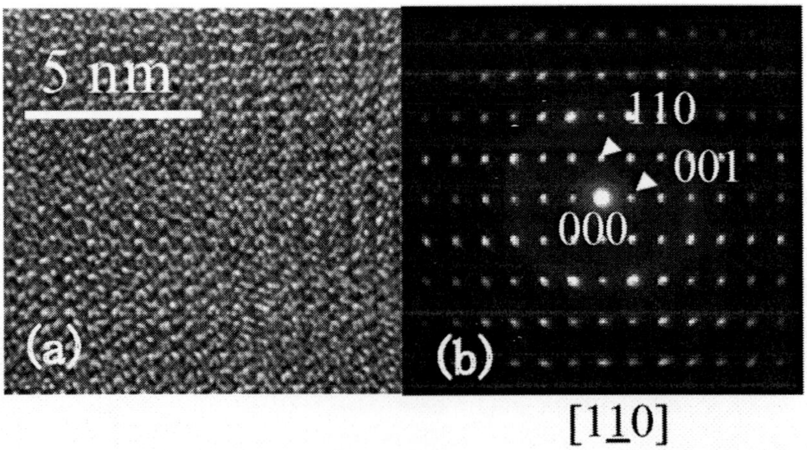

Figure 2. (a) TEM micrograph of the Li_2Pd_3B grain taken along the [1 1 0]. (b) Corresponding electron diffraction pattern for the image of (a).

The XRD patterns of the reacted materials were checked for the identification of the $Li_2(Pd/Pt)_3B$ phase. Crystallographic data on these materials have been reported by Eibenstein et al. [20]. Figure 2(a) and (b) show the TEM micrograph and diffraction pattern of the Li_2Pd_3B grain taken along the [1 1 0] axis. The index of the diffraction pattern and lattice parameter of a=0.67 nm illustrated in Fig. 2(b) is consistent with the XRD results and the report of a single-crystal structure analysis [20].

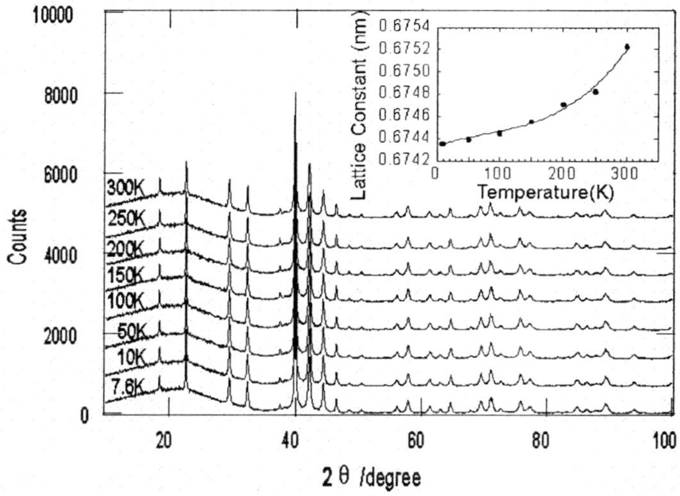

Figure 3. Temperature dependence of X-ray powder diffraction patterns and lattice parameters (inset) for the Li_2Pd_3B from 300 K down to 7.6 K.

The temperature dependence of the XRD patterns for the arc-melted Li_2Pd_3B is illustrated in Fig. 3. A broad peak of around 20 degrees/2θ was attributed to the grease (APIEZON N) for keeping samples in a copper holder. No new or additional peaks were observed in the temperature range between 300 K and 7.6 K. Even in the range near T_C, no significant change was revealed through the transition temperature. The shrinking of the lattice parameter calculated by using a Rietveld fitting program (FullProf) is shown in the inset of Fig. 3. These results proved no phase transition in the temperatures down to T_C from room temperature and was also supported by the neutron diffraction study [21].

The magnetic properties of the Li_2Pd_3B material synthesized using the three synthesizing processes are shown in comparison with the data of a Nb single crystal in Fig.4. Three samples of Li_2Pd_3B were analyzed by the ICP method, and their compositional ratio coincided with the stoichiometric one within 4 % for Li/Pd and 2 % for B/Pd. Those data of the magnetization were normalized by the volume (cm^3). Three Li_2Pd_3B materials and Nb showed the same magnitude of volume magnetization under 20 Oe by zero-field cooling (ZFC). The shielding effect is consistent with the full volume of superconductivity. This fact supports the bulk superconductivity in Li_2Pd_3B.The signals measured under field cooling (FC) were much smaller than those under ZFC. The signal of the material synthesized under the conventional heating process at 700 °C for 48 hours was relatively larger than that of the others. This suggests that the materials of Li_2Pd_3B by arc- and RF-heating contain many effective pinning centers to trap the magnetic flux.

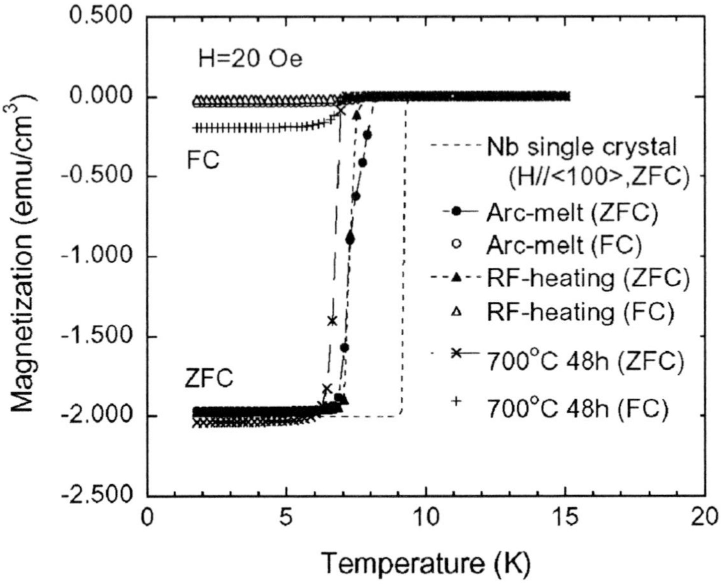

Figure 4. Magnetization vs. temperature curves normalized by unit volume ($1cm^3$) for the three samples (arc-melted, RF-heated, and heated at 700 °C for 48 hours) compared to a Nb single crystal.

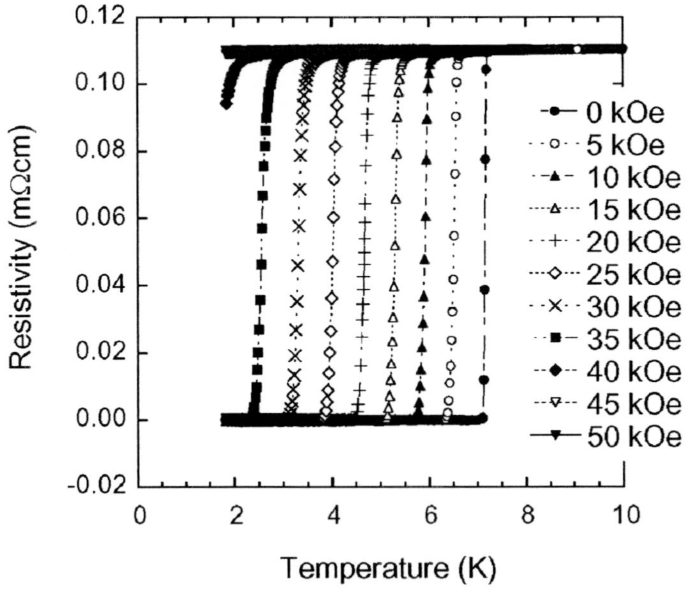

Figure 5. Electrical resistivity of the Li_2Pd_3B sample in dependence with an applied magnetic field up to 50 kOe.

We have measured the resistivity of Li_2Pd_3B and have reported before in ref. [22]. The H_{C2}(T) could be well described by the simple expression $H_{C2}(T) = H^*_{C2}(1-T/T_C)^{1+\alpha}$ [23] in a wide temperature range. The parameter α ($0<\alpha<1$) measures the positive curvature of $H_{C2}(T)$ and H^*_{C2}~1.1-1.2 $H_{C2}(0)$ gives an upper bound for the true value of $H_{C2}(0)$ at T=0.

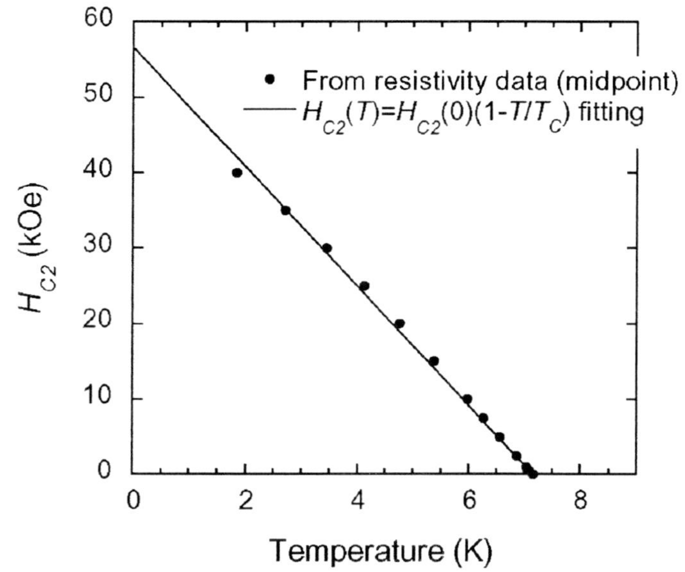

Figure 6. $H_{C2}(T)$ vs. temperature plot for the Li_2Pd_3B defined by the midpoint T_c. The solid line is illustrated by the equation of $H_{C2}(T)=H_{C2}(0)(1-T/T_C)$.

Consequently, the $H_{C2}(0)$ was estimated at 62 kOe. The positive curvature is similar to the polycrystalline borocarbides [24] and MgB_2 [25]. Recently, we tried to confirm the curvature of the H_{C2} near T_C using the sample which was used in the specific heat measurements. Figure 5 shows the electrical resistivity of the annealed Li_2Pd_3B sample in dependence with applied magnetic field. The upper critical field H_{C2} curve vs. temperature shown in Fig. 6 was derived from the data of the resistivity. We found that H_{C2}(T) was well fitted by the linear equation of $H_{C2}(T)=H_{C2}(0)(1-T/T_C)$ and it gave $H_{C2}(0)$=57 kOe. No positive curvature was seen in the plot for Li_2Pd_3B. According to the Ginzburg-Landau(G-L) theory, H_{C2} near T_C is given by the equation:

$$H_{C2}(T) = \frac{\Phi_0}{2\pi\xi^2} = \frac{\Phi_0}{2\pi\xi(0)^2}(1-\frac{T}{T_C}),$$

where Φ_0 :fluxoid quantum, ξ :G-L coherence length, and T_C: superconducting transition temperature. ξ (0) is estimated to be 7.6 nm for this material of Li_2Pd_3B by plotting H_{C2} vs. $(1-T/T_c)$ near T_C. The electrical resistivity of the annealed Li_2Pt_3B sample was also measured and is shown in Fig. 7. The H_{C2} curve vs. temperature was plotted in Fig. 8 in the same way of Fig. 6 for Li_2Pd_3B. It was also linearly fitted by the simple equation of $H_{C2}(T)=H_{C2}(0)(1-T/T_C)$ and it gave $H_{C2}(0)$=24 kOe. Then, ξ (0) is estimated to be 11.8 nm.

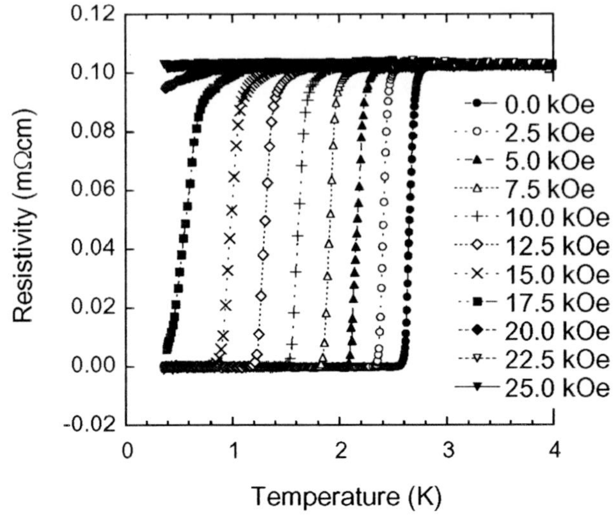

Figure 7. Electrical resistivity of the Li_2Pt_3B sample in dependence with an applied magnetic field up to 25 kOe.

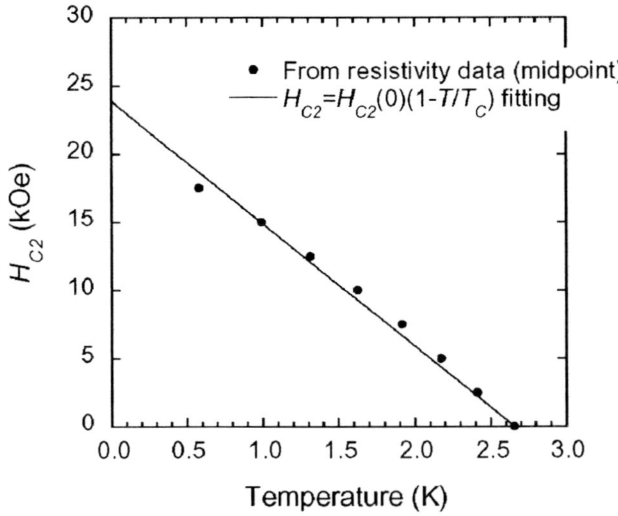

Figure 8. $H_{C2}(T)$ vs. temperature plot for the Li_2Pt_3B defined by the midpoint T_C. The solid line is illustrated by the equation of $H_{C2}(T)=H_{C2}(0)(1-T/T_C)$.

The specific heat of Li_2Pd_3B is shown in Fig.9. The normal-state properties of Li_2Pd_3B were revealed after quenching the superconductivity by applying a magnetic field of 50 kOe over H_{c2} (T>0.4K). The normal-state specific heat was described using Sommerfeld (γT) and Debye (βT^3) terms as a conventional formula, $C_n=\gamma T+\beta T^3$. According to the fitting formula, γ=8.3 mJmol^{-1}K^{-2} and β=1.4 mJmol^{-1}K^{-1} for Li_2Pd_3B were given in the temperature range of

$T<T_C$. The Debye temperature was θ_D=202 K. A superconducting jump appeared at 7.0 K in a zero magnetic field. The value of $\Delta C/\gamma T$ at T_C was calculated to be 2.1. Figure 10 illustrates the specific heat of Li_2Pt_3B under a zero field and a magnetic field of 15 kOe (>H_{C2}, T>0.4K). The γ and β values were 9.6 mJmol^{-1}K^{-2} and 0.85 mJmol^{-1}K^{-2} respectively estimated by the fitting formula of $C_n=\gamma T+\beta T^3$ ($T<T_C$). θ_D=240 K, T_C =2.6 K, and $\Delta C/\gamma T_C$=0.75 were obtained as well as given for Li_2Pd_3B. Since $\Delta C/\gamma T_C$ in the weak-coupling limit by the BCS theory is 1.43, the value of 2.1 for Li_2Pd_3B suggested the possibility of strong-coupling corrections, while the value of 0.75 for Li_2Pt_3B was much lower. The above-mentioned physical parameters of Li_2Pd_3B and Li_2Pt_3B are listed in Table 1.

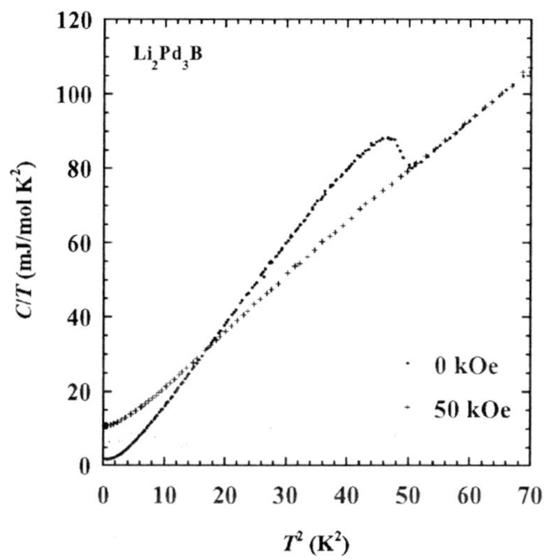

Figure 9. Specific heat divided by the temperature (C/T) of Li_2Pd_3B vs. T^2 under a zero and a 50 kOe magnetic field.

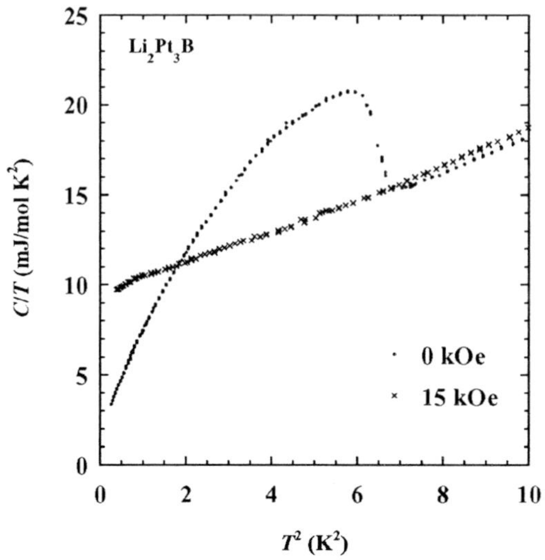

Figure 10. Specific heat divided by the temperature (C/T) of Li_2Pt_3B vs. T^2 under a zero and a 15 kOe magnetic field.

Table 1. Normal and superconducting parameters of Li_2Pd_3B and Li_2Pt_3B

Compound	T_C [K]	γ [mJ/mol K^2]	θ_D [K]	$\Delta C/\gamma T_C$ [none]
Li_2Pd_3B	7.2	8.3	202	2.1
Li_2Pt_3B	2.6	9.6	240	0.75

The γ of Li_2Pt_3B was almost 16% higher than that of Li_2Pd_3B, and the θ_D was also 19% higher. γ is related to the electron density state of $N(E_F)$ expressed by $\gamma=(2/3)\pi^2 k_B^2(1+\lambda)N(E_F)$, whereas k_B and λ denote the Boltzmann's constant and the electronic mass-enhancement. T_C is described by $T_C=0.85\theta_D\exp(-1/N(E_F)V)$, V: BCS interaction parameter in the BCS model [26]. Therefore, assuming the same values of λ and V for the two compounds, the T_C of Li_2Pt_3B was calculated to be 13 K on the basis of the measured T_C=7.0 K for Li_2Pd_3B. The measured T_C for Li_2Pt_3B was only 2.6 K, much lower than T_C of Li_2Pd_3B. The electronic structures of these isotypic compounds have been calculated in several studies [27-29]. According to the calculations, the density of states at the Fermi level is higher in Li_2Pt_3B than in Li_2Pd_3B. This result is consistent with the γ from the specific heat measurements.

It is difficult to control the Li stoichiometry accurately because of the loss during arc melting. Since there is a strong dependence of T_C on the stoichiometry of the Li content, studying the boron isotope effect on these materials is discouraged. Paying attention to the isotypic structure and the similarity of the electronic configurations between Pd and Pt, their pseudo-isotope effect on superconductivity was evaluated. Since the isotope effect is described as $T_C M^{\alpha}$=constant (α~1/2), the relation $(T_{C,\mathrm{Pt}}\sqrt{M_{Pt}})\ /\ (T_{C,\mathrm{Pd}}\sqrt{M_{Pd}}) = 0.5$ was 50% away from the BCS standard model. The T_C of Li_2Pt_3B from the isotope effect should be 5.2

K when extrapolating from the T_C of Li_2Pd_3B. The measured T_C for Li_2Pt_3B, 2.6 K, was much lower. Therefore, the difference in T_C between Li_2Pd_3B and Li_2Pt_3B cannot be attributed solely to a modification of the lattice vibrations and the density of states at the Fermi level.

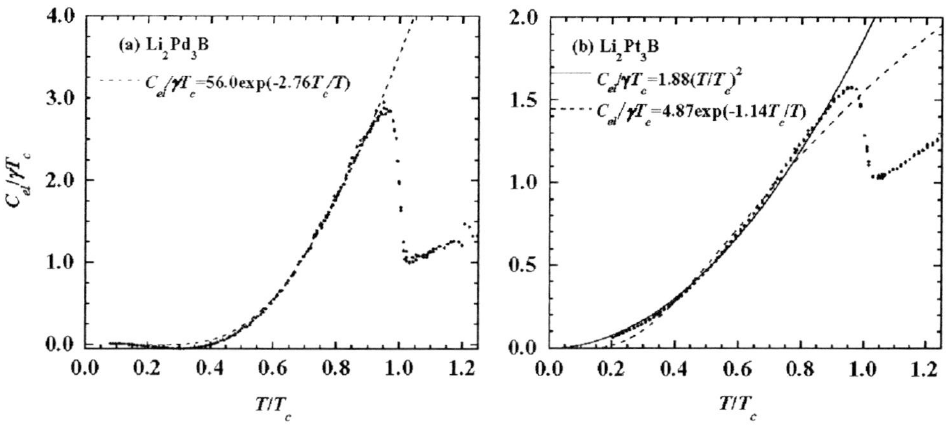

Figure 11. Electronic specific heat ($C_{el}/\gamma T_C$) vs. temperature (T/T_C) of (a) Li_2Pd_3B and (b) Li_2Pt_3B with fitting curves.

Figures 11 (a) and (b) show the electronic specific heat (C_{el}) values of both compounds. They are the electron contribution estimated by subtracting the lattice one, βT^3, from the entire specific heat. The obtained C_{el} data below T_C in Li_2Pd_3B gives clearly exp(-1/T) dependence following the fitting curve in the figure as predicted by the BCS theory for the superconductivity with isotropic s-wave symmetry. On the other hand, the C_{el} of Li_2Pt_3B seems not to reflect an exponential behavior as seen in that of Li_2Pd_3B, but a power function of T^2. This result is not an enough evidence of unconventional behavior, but Li_2Pt_3B did not totally follow conventional superconducting behavior in contract to Li_2Pd_3B. According to the penetration depth study [30], they claimed that Li_2Pt_3B has line nodes in the energy gap.

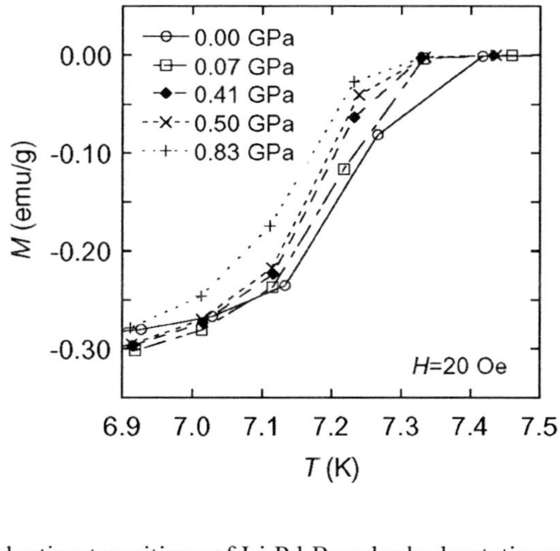

Figure 12. Superconducting transitions of Li_2Pd_3B under hydrostatic pressure up to 0.83 GPa.

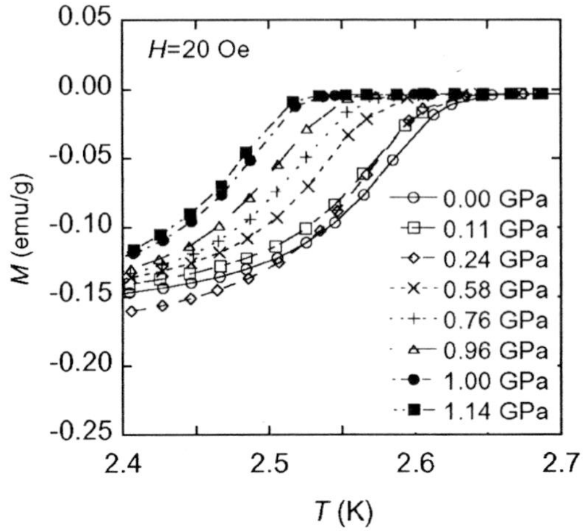

Figure 13. Superconducting transitions of Li_2Pt_3B under hydrostatic pressure up to 1.14 GPa.

Hydrostatic pressure effect was measured with a SQUID magnetometer using a low temperature hydrostatic micro pressure cell. Figures 12 and 13 are the pressure dependence of the superconducting transitions for Li_2Pd_3B and Li_2Pt_3B respectively at pressures up to around 1 GPa. The superconducting transition temperatures of both Li_2Pd_3B and Li_2Pt_3B decreased monotonically with pressure. The pressure dependence, dT_C/dP, were 0.091 and 0.090 K/GPa, respectively. These values are comparable to those of conventional metallic superconductors, but much smaller than the value, 1.03 K/GPa, of MgB_2. In the equation, $T_C=0.85\theta_D\exp(-1/N(E_F)V)$, described above, T_C is dependent on θ_D, $N(E_F)$ and V. Usually, the pressure effect is expected on the increase of θ_D, but the behaviors of $N(E_F)$ and V are not simple. According to ref.[26], the decrease in T_C is caused by the variation in V with an approximately constant value of $N(E_F)$ for nontransition metals. On the other hand, it is explained that the change in T_C with pressure in the transition metals is mainly caused by a change in $N(E_F)$. Since d-bands of Pt and Pd of these superconductors are mainly responsible for the Fermi surface[27], T_C dependence of pressure effect is negative as commonly seen in transition metals[26]. Further experiments of specific heat measurements with applying pressure and theoretical calculations for the pressure effect on $N(E_F)$ and V are expected for the clarification of the pressure effect on these new superconductors.

4. Conclusion

The two superconductors of Li_2Pd_3B and Li_2Pt_3B have been newly found and their physical properties were studied. They do not have symmetries of inversion and mirror in the crystal structure of $P4_332$. Pd and Pt are metals of the platinum family. Therefore, their physical properties can be expected by the analogy between Pd and Pt at the first simple approximation. However, the similarity in the two compounds was not simple, as there was considerable difference in their specific heat behaviors. The electronic density state γ of Li_2Pt_3B was almost 16% higher than that of Li_2Pt_3B, while θ_D was 19% higher. Chandra et al. calculated their electronic structures [24]. They claimed that the density of states at Fermi level is larger in Li_2Pt_3B than that in Li_2Pd_3B. This result is consistent with the BCS scenario, but those parameters could not explain the difference in T_C between Li_2Pd_3B and Li_2Pt_3B well. The obtained C_{el} data below T_C in Li_2Pd_3B and Li_2Pt_3B were compared. The former clearly followed an exponential dependence, while the latter did not follow a clear exponential function, but a power one. The superconducting behavior of Li_2Pt_3B is likely to be unconventional.

We are now measuring NMR, μSR and other physical properties at lower temperatures, especially for Li_2Pt_3B. In addition, we also hope to perform the pressure dependence in higher pressure range for both compounds and to know the behaviors of dT_c/dP and $dN(E_F)/dP$ for both compounds. It seems to be difficult to grow the single crystals of Li_2Pd_3B and Li_2Pt_3B because of the volatility of Li, but it is necessary to better understand the nature of its superconductivity.

Acknowledgements

The author would like to thank Dr. S. Kasahara, Dr. M. El Massalami, Dr. K. Hirata Dr. T. Mochiku, Dr. H. Fujii and Dr. K. Togano for assistance and discussion in the course of this study. This work was partly supported by the Japanese Society of the Promotion of Science and the Japan Space Forum.

References

[1] G. Profeta, A. Continenza, F. Bernardini, S. Massidda, *Phys. Rev. B* **65**, 054502 (2002) .

[2] H. Rosner, A.Kitaigorodsky, W. E. Pickett, *Phys. Rev. Lett.* **88**, 127001 (2002).

[3] A.S. Cooper, E. Corenzwit, L.D. Longinotti, B.T. Matthias, WH. Zachariasen, *Proc. National Academy of Sciences*, Vol.67, No.1, pp.313-319, September 1970.

[4] D. Kaczorowski, A.J. Zaleski, O.J. Zogal, J. Klamut, *cond-mat/0103571.*

[5] V.A. Gasparov, N.S. Sidorov, I.I. Zever'kova, M.P. Kulakov, *JTEP Lett.* **73**, 532 (2001).

[6] H. Takeya,A. Matsumoto, K.Hirata, Y.S.Sung, K.Togano, *Physica C* 412-414, 111 (2004).

[7] H.Takeya, K.Togano, Y.S.Sung, T.Mochiku, K.Hirata, *Physica C* 408-410, 144 (2004).

[8] H. Takeya, H. Takei, Y. Koike, T. Fukase, T. Shishido, N. Toyota, Y. Syono, *Jpn. J. Appl. Phys.*, **25**, 1945 (1986).

[9] Y. Koike, T. Fukase, H. Takeya, H. Takei, T. Shishido, N. Toyota, Y. Syono, *Physica B*, **148**, 126 (1987).

[10] T. He, Q. Huang, A. P. Ramirez, Y. Wang, K. A. Regan, N. Rogado, M. A. Hayward, M. K. Haas, J. S. Slusky, K. Inumara, H. W. Zandbergen, N. P. Ong, R. J. Cava, *Nature* **411**, 54 (2001).

[11] G. Amano, S. Akutagawa, T. Muranaka, Y. Zenitani, J. Akimitsu, *J. Phys. Soc. Jpn.* **73**, 530 (2004).

[12] K. Kawashima, A. Kawano, T. Muranaka, J. Akimitsu, *J. Phys. Soc. Jpn.* **74**, 700 (2005).

[13] H. Fujii, T. Mochiku, H. Takeya, A. Sato, *Phys. Rev. B* **72**, 214520 (2005).

[14] E. A. Ekimov, V. A. Sidorov, E. D. Bauer, N. N. Mel'nik, N. J. Curro, J. D. Thompson, S. M. Stishov, *Nature* **428**, 542 (2004).

[15] K. Togano, P. Badica, Y. Nakamori, S. Orimo, H. Takeya, K. Hirata, *Phys. Rev. Lett.* **93**, 247004 (2004).

[16] F. Izumi, R. A. Dilanian, Commission on Powder Diffraction, *IUCr Newslett.*, No. **32**, 2005, pp. 59-63.

[17] E. Bauer, G. Hilscher, H. Michor, C. Paul, E. W. Scheidt, A. Gribanov, Y. Seropegin, H. Noël, M. Sigrist, P. Rogl, *Phys. Rev. Lett.* **92**, 027003 (2004).

[18] J. Diederichs, A. K. Gangopadhyay, J. S. Schilling, *Phys. Rev. B*, **54**, R9662 (1996).

[19] T.F. Smith, C.W. Chu, *Phys. Rev.*, **159**, 353 (1967) .

[20] U. Eibenstein, W. Jung, *J. Solid State Chem.* **133**, 21 (1997).

[21] H. Takeya, K. Hirata, K. Yamaura, K. Togano, M. El Massalami, R. Rapp, F. A. Chaves, B. Ouladdiaf, *Phys. Rev. B* **72**, 104506 (2005).

[22] H. Takeya, K. Yamada, K. Yamaura, T. Mochiku, H. Fujii, T. Furubayashi, K. Hirata, K. Togano, *Physica C*, 426–431, 411 (2005).

[23] G. Fuchs, K.-H. Müller, S.-L. Drechsler, S. Shulga, K. Nenkov, J. Freudenberger, G. Behr, D. Souptel, A. Handstein, A. Wälte, D. Lipp, L.C. Gupta, *Physica C* 408-410, 107 (2004).

[24] Y. Takano, H. Takeya, H. Fujii, H. Kumakura, T. Hatano, K. Togano, H. Kito and H. Ihara: Appl. *Phys. Lett.* **78**, 2914 (2001).

[25] H. Takagi, R. J. Cava, H. Eisaki, J. O. Lee, K. Mizuhashi, B. Batlogg, S. Uchida, J. J. Krajewski and W. F. Peck, Jr., *Physica C* **228**, 389 (1994).

[26] R. Meservey, B. B. Schwartz, "Equilibrium Properties," in *Superconductivity*, edited by R. D. Parks, New York: Marcel Dekker, Inc., 1969, pp. 117-191.

[27] S. Chandra, S. Mathi Jaya, M. C. Valsakumar, *Physica C* **432**, 116 (2005).

[28] S. K. Bose, E. S. Zijlstra, *Physica C* **432**, 173 (2005).

[29] K.-W. Lee, W. E. Pickett, *Phys. Rev. B* **72**, 174505 (2005).

[30] H. Q. Yuan, D. F. Agterberg, N. Hayashi, P. Badica, D. Vandervelde, K. Togano, M. Sigrist, M. B. Salamon, *Phys. Rev. Lett.* **97**, 017006 (2006).

In: New Research on Superconductivity
Editor: Barry P. Martins, pp. 187-215
ISBN 978-1-59454-197-1

Chapter 7

APPLICATIONS OF CONFINED QUANTUM FIELD THEORY TO CONDENSED MATTER SYSTEMS: SUPERCONDUCTING PHASE TRANSITIONS IN FILMS, WIRES AND GRAINS

***A. P. C. Malbouisson*[1,*], *J. M. C. Malbouisson*[2,†], *A. E. Santana*[3,‡]**
[1]Centro Brasileiro de Pesquisas Físicas (CBPF/MCT),
Rua Dr. Xavier Sigaud, 150, Rio Janeiro, 222290-180, RJ, Brazil
[2]Instituto de Física, Universidade Federal da Bahia,
Salvador, 40210-340, BA, Brazil
[3]Instituto de Física, Universidade de Brasília,
Brasília, 70910-900, Brazil

Abstract

We consider the $\lambda\phi^4$ theory in the D-dimensional Euclidean space with d ($d \leq D$) compactified dimensions. Parameterizing the mass term in the form $m^2(T) = \alpha(T - T_0)$, we retrieve Ginzburg-Landau (GL) models with the order parameter confined in d spatial dimensions. We use methods of Quantum Field Theory to discuss second-order phase transitions in such framework. Specifically, we consider the 3-dimensional GL model compactified in $d = 1, 2$ and 3 space coordinates describing a film, a wire and a grain, respectively. We determine the dependence of the transition temperature with the relevant size of the system for these cases. In particular, we address the question of the existence of a minimal size of the system for sustaining the superconducting phase. We also consider the critical behavior of type-II superconducting films in a magnetic field. The results are in qualitative agreement with the experimental measures for the case of superconducting transitions.

Keywords: $\lambda\phi^4$ theory; Ginzburg-Landau model; Second-order phase transitions; Superconducting films, wires and grains

*E-mail address: adolfo@cbpf.br
†E-mail address: jmalboui@ufba.br
‡E-mail address: asantana@fis.unb.br

1. Introduction: Phase Transitions in Confined Systems

The idea of describing thermodynamical phases through classical fields, the order parameters, was first introduced by Landau [1]. The free energy of a system is written as a functional of the order parameter and the equilibrium state is obtained from a variational principle. Phase transitions of first- and second-order type can be described by choosing an appropriated expansion of the free energy functional around the critical point.

For a second-order phase transition, the free energy can be written (in natural units, $\hbar = c = k_B = 1$) as [2]

$$F[\phi(\mathbf{r})] = \int d\mathbf{r}\, \mathcal{F}(\phi(\mathbf{r}), \nabla\phi(\mathbf{r}))\,, \tag{1}$$

with the free energy density given by

$$\mathcal{F}(\phi(\mathbf{r}), \nabla\phi(\mathbf{r})) = |\nabla\phi(\mathbf{r})|^2 + a(T)\,|\phi(\mathbf{r})|^2 + \frac{b}{2}\,|\phi(\mathbf{r})|^4\,, \tag{2}$$

where $a(T) = a_0(T - T_0)$, T_0 being the critical temperature, and $b > 0$ is independent of the temperature T. The truncated expansion (2) is supposed to be correct near to the transition point, where the order parameter is close to zero. In the equilibrium, the order parameter must then satisfy the equation

$$\nabla^2\phi(\mathbf{r}) - a(T)\phi(\mathbf{r}) - b|\phi(\mathbf{r})|^2\phi(\mathbf{r}) = 0. \tag{3}$$

Neglecting spatial variations of ϕ, the possible solutions are $\phi_0 = 0$ and $\phi_0 = -a(T)/b$; the null solution is stable when $T > T_0$, while the other one gives the true minimum of free energy for $T < T_0$, corresponding to the ordered phase. We also see from Eq. (3) that the spatial changes of the order parameter $\phi(\mathbf{r})$ are characterized by the GL coherence length $\eta^2 = 1/|a(T)|$.

In this chapter, our aim is to address the question of second-order phase transitions for systems with compactified dimensions (that is, in films, wires and grains), using methods of the quantum theory of fields.

Under a field theoretical point of view, the Ginzburg-Landau free energy density in absence of external fields, Eq. (2), can be looked at as the Hamiltonian density for the Euclidean self-interacting scalar field theory, the $\lambda\phi^4$ model. Therefore, we can take advantage of this similarity to explore the powerful methods of quantum field theory to treat fluctuations of the order parameter in the GL model. In this Chapter, we shall use these methods to discuss second-order phase transitions in confined systems, that is, in situations where the order parameter is restricted to exist in limited regions of space, for example in films, wires and grains.

In last decades, a large amount of work has been done on the Ginzburg-Landau model applied to the study of the superconducting transition, both in the single component and in the N- component versions of the model, using the renormalization group approach. The state of the subject, for type-I and type-II superconductors and related topics, can be found for instance in Refs. [3, 4, 5, 6, 7, 8]. Questions concerning stability and the existence of phase transitions may also be raised if one considers field theories in the presence of spacial boundaries. Studies on confined field theory have been done in the literature since a long time ago. In particular, an analysis of the renormalization group in finite size geometries can be found in Refs. [9, 10]. Those studies have been performed to take into account boundary effects on scaling laws. The existence of phase transitions would be in this case associated to some spatial parameters describing the breaking of translational invariance, for instance the distance L between planes confining the system.

Here, we shall take a distinct route, analyzing directly the effects of boundaries on the transition. Such confined systems will be obtained by compactifying spatial dimensions. Compactification will be engendered here as a generalization of the Matsubara (imaginary-time) prescription to account for constraints on spatial coordinates. In the original Matsubara formalism, the time is rotated to the imaginary axis, $t \to i\tau$ where τ (the Euclidean time) is limited to the interval $0 \leq \tau \leq \beta$, with $\beta = 1/T$ standing for the inverse temperature. The fields then fulfill periodic (bosons) or anti-periodic (fermions) boundary conditions and are compactified on the τ-axis in a S^1-topology, a circumference of length β. Such formalism leads to the description of a system in thermal equilibrium at the temperature β^{-1}. Since in an Euclidean field theory space and time are treated on the same footing, one can envisage a generalization of the Matsubara approach to any set of spatial coordinates as well; some general aspects of such approach are discussed in Refs. [11, 12, 13, 14]. In such a case, however, the interpretation of this prescription is quite different: it provides a general and practical way to account for systems confined in limited regions of space at finite temperature. Distinctly, here, we will be concerned with stationary field theories and shall employ the generalized Matsubara prescription to implement spatial compactification [15, 16]. No imaginary-time compactification will be done, the temperature will be introduced through the mass parameter in the hamiltonian as done in the GL model.

For a system described by an N-component bosonic field, $\varphi_a(\mathbf{r})$ with $a = 1, 2, ..., N$, living in a D-dimensional Euclidean space with d compactified coordinates, the partition function (the generating functional for the correlation functions in field theory) is given by

$$\begin{aligned} \mathcal{Z}[\varphi_a] &= \int \mathcal{D}\varphi_1 \cdots \mathcal{D}\varphi_N \times \exp\left[- \int_0^{L_1} dx_1 \right. \\ & \left. \cdots \int_0^{L_d} dx_d \int d^{D-d}\mathbf{z}\, \mathcal{H}(\varphi(\mathbf{r}), \nabla\varphi(\mathbf{r})) \right], \end{aligned} \tag{4}$$

where $\mathcal{H}(\varphi(\mathbf{r}), \nabla\varphi(\mathbf{r}))$ is the Hamiltonian density, $\mathbf{r} = (x_1, ..., x_d, \mathbf{z})$ with $\mathbf{z}$ being a $(D - d)$-dimensional vector; the corresponding momentum is $\mathbf{k} = (k_1, ..., k_d, \mathbf{q})$, $\mathbf{q}$ referring to a $(D - d)$ -dimensional vector in momentum space. In this situation, the Feynman rules should be modified according to the generalized Matsubara prescription,

$$\int \frac{dk_i}{2\pi} \to \frac{1}{L_i} \sum_{n_i=-\infty}^{+\infty} ; \qquad k_i \to \frac{2n_i\pi}{L_i}, \quad i = 1, 2..., d. \tag{5}$$

Thus, for $D = 3$ and $d = 1, 2, 3$ we have respectively the system constrained to a slab of thickness L_1, to a wire of rectangular section $L_1 \times L_2$ and to a parallelepiped of volume $L_1 \times L_2 \times L_3$.

Studies using this approach have been recently performed [15, 16] concerning with the spontaneous symmetry breaking in the $\lambda\phi^4$ theory. Also, if one considers the Ginzburg-Landau model confined to limited regions of space, the question of how the critical temperature depends on the relevant lengths of the system can be raised; such investigation has been carried out for films, wires and grains [17, 18, 19, 20, 21, 22, 23], and will be revised here.

We consider the Ginzburg-Landau model in its N-component version, firstly compactified in one of the spatial dimensions. A general formalism describing the compactfication in an arbitrary dimension is then introduced. In practical applications, we investigate the critical behavior of the N -component Ginzburg-Landau model at the large-N limit in three situations: confined between two parallel planes a distance L apart from one another, confined to an infinitely long cylinder having a square cross-section of area L^2 and to a cubic box of volume L^3. These situations are supposed to describe samples of a material undergoing a phase transition, respectively in the form of a film, a wire and a grain. We obtain equations for the critical temperature as a function of L, and determine the limiting sizes sustaining the transition. For type-II superconducting films, we can show that each value of L defines a critical line on the $H \times T$ plane, where H is the magnetic field and T is the temperature. This set of critical lines also suggests the existence of a minimal value for the thickness L below which superconductivity is suppressed.

2. Superconducting Transition in Films

In this section our goal is to present, in a field theoretical framework, effects associated to spatial confinement for the large-N Ginzburg-Landau model. We consider the N-component Euclidean $\lambda\varphi_D^4$ theory at leading order in $1/N$, the system being submitted to the constraint of confinement between two parallel planes a distance L apart from one another. From a physical point of view, for $D = 3$ and introducing temperature by means of the mass term in the Hamiltonian, this corresponds to a film-like material described by the N-component Ginzburg-Landau model. The large-N limit allows us to incorporate in the model the L-corrections to the coupling constant in a non-perturbative manner. We investigate how the physically relevant quantities, as the coupling constant, the mass, and in particular the critical temperature, are affected by the compactification of one of the spatial dimensions. After a mass renormalization, the critical curve relating the transition temperature and the film thickness is established and a comparison is made with the results for the standard Ginzburg-Landau model without coupling-constant corrections.

2.1. The Effective Potential for the GL Model with One Compactified Dimension

We consider the N-component vector model described by the Ginzburg-Landau Hamiltonian density,

$$\mathcal{H} = \nabla\varphi_a \cdot \nabla\varphi_a + \bar{m}_0^2\varphi_a\varphi_a + u\,(\varphi_a\varphi_a)^2\,, \tag{6}$$

in Euclidean D-dimensional space, where u is the coupling constant and $\bar{m}_0^2$ is an L-modified mass parameter such that

$$\lim_{L\to\infty} \bar{m}_0^2(L,T) = m_0^2(T) \equiv \alpha\,(T - T_0)\,, \tag{7}$$

$m_0^2(T)$ being the constant mass parameter present in the usual free-space Ginzburg-Landau model and T_0 the bulk transition temperature. Summation over repeated indices a is assumed. In the following we will consider the model described by the Hamiltonian (6) with N components and take the large N limit with $Nu = \lambda$ *fixed*. We take the system confined between two parallel planes, normal to the x-axis, a distance L apart from one another and use Cartesian coordinates $\mathbf{r} = (x, \mathbf{z})$, where $\mathbf{z}$ is a $(D-1)$-dimensional vector, with corresponding momenta $\mathbf{k} = (k_x, \mathbf{q})$, $\mathbf{q}$ being a $(D-1)$-dimensional vector in momentum space. In this case, the model is supposed to describe a film of thickness L.

Under these conditions the generating functional of the correlation functions is written as,

$$\mathcal{Z} = \int \mathcal{D}\varphi \exp\left[-\int_0^L dx \int d^{D-1}\mathbf{z}\, \mathcal{H}(\varphi, \nabla\varphi)\right], \tag{8}$$

with the field $\varphi(x, \mathbf{z})$ satisfying the condition of confinement along the x-axis, $\varphi(x = 0, \mathbf{z}) = \varphi(x = L, \mathbf{z}) = 0$. Then the field should have a mixed series-integral Fourier expansion of the form,

$$\varphi(x, \mathbf{z}) = \sum_{n=-\infty}^{\infty} \int d^{D-1}\mathbf{q}\; e^{-i\omega_n x\; -i\mathbf{q}\cdot\mathbf{z}}\tilde{\varphi}(\omega_n, \mathbf{q}), \tag{9}$$

where $\omega_n = 2\pi n/L$. According to Eq. (5), for the particular case of $d = 1$, the Feynman rules are modified following the prescription,

$$\int \frac{dk_x}{2\pi} \to \frac{1}{L}\sum_{n=-\infty}^{+\infty}, \qquad k_x \to \frac{2n\pi}{L} \equiv \omega_n. \tag{10}$$

We start from the well known expression for the one-loop contribution to the effective potential [24],

$$U_1(\varphi_0) = \sum_{s=1}^{\infty} \frac{(-1)^{s+1}}{2s}\left[12\lambda\varphi_0^2\right]^s \int \frac{d^Dk}{(2\pi)^D}\frac{1}{(k^2+m^2)^s}. \tag{11}$$

where m is the physical mass and φ_0 is the normalized vacuum expectation value of the field (the classical field). To deal with dimensionless quantities in the regularization procedures, we introduce parameters

$$c = \frac{m}{2\pi\mu}, \quad b = \frac{1}{L\mu}, \quad g = \frac{\lambda}{4\pi^2\mu^{4-D}}, \quad \phi_0^2 = \frac{\varphi_0^2}{\mu^{D-2}}, \tag{12}$$

where μ is a mass scale. In terms of these parameters and performing the replacements (10), the one-loop contribution to the effective potential can be written in the form,

$$U_1(\phi_0, b) = \mu^D\, b \sum_{s=1}^{\infty} \frac{(-1)^s}{2s}\left[12g\phi_0^2\right]^s \sum_{n=-\infty}^{+\infty} \int \frac{d^{D-1}q'}{(b^2n^2 + c^2 + \mathbf{q}'^2)^s}, \tag{13}$$

where $\mathbf{q}' = \mathbf{q}/2\pi\mu$ is dimensionless.

Using the well-known dimensional regularization formula [9]

$$\int \frac{d^dp}{(2\pi)^d}\frac{1}{(p^2+M)^s} = \frac{\Gamma\left(s-\frac{d}{2}\right)}{(4\pi)^{\frac{d}{2}}\Gamma(s)}\frac{1}{M^{s-\frac{d}{2}}}, \tag{14}$$

Eq. (13) can be cast in the form

$$U_1(\phi_0, b) = \mu^D b \sum_{s=1}^{\infty} f(D,s) \left[12 g \phi_0^2\right]^s A_1^{c^2}\left(s - \frac{D-1}{2}; b\right), \tag{15}$$

where

$$f(D,s) = \pi^{(D-1)/2} \frac{(-1)^{s+1}}{2s\Gamma(s)} \Gamma(s - \frac{D-1}{2}) \tag{16}$$

and $A_1^{c^2}\left(s - \frac{D-1}{2}; b\right)$ is one of the Epstein-Hurwitz $zeta$-functions defined by

$$A_K^{c^2}(\nu; b_1, ..., b_K) = \sum_{n_1,...,n_K=-\infty}^{+\infty} (b_1^2 n_1^2 + \cdots + b_K^2 n_K^2 + c^2)^{-\nu}, \tag{17}$$

which is regular for $Re(\nu) > K/2$ (in our case $Re(s) > D/2$).

The function $A_1^{c^2}(\nu; b)$ can be analytically extended to the whole complex ν-plane [25] as follows: using the identity [26]

$$\frac{1}{\Delta^\nu} = \frac{1}{\Gamma(\nu)} \int_0^\infty dt\, t^{\nu-1} e^{-\Delta t}, \tag{18}$$

one can write

$$\begin{aligned} A_1^{c^2}(\nu; b) &= \sum_{n=-\infty}^{+\infty} \frac{1}{(b^2 n^2 + c^2)^\nu} = \frac{1}{c^{2\nu}} + 2 \sum_{1}^{\infty} \frac{1}{(b^2 n^2 + c^2)^\nu} \\ &= \frac{1}{\Gamma(\nu)} \int_0^\infty dt\, t^{\nu-1} e^{-c^2 t} \left[1 + 2 T_1(t,b)\right], \end{aligned} \tag{19}$$

where the function $T_1(t, b)$ is given by

$$T_1(t,b) = \sum_{n=1}^{\infty} e^{-b^2 n^2 t}, \tag{20}$$

and possesses the property

$$T_1(t,b) = -\frac{1}{2} + \sqrt{\frac{\pi}{b^2 t}} \left[\frac{1}{2} + S(\frac{\pi^2}{b^2 t})\right], \tag{21}$$

with $S(x)$ defined by

$$S(x) = \sum_{n=1}^{\infty} e^{-n^2 x}. \tag{22}$$

Next, considering the representation for Bessel functions of the third kind, K_ν, [26]

$$2(a/b)^{\frac{\nu}{2}} K_\nu(2\sqrt{ab}) = \int_0^\infty dx\, x^{\nu-1} e^{-(a/x)-bx}, \tag{23}$$

it is found that

$$\begin{aligned} A_1^{c^2}(\nu;b) &= \frac{2^{\nu+\frac{1}{2}}\pi^{2\nu-\frac{1}{2}}}{b\,\Gamma(\nu)}\left[2^{\nu-\frac{3}{2}}\Gamma\left(\nu-\frac{1}{2}\right)(2\pi c)^{1-2\nu}\right. \\ &\left. + 2\sum_{n=1}^{\infty}\left(\frac{n}{2\pi cb}\right)^{\nu-\frac{1}{2}}K_{\nu-\frac{1}{2}}\left(\frac{2\pi cn}{b}\right)\right]. \end{aligned} \tag{24}$$

Taking $\nu = s-(D-1)/2$ in Eq. (24) and inserting it in Eq. (15), we obtain the one-loop correction to the effective potential in D dimensions, with one compactified dimension, in the form (recovering the dimensional parameters)

$$\begin{aligned} U_1(\varphi_0, L) &= \sum_{s=1}^{\infty}\left[12\lambda\varphi_0^2\right]^s h(D,s)\left[2^{s-\frac{D}{2}-2}\Gamma(s-\frac{D}{2})m^{D-2s}\right. \\ &\left. + \sum_{n=1}^{\infty}\left(\frac{m}{Ln}\right)^{\frac{D}{2}-s}K_{\frac{D}{2}-s}(mLn)\right], \end{aligned} \tag{25}$$

where

$$h(D,s) = \frac{1}{2^{D/2+s-1}\pi^{D/2}}\frac{(-1)^{s+1}}{s\Gamma(s)}. \tag{26}$$

The above one-loop contribution to the effective potential has been obtained by the concurrent use of dimensional and *zeta*-function analytic regularizations, in evaluating formally the integral over the continuous momenta and the summation over discrete frequencies. We get sums of polar (L-independent) terms plus L-dependent analytic corrections. These polar terms are proportional to $\Gamma(s-D/2)$, $s = 1, 2, \ldots$, and thus may lead to divergent contributions to the mass and the coupling constant when D is even; renormalized quantities are obtained by subtraction of these divergent (polar) terms. In order to have a coherent renormalization scheme in any dimension, these subtractions should be performed even in the case of odd dimensions, where no poles of Γ-functions are present. In these cases a finite renormalization is performed.

The mass and the coupling constant renormalizations are implemented by using the renormalization conditions, which here will be taken in the form

$$\left.\frac{\partial^2}{\partial\varphi_0^2}U(D,L)\right|_{\varphi_0=0} = m^2, \tag{27}$$

$$\left.\frac{\partial^4}{\partial\varphi_0^4}U(D,L)\right|_{\varphi_0=0} = \lambda, \tag{28}$$

where U is the sum of the tree-level and one-loop contributions to the effective potential. This will be done in the next subsections.

2.2. Mass Renormalization and Transition Temperature

Initially, we shall neglect corrections to the coupling constant, that is, we assume that λ corresponds to the renormalized coupling constant. Then, from Eqs. (25) and (27), subtracting the polar term proportional to $\Gamma(1 - D/2)$, we get the renormalized mass

$$m^2(L,T) = \bar{m}_0^2(L,T) + \frac{24\lambda}{(2\pi)^{D/2}} \sum_{n=1}^{\infty} \left(\frac{m}{nL}\right)^{\frac{D}{2}-1} K_{\frac{D}{2}-1}(nLm)\,, \tag{29}$$

where L is the separation between the planes, the film thickness.

We shall now obtain the critical curve and determine the dependence of the critical temperature, T_c, with the thickness of a film. If we start in the ordered phase with a negative squared mass, the model exhibits spontaneous symmetry breaking of the $O(N)$ symmetry to $O(N-1)$, but for sufficiently small values of T^{-1} and L the symmetry is restored. We can define the critical curve $C(T_c, L) = 0$ as the line in the $T \times L$ plane for which the inverse squared correlation length, $\xi^{-2}(T, L, \phi_0)$, vanishes in the large-N gap equation

$$\begin{aligned} \xi^{-2}(T,L,\varphi_0) &= \bar{m}_0^2(L,T) + 12\lambda\varphi_0^2 \\ &+ \frac{24\lambda}{L} \sum_{n=-\infty}^{+\infty} \int \frac{d^{D-1}q}{(2\pi)^{D-1}} \frac{1}{\mathbf{q}^2 + \omega_n^2 + \xi^{-2}(T,L,\varphi_0)}\,, \end{aligned} \tag{30}$$

where φ_0 is the normalized vacuum expectation value of the field (different from zero in the ordered phase).

In the disordered phase, in particular in the neighborhood of the critical curve, φ_0 vanishes and the gap equation reduces to the L-dependent Dyson-Schwinger equation for the mass given by Eq. (29). This equation can not be solved in closed form but if we limit ourselves to the neighborhood of criticality ($m^2 \approx 0$) and consider L finite and sufficiently small, we may use the asymptotic formula for small values of the argument of Bessel functions,

$$K_\nu(z) \approx \frac{1}{2}\Gamma(\nu)\left(\frac{z}{2}\right)^{-\nu} \qquad (z \sim 0\,;\quad Re(\nu) > 0)\,, \tag{31}$$

and Eq. (29) reduces to

$$m^2(L,T) \approx \bar{m}_0^2(L,T) + \frac{6\lambda}{\pi^{D/2}L^{D-2}}\Gamma\left(\frac{D}{2}-1\right)\zeta(D-2) \tag{32}$$

where

$$\zeta(z) = \sum_{n=1}^{\infty} \frac{1}{n^z} \tag{33}$$

is the Riemann *zeta*-function, which is a meromorphic function having only a simple pole at $z = 1$.

For $D = 3$, which is the dimensionality of the ordinary space, Eq.(32) is not well defined due to the pole of $\zeta(D-2)$. However, it can be made physically meaningful by the following regularization procedure: from the Laurent expansion of $\zeta(z)$,

$$\zeta(z) = \frac{1}{z-1} + \gamma_0 - \gamma_1(z-1) + \cdots\,, \tag{34}$$

where $\gamma_0 \simeq 0.577$ and $\gamma_1 \simeq 0.0728$ (the Euler-Mascheroni and the first Stieltjes constants, respectively), we define the L-dependent bare mass in such a way that the pole at $D = 3$ in Eq.(32) is suppressed, that is, we take

$$\bar{m}_0^2(L,T) \approx M - \frac{1}{(D-3)}\frac{6\lambda}{\pi L}, \tag{35}$$

where M is independent of D. To fix the finite term, we make the simplest choice satisfying (7)

$$M = m_0^2(T) = \alpha\,(T - T_0)\ , \tag{36}$$

T_0 being the bulk critical temperature. In this case, using Eq. (35) in Eq. (32) and taking the limit as $D \to 3$, the L-dependent renormalized mass term in the vicinity of criticality becomes

$$m^2(L,T) \approx \alpha\,(T - T_c(L))\ , \tag{37}$$

where the modified, L-dependent, transition temperature is given by

$$T_c(L) = T_0 - C_1\frac{\lambda}{\alpha L}\ , \tag{38}$$

L being the thickness of the film, with the constant C_1 given by

$$C_1 = \frac{6\gamma}{\pi} \approx 1.1024\ . \tag{39}$$

From this equation, we see that for L smaller than

$$L_{\min} = C_1\frac{\lambda}{\alpha T_0}\ , \tag{40}$$

$T_c(L)$ becomes negative, meaning that the transition does not occurs. In other words, superconductivity is suppressed in films with thickness smaller than $L_{\min}$.

2.3. Effect of the Coupling-Constant Correction on $T_c(L)$

We now consider the effect of the L-correction of the coupling constant on critical temperature of films. The renormalized coupling constant will be defined in terms of the four-point function at zero external momenta which, at leading order in $\frac{1}{N}$, is given by the sum of all chains of one-loop subdiagrams of the type shown in Fig. 1. This is actually a geometric series whose sum is given by

$$\Gamma_D^{(4)}(\mathbf{p}=0,m,L) = \frac{u}{1+Nu\Pi(D,m,L)}, \tag{41}$$

where $\Pi(D,m,L) = \Pi(\mathbf{p}=0,D,m,L)$ corresponds to the one-loop subdiagram,

$$\Pi(D,m,L) = \frac{1}{L}\sum_{n=-\infty}^{\infty}\int\frac{d^{D-1}q}{(2\pi)^{D-1}}\frac{1}{[\mathbf{q}^2+\omega_n^2+m^2]^2}\ . \tag{42}$$

This subdiagram can be calculated with the technique used before or from the renormalization condition (28). However, comparing Eq. (42) with Eq. (13), we see that $\Pi(D,m,L)$

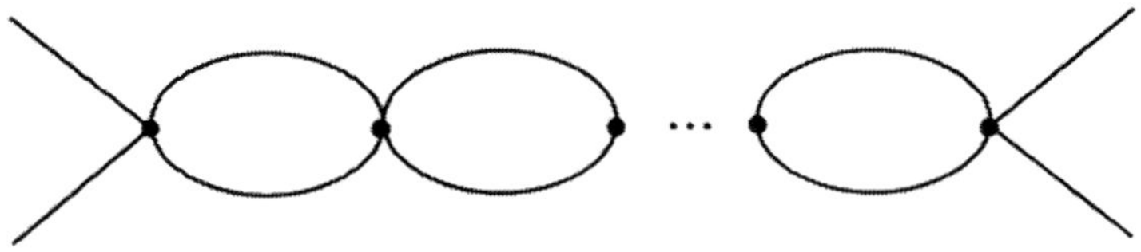

Figure 1. Chain of one-loop subdiagrams contributing to the large- N four-point function.

can be directly obtained from the $s = 2$ term in the expression (25) for the one-loop contribution to the effective potential. Again, suppressing the polar term proportional to $\Gamma(2 - D/2)$, we obtain the renormalized one-loop subdiagram

$$\Pi_R(D, m, L) = \frac{1}{(2\pi)^{D/2}} \sum_{n=1}^{\infty} \left[\frac{m}{nL}\right]^{(D-4)/2} K_{\frac{D-4}{2}}(nLm) \, . \tag{43}$$

From the properties of Bessel functions, it can be seen from Eq. (43) that for any dimension D, $\Pi_R(D, m, L)$ satisfies the conditions

$$\lim_{L\to\infty} \Pi_R(D, m, L) = 0 \, , \qquad \lim_{L\to 0} \Pi_R(D, m, L) \to \infty \, . \tag{44}$$

We also conclude that $\Pi_R(D, m, L)$ is positive for all values of D and L.

Let us define the L-dependent renormalized coupling constant $u_R(m, D, L)$, at the leading order in $1/N$, as

$$\Gamma^{(4)}_{D,R}(\mathbf{p} = 0, m, L) \equiv u_R(D, m, L) = \frac{u}{1 + Nu\Pi_R(D, m, L)} \tag{45}$$

and the renormalized coupling constant in the absence of boundaries as

$$u_R(D, m) = \lim_{L\to\infty} \Gamma^{(4)}_{D,R}(\mathbf{p} = 0, m, L) \, . \tag{46}$$

From Eqs. (45) and (44) we get simply $u_R(D, m) = u$. In other words we have done a choice of renormalization scheme such that the constant u introduced in the Hamiltonian corresponds to the *renormalized* coupling constant. From Eq. (45) we obtain the L-dependent renormalized coupling constant

$$Nu_R(D, m, L) \equiv \lambda_R(D, m, L) = \frac{\lambda}{1 + \lambda\Pi_R(D, m, L)} \, , \tag{47}$$

where we have defined the *fixed* coupling constant $\lambda = Nu$ (see comments after Eq. (6)).

Considering the L-corrections to the coupling constant, the Dyson-Schwinger equation for the mass (29) becomes

$$\begin{aligned} m^2(L, T) &= \bar{m}_0^2(L, T) + \frac{\lambda_R(D, m, L)}{2(2\pi)^{D/2}} \\ &\times \sum_{n=1}^{\infty} \left[\frac{m}{nL}\right]^{\frac{D-2}{2}} K_{\frac{D-2}{2}}(nLm) \, ; \end{aligned} \tag{48}$$

this equation, together with Eqs. (43) and (47), form a very complicate set of equations since in their right hand sides $m = m(D, T, L)$. Nevertheless, limiting ourselves to the neighborhood of criticality, $m^2(D, T, L) \approx 0$, we may investigate the behavior of the system by using in Eqs. (43) and (48) the asymptotic formula (31) for small values of the argument of the Bessel function. Making such an approximation, Eq. (48) reduces to Eq. (32) with λ replaced by $\lambda_R(D, m, L)$. So it remains to calculate the renormalized value of the coupling constant close to criticality.

Inserting (31) in (43), we obtain

$$\Pi_R(D, L) \approx \frac{1}{8\pi^{D/2}} L^{4-D} \Gamma\left(\frac{D-4}{2}\right) \zeta(D-4), \tag{49}$$

which shows that the renormalized one-loop subdiagram is independent of the mass close to criticality. This result leads to the renormalized coupling constant, for $m^2 \approx 0$, given by

$$\lambda_R(D, L) \approx \frac{\lambda}{1 + \lambda C(D) L^{4-D} \zeta(D-4)}, \tag{50}$$

where $C(D) = \frac{1}{8\pi^{D/2}} \Gamma(\frac{D-4}{2})$. Notice that, since $\Gamma(-1/2) = -2\sqrt{\pi}$ and $\zeta(-1) = -1/12$, we find that $\lambda_R(D, L)$ is finite and positive for $D = 3$.

The mass, close to criticality and for generic dimension D, thus becomes

$$m^2(L, T) \approx \bar{m}_0^2(L, T) + \frac{6\lambda_R(D, L)}{\pi^{D/2} L^{D-2}} \Gamma\left(\frac{D}{2} - 1\right) \zeta(D-2), \tag{51}$$

with $\lambda_R(D, L)$ given by Eq. (50). Although $\lambda_R(D, L)$ remains finite as $D \to 3^+$, we have again a singularity in the mass, due to the pole of $\zeta(D-2)$. As in preceding subsection, a regularization procedure is needed. Thus, following the same steps which led from Eq. (32) to Eq. (37) with λ replaced by $\lambda_R(3, L)$, we obtain

$$T_c(L) = T_0 - \frac{48\pi C_1 \lambda}{48\pi\alpha L + \alpha\lambda L^2}. \tag{52}$$

It is interesting to compare this result with the critical curve for a film deduced from the Ginzburg-Landau model in which the L-correction to the coupling constant is neglected, obtained from Eq. (51) by taking $\lambda_R = \lambda$. In this lowest level of approximation, the critical temperature is simply a linear decreasing function of $1/L$, which is plotted in Fig. 2, together with the critical curve (52), for comparison.

We find that the critical temperature (52) decreases from T_0 (the bulk transition temperature) as L diminishes reaching zero for a minimal thickness $L^{(0)}$, below which the transition is suppressed. This minimal thickness is given by

$$L'_{\text{min}} = \frac{24\pi}{\lambda} \left[\sqrt{1 + \frac{\lambda L_{\text{min}}}{12\pi}} - 1\right], \tag{53}$$

where $L_{\text{min}} = C_1\lambda/(\alpha T_0)$ is the minimal thickness for the existence of the ordered phase in the bare approximation of neglecting L-corrections to the coupling constant, Eq. (40). This minimal thickness (not considering coupling constant corrections) coincides with the result

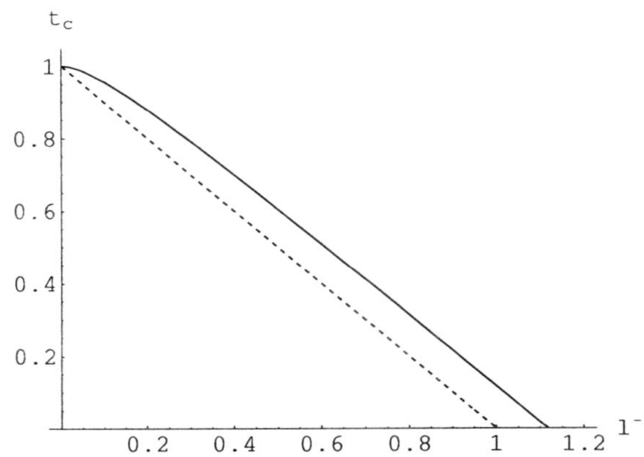

Figure 2. Reduced transition temperature, $t_c = T_c/T_0$, for films as a function of the inverse of the reduced thickness, $l^{-1} = L_{\min}/L$ (with $L_{\min} = C_1\lambda/(\alpha T_0)$), fixing $\lambda L_{\min} = 20$. The dashed line corresponds to $t_c(l) = 1 - l^{-1}$, obtained without considering L-corrections to the coupling constant in the Ginzburg-Landau model.

for the standard (two-component) Ginzburg-Landau model [18], except for a simple symmetry factor. We also find that the predicted minimal film thickness, for the N-component model including the L-correction to the coupling constant ($L'_{\min}$), is lower than the value $L_{\min}$ it would have without finite thickness corrections to the coupling constant, but the general behavior of the curves $t_c = t_c(l)$, in both cases, is very similar. From Fig. 2 we see that, for the chosen values of the parameters, the inclusion of L-corrections to the coupling constant leads to a decrease of the minimum thickness for maintaining superconductivity of about 10%.

Notice that the results obtained here might be applicable to any physical system undergoing a second-order phase transition described by the Ginzburg-Landau model. For example, the decrease of the transition temperature with the inverse of the film thickness (as described above) was experimentally observed for superconductors more than thirty years ago [28]. In fact, our results do not depend on particular physical systems, appearing only as a topological consequence of the compactification in one spatial dimension of the Ginzburg-Landau model.

3. Superconducting Transition Temperature in Wires and Grains

We now generalize the procedure presented before to account for the transition on wires and grains. To do so, we have to extend the compactification process to more than one spatial dimensions. For sake of generality, we consider initially the D-dimensional GL model with d ($\leq D$) compactified coordinates. Fixing $D = 3$, the cases $d = 2$ and $d = 3$ will correspond to wires and grains respectively.

3.1. Compactification of a d-dimensional Subspace

For such a general situation, the partition function is given by Eq. (4), with the hamiltonian density (6), and the generalized Matsubara prescription must be taken as in (5). We consider λ to be the *renormalized* coupling constant (i.e., we will work in the approximation of neglecting boundary corrections to the coupling constant) and take $\bar{m}_0^2$ as the boundary-modified mass parameter depending on $\{L_i\}$ $i = 1, 2...d$, in such a way that

$$\lim_{\{L_i\}\to\infty} \bar{m}_0^2(L_1, ..., L_d, T) = m_0^2(T) \equiv \alpha\,(T - T_0)\,, \tag{54}$$

$m_0^2(T)$ being the bulk mass parameter. In this case, the one-loop contribution for the effective potential, Eq. (11), becomes

$$\begin{aligned} U_1(\phi_0, b_1, ..., b_d) &= \mu^D b_1 \cdots b_d \sum_{s=1}^{\infty} \frac{(-1)^s}{2s} [12g\phi_0^2]^s \\ &\quad \sum_{n_1,...,n_d=-\infty}^{+\infty} \int \frac{d^{D-d}q}{(b_1^2 n_1^2 + \cdots + b_d^2 n_d^2 + c^2 + \mathbf{q}^2)^s}, \end{aligned} \tag{55}$$

where we used the dimensionless quantities (12), now with $b_i = (L_i\mu)^{-1}$ for $i = 1, 2, \ldots d$.

Using the dimensional regularization formula (14) to perform the integration over the $D - d$ non-compactified momentum variables, the above expression can be written as

$$\begin{aligned} U_1(\phi_0, b_1, ..., b_d) &= \mu^D b_1 \cdots b_d \sum_{s=1}^{\infty} f(D, d, s) \\ &\quad \times [12g\phi_0^2]^s \, A_d^{c^2}\left(s - \frac{D-d}{2}; b_1, ..., b_d\right), \end{aligned} \tag{56}$$

with

$$f(D, d, s) = \pi^{(D-d)/2} \frac{(-1)^{s+1}}{2s\Gamma(s)} \Gamma(s - \frac{D-d}{2}), \tag{57}$$

and

$$\begin{aligned} A_d^{c^2}(\nu; b_1, ..., b_d) &= \sum_{\{n_i\}=-\infty}^{+\infty} (b_1^2 n_1^2 + \cdots + b_d^2 n_d^2 + c^2)^{-\nu} \\ &= \frac{1}{c^{2\nu}} + 2\sum_{i=1}^{d} \sum_{n_i=1}^{\infty} (b_i^2 n_i^2 + c^2)^{-\nu} \\ &\quad + 4\sum_{i<j=1}^{d} \sum_{n_i,n_j=1}^{\infty} (b_i^2 n_i^2 + b_j^2 n_j^2 + c^2)^{-\nu} + \cdots \\ &\quad + 2^d \sum_{n_1,...,n_d=1}^{\infty} (b_1^2 n_1^2 + \cdots + b_d^2 n_d^2 + c^2)^{-\nu}. \end{aligned} \tag{58}$$

The analytical extension of the multi-parameter Epstein-Hurwitz function has been constructed in Ref. [16]. Using (18), one gets

$$\begin{aligned} A_d^{c^2}(\nu; b_1, ..., b_d) &= \frac{1}{\Gamma(\nu)} \int_0^\infty dt\, t^{\nu-1} e^{-c^2 t} \left[1 + 2\sum_{i=1}^{d} T_1(t, b_i) \right. \\ &\quad \left. + 4 \sum_{i<j=1}^{d} T_2(t, b_i, b_j) + \cdots + 2^d\, T_d(t, b_1, ..., b_d) \right], \end{aligned} \tag{59}$$

where, the function T_1 is given by (20) and

$$T_j(t, b_1, ..., b_j) = \prod_{l=1}^{j} T_1(t, b_l) \quad , \qquad j = 2, ..., d. \tag{60}$$

Considering the property (21) of the functions T_1, we notice that the surviving terms in Eq. (59) are those proportional to $(b_1 \cdots b_d)^{-1}$, and we find

$$\begin{aligned} A_d^{c^2}(\nu; b_1, ..., b_d) &= \frac{\pi^{\frac{d}{2}}}{b_1 \cdots b_d} \frac{1}{\Gamma(\nu)} \int_0^\infty dt\, t^{(\nu - \frac{d}{2})-1} e^{-c^2 t} \left[1 + 2\sum_{i=1}^{d} S(\frac{\pi^2}{b_i^2 t}) \right. \\ &\quad \left. + 4 \sum_{i<j=1}^{d} S(\frac{\pi^2}{b_i^2 t}) S(\frac{\pi^2}{b_j^2 t}) + \cdots + 2^d \prod_{i=1}^{d} S(\frac{\pi^2}{b_i^2 t}) \right], \end{aligned} \tag{61}$$

with $S(x)$ given by (22). Using again the representation (23) for the Bessel function K_ν, we find

$$\begin{aligned} A_d^{c^2}(\nu; b_1, ..., b_d) &= \frac{2^{\nu - \frac{d}{2}+1} \pi^{2\nu - \frac{d}{2}}}{b_1 \cdots b_d\, \Gamma(\nu)} \left[2^{\nu - \frac{d}{2} - 1} \Gamma\left(\nu - \frac{d}{2}\right) (2\pi c)^{d-2\nu} \right. \\ &\quad + 2 \sum_{i=1}^{d} \sum_{n_i=1}^{\infty} \left(\frac{n_i}{2\pi c b_i} \right)^{\nu - \frac{d}{2}} K_{\nu - \frac{d}{2}} \left(\frac{2\pi c n_i}{b_i} \right) + \cdots \\ &\quad + 2^d \sum_{n_1, ..., n_d = 1}^{\infty} \left(\frac{1}{2\pi c} \sqrt{\frac{n_1^2}{b_1^2} + \cdots + \frac{n_d^2}{b_d^2}} \right)^{\nu - \frac{d}{2}} \\ &\quad \left. \times K_{\nu - \frac{d}{2}} \left(2\pi c \sqrt{\frac{n_1^2}{b_1^2} + \cdots + \frac{n_d^2}{b_d^2}} \right) \right]. \end{aligned} \tag{62}$$

Taking $\nu = s - (D-d)/2$ in Eq. (62) and inserting it in Eq. (56), we obtain the one-loop correction to the effective potential in D dimensions with a compactified d-dimensional subspace in the form (recovering the dimensional parameters)

$$\begin{aligned}
U_1(\varphi_0, L_1, ..., L_d) &= \sum_{s=1}^{\infty} \left[12g\varphi_0^2\right]^s h(D,s) \left[2^{s-\frac{D}{2}-2}\Gamma(s-\frac{D}{2})m^{D-2s}\right. \\
&+ \sum_{i=1}^{d}\sum_{n_i=1}^{\infty}\left(\frac{m}{L_i n_i}\right)^{\frac{D}{2}-s} K_{\frac{D}{2}-s}(mL_i n_i) \\
&+2\sum_{i<j=1}^{d}\sum_{n_i,n_j=1}^{\infty}\left(\frac{m}{\sqrt{L_i^2 n_i^2+L_j^2 n_j^2}}\right)^{\frac{D}{2}-s} \\
&\times K_{\frac{D}{2}-s}\left(m\sqrt{L_i^2 n_i^2+L_j^2 n_j^2}\right)+\cdots \\
&+ 2^{d-1}\sum_{n_1,\ldots,n_d=1}^{\infty}\left(\frac{m}{\sqrt{L_1^2 n_1^2+\cdots+L_d^2 n_d^2}}\right)^{\frac{D}{2}-s} \\
&\left.\times K_{\frac{D}{2}-s}\left(m\sqrt{L_1^2 n_1^2+\cdots+L_d^2 n_d^2}\right)\right], \qquad (63)
\end{aligned}$$

with

$$h(D,s) = \frac{1}{2^{D/2+s-1}\pi^{D/2}}\frac{(-1)^{s+1}}{s\Gamma(s)}\,. \qquad (64)$$

Criticality is attained when the inverse squared correlation length, ξ^{-2}, vanishes in the large-N gap equation,

$$\begin{aligned}
\xi^{-2}(T,\{L_i\},\varphi_0) &= \bar{m}_0^2(\{L_i\},T) + 12\lambda\varphi_0^2 + \frac{24\lambda}{L_1\cdots L_d}\sum_{\{n_i\}=-\infty}^{\infty} \\
&\times \int\frac{d^{D-d}q}{(2\pi)^{D-d}}\frac{1}{\mathbf{q}^2+\sum_{i=1}^{d}(\frac{2\pi n_i}{L_i})^2+\xi^{-2}(T,\{L_i\},\varphi_0)}, \qquad (65)
\end{aligned}$$

where φ_0 is the normalized vacuum expectation value of the field (different from zero in the ordered phase). In the disordered phase, φ_0 vanishes and the inverse correlation length equals the physical mass, given below by Eq. (66). Recalling the renormalization condition (27) (remembering that at the large-N limit it is enough to take the one-loop contribution to the mass), we obtain

$$
\begin{aligned}
m^2(L_1,...,L_d,T) &= \bar{m}_0^2(L_1,...,L_d,T) \\
&+ \frac{24\lambda}{(2\pi)^{D/2}}\left[\sum_{i=1}^{d}\sum_{n_i=1}^{\infty}\left(\frac{m}{L_i n_i}\right)^{\frac{D}{2}-1} K_{\frac{D}{2}-1}(mL_i n_i)\right. \\
&+ 2\sum_{i<j=1}^{d}\sum_{n_i,n_j=1}^{\infty}\left(\frac{m}{\sqrt{L_i^2 n_i^2 + L_j^2 n_j^2}}\right)^{\frac{D}{2}-1} \\
&\times K_{\frac{D}{2}-1}\left(m\sqrt{L_i^2 n_i^2 + L_j^2 n_j^2}\right) + \cdots \\
&+ 2^{d-1}\sum_{n_1,\ldots,n_d=1}^{\infty}\left(\frac{m}{\sqrt{L_1^2 n_1^2 + \cdots + L_d^2 n_d^2}}\right)^{\frac{D}{2}-1} \\
&\left.\times K_{\frac{D}{2}-1}\left(m\sqrt{L_1^2 n_1^2 + \cdots + L_d^2 n_d^2}\right)\right]. \qquad (66)
\end{aligned}
$$

Notice that, in writing Eq. (66), we have suppressed the parcel $2^{-\frac{D}{2}-1}\Gamma(1-\frac{D}{2})m^{D-2}$ from its square bracket, the parcel that emerges from the first term in the square bracket of Eq. (63). This expression, which does not depend explicitly on L_i, diverges for D even due to the poles of the gamma function; in this case, this parcel is subtracted to get a renormalized mass equation. For D odd, $\Gamma\left(1-\frac{D}{2}\right)$ is finite but we also subtract this term (corresponding to a finite renormalization) for sake of uniformity; besides, for $D \geq 3$, the factor m^{D-2} does not contribute in the criticality.

The vanishing of Eq. (66) defines criticality for our compactified system. Notice that the parameter m in the right hand side of Eq. (66) is the boundary-modified mass $m(L_1,...,L_d,T)$, which means that (66) is a self-consistency equation, a very complicated modified Schwinger-Dyson equation for the mass, not soluble by algebraic means. Nevertheless, as we will see in the next sections, a solution is possible at criticality, which allows us to obtain a closed formula for the boundary-dependent critical temperature.

One finds immediately that Eq. (66) reduces to Eq. (29) when $d = 1$, corresponding to the case of a film. We claim that, taking $d = 2$ and $d = 3$, with $D = 3$, we are able to describe respectively the critical behavior of samples of materials in the form square wires and cubic grains. We will derive equations for the critical temperature as a function of the confining dimensions and calculate the minimal system size (cross-section area or volume) below which the phase transition does not take place; this is the subject of the following subsections.

3.2. Critical Behavior for Wires

We now focus on the situation where two spatial dimensions are compactified. From Eq. (66), taking $d = 2$, we get (in the disordered phase)

$$\begin{aligned} m^2(L_1, L_2, T) &= \bar{m}_0^2(L_1, L_2, T) + \frac{24\lambda}{(2\pi)^{D/2}} \\ &\times \left[\sum_{i=1}^{2} \sum_{n_i=1}^{\infty} \left(\frac{m}{n_i L_i}\right)^{\frac{D}{2}-1} K_{\frac{D}{2}-1}(n_i L_i m) \right. \\ &+ 2 \sum_{n_1,n_2=1}^{\infty} \left(\frac{m}{\sqrt{L_1^2 n_1^2 + L_2^2 n_2^2}}\right)^{\frac{D}{2}-1} \\ &\left. \times K_{\frac{D}{2}-1}\left(m\sqrt{L_1^2 n_1^2 + L_2^2 n_2^2}\right) \right]. \end{aligned} \quad (67)$$

If we limit ourselves to the neighborhood of criticality, $m^2 \approx 0$, and taking both L_1 and L_2 finite and sufficiently small, we may use Eq.(23) to rewrite Eq. (67) as

$$\begin{aligned} m^2(L_1, L_2, T) &\approx \bar{m}_0^2(L_1, L_2, T) + \frac{6\lambda}{\pi^{D/2}} \Gamma\left(\frac{D}{2} - 1\right) \\ &\times \left[\left(\frac{1}{L_1^{D-2}} + \frac{1}{L_2^{D-2}} \right) \zeta(D-2) \right. \\ &\left. + 2\, E_2\left(\frac{D-2}{2}; L_1, L_2\right) \right], \end{aligned} \quad (68)$$

where $E_2\left(\frac{D-2}{2}; L_1, L_2\right)$ is the generalized (two-dimensional) Epstein $zeta$-function defined by

$$E_2\left(\frac{D-2}{2}; L_1, L_2\right) = \sum_{n_1, n_2=1}^{\infty} \left[L_1^2 n_1^2 + L_2^2 n_2^2\right]^{-\left(\frac{D-2}{2}\right)}, \quad (69)$$

defined for $\mathrm{Re}\{D\} > 3$.

As mentioned before, the Riemann $zeta$-function $\zeta(D-2)$ has an analytical extension to the whole complex D-plane, having an unique simple pole (of residue 1) at $D = 3$. One can also construct analytical continuations (and recurrence relations) for the multi-dimensional Epstein functions which permit to write them in terms Kelvin and Riemann $zeta$ functions. To start one considers the analytical continuation of the Epstein-Hurwitz $zeta$-function given by [25]

$$\begin{aligned} \sum_{n=1}^{\infty} \left(n^2 + p^2\right)^{-\nu} &= -\frac{1}{2} p^{-2\nu} + \frac{\sqrt{\pi}}{2 p^{2\nu-1} \Gamma(\nu)} \\ &\times \left[\Gamma\left(\nu - \frac{1}{2}\right) + 4 \sum_{n=1}^{\infty} (\pi p n)^{\nu - \frac{1}{2}} K_{\nu - \frac{1}{2}}(2\pi p n) \right]. \end{aligned} \quad (70)$$

Using this relation to perform one of the sums in (69) leads immediately to the question of which sum is firstly evaluated. As it is done in Ref. [27], whatever the sum one chooses to perform firstly, the manifest $L_1 \leftrightarrow L_2$ symmetry of Eq. (69) is lost. To overcome such an obstacle, in order to preserve this symmetry, we adopt here a symmetrized summation generalizing the prescription introduced in [16] for the case of many variables.

To derive an analytical continuation and symmetrized recurrence relations for the multidimensional Epstein functions, we start by taking these functions defined as the symmetrized summations

$$E_d\left(\nu; L_1, ..., L_d\right) = \frac{1}{d!} \sum_{\sigma} \sum_{n_1=1}^{\infty} \cdots \sum_{n_d=1}^{\infty} \left[\sigma_1^2 n_1^2 + \cdots + \sigma_d^2 n_d^2\right]^{-\nu}, \tag{71}$$

where $\sigma_i = \sigma(L_i)$, with σ running in the set of all permutations of the parameters $L_1, ..., L_d$, and the summations over $n_1, ..., n_d$ being taken in the given order. Applying (70) to perform the sum over n_d, one gets

$$\begin{aligned} E_d\left(\nu; L_1, ..., L_d\right) &= -\frac{1}{2d} \sum_{i=1}^{d} E_{d-1}\left(\nu; ..., \widehat{L_i}, ...\right) \\ &+ \frac{\sqrt{\pi}}{2d\,\Gamma(\nu)} \Gamma\left(\nu - \frac{1}{2}\right) \sum_{i=1}^{d} \frac{1}{L_i} E_{d-1}\left(\nu - \frac{1}{2}; ..., \widehat{L_i}, ...\right) \\ &+ \frac{2\sqrt{\pi}}{d\,\Gamma(\nu)} W_d\left(\nu - \frac{1}{2}, L_1, ..., L_d\right), \end{aligned} \tag{72}$$

where the hat over the parameter L_i in the functions E_{d-1} means that it is excluded from the set $\{L_1, ..., L_d\}$ (the others being the $d-1$ parameters of E_{d-1}), and

$$\begin{aligned} W_d\left(\eta; L_1, ..., L_d\right) &= \sum_{i=1}^{d} \frac{1}{L_i} \sum_{n_1, ..., n_d=1}^{\infty} \left(\frac{\pi n_i}{L_i \sqrt{(\cdots + \widehat{L_i^2 n_i^2} + \cdots)}}\right)^{\eta} \\ &\times K_\eta\left(\frac{2\pi n_i}{L_i} \sqrt{(\cdots + \widehat{L_i^2 n_i^2} + \cdots)}\right), \end{aligned} \tag{73}$$

with $(\cdots + \widehat{L_i^2 n_i^2} + \cdots)$ representing the sum $\sum_{j=1}^{d} L_j^2 n_j^2 - L_i^2 n_i^2$. In particular, noticing that $E_1\left(\nu; L_j\right) = L_j^{-2\nu} \zeta(2\nu)$, one finds

$$\begin{aligned} E_2\left(\frac{D-2}{2}; L_1^2, L_2^2\right) &= -\frac{1}{4}\left(\frac{1}{L_1^{D-2}} + \frac{1}{L_2^{D-2}}\right) \zeta(D-2) \\ &+ \frac{\sqrt{\pi}\,\Gamma(\frac{D-3}{2})}{4\Gamma(\frac{D-2}{2})} \left(\frac{1}{L_1 L_2^{D-3}} + \frac{1}{L_1^{D-3} L_2}\right) \zeta(D-3) \\ &+ \frac{\sqrt{\pi}}{\Gamma(\frac{D-2}{2})} W_2\left(\frac{D-3}{2}; L_1, L_2\right), \end{aligned} \tag{74}$$

Using the above expression, Eq. (68) can be rewritten as

$$\begin{aligned} m^2(L_1,L_2,T) &\approx \bar{m}_0^2(L_1,L_2,T)+\frac{3\lambda}{\pi^{D/2}} \\ &\left[\left(\frac{1}{L_1^{D-2}}+\frac{1}{L_2^{D-2}}\right)\Gamma\left(\frac{D-2}{2}\right)\zeta(D-2)\right. \\ &+\sqrt{\pi}\left(\frac{1}{L_1L_2^{D-3}}+\frac{1}{L_1^{D-3}L_2}\right)\Gamma\left(\frac{D-3}{2}\right)\zeta(D-3) \\ &\left.+\,2\sqrt{\pi}W_2\left(\frac{D-3}{2};L_1,L_2\right)\right]. \end{aligned} \tag{75}$$

For $D=3$, the first and second terms between brackets of Eq. (75) are divergent due to the ζ-function and Γ-function, respectively. We can deal with these divergences remembering the Laurent expansion of $\zeta(z)$, Eq. (34), considering the expansion of $\Gamma(z)$ around $z=0$,

$$\Gamma(z)=\frac{1}{z}-\gamma_0+\left(\gamma_0+\frac{\pi^2}{6}\right)z+\cdots, \tag{76}$$

and using the fact that $\zeta(0)=-\frac{1}{2}$ and $\Gamma(\frac{1}{2})=\sqrt{\pi}$.

We notice however that, differently from the case treated in the previous section (for a film), where a renormalization procedure was needed, here the two divergent terms generated by the use of formulas (34) and (76) cancel exactly between them. No renormalization is needed. Thus, for $D=3$, taking the bare mass given by $\bar{m}_0^2(L_1,L_2,T)=\alpha\,(T-T_0)$, we obtain the physical boundary-dependent mass in the form,

$$m^2(L_1,L_2,T)\approx\alpha\,(T-T_c(L_1,L_2))\,, \tag{77}$$

with the boundary-dependent critical temperature given by

$$T_c(L_1,L_2)=T_0-\frac{9\lambda\gamma}{2\pi\alpha}\left(\frac{1}{L_1}+\frac{1}{L_2}\right)-\frac{6\lambda}{\pi\alpha}W_2(0;L_1,L_2), \tag{78}$$

where

$$W_2(0;L_1,L_2)=\sum_{n_1,n_2=1}^{\infty}\left\{\frac{1}{L_1}K_0\left(2\pi\frac{L_2}{L_1}n_1n_2\right)+\frac{1}{L_2}K_0\left(2\pi\frac{L_1}{L_2}n_1n_2\right)\right\}. \tag{79}$$

The quantity $W_2(0;L_1,L_2)$, appearing in Eq.(78), involves complicated double sums, very difficult to handle for $L_1\neq L_2$; in particular, it is not analitically possible to take limits such as $L_i\to\infty$. For this reason we will restrict ourselves to the case $L_1=L_2$. For a wire with square cross-section, we have $L_1=L_2=L=\sqrt{A}$ and Eq. (78) reduces to

$$T_c(A)=T_0-C_2\frac{\lambda}{\alpha\sqrt{A}}, \tag{80}$$

where C_2 is a constant given by

$$C_2=\frac{9\gamma}{\pi}+\frac{12}{\pi}\sum_{n_1,n_2=1}^{\infty}K_0(2\pi n_1n_2)\approx 1.6571\,. \tag{81}$$

We see that the critical temperature of the square wire depends on the bulk critical temperature and the Ginzburg-Landau parameters α and λ (which are characteristics of the material constituting the wire), and also on the area of its cross-section. Since T_c decreases linearly with the inverse of the side of the square, this suggests that there is a minimal area for which $T_c(A_{\text{min}}) = 0$,

$$A_{\text{min}} = \left(C_2 \frac{\lambda}{\alpha T_0} \right)^2 ; \tag{82}$$

for square wires of cross-section areas smaller than this value, in the context of our model the transition should be suppressed. On topological grounds, we expect that (apart from appropriate coefficients) our result should be independent of the cross-section shape of the wire, at least for cross-sectional regular polygons.

3.3. Critical Behavior for Grains

We now turn our attention to the case where all three spatial dimensions are compactified, corresponding to the system confined in a box of sides L_1, L_2, L_3. Taking $d = 3$ in Eq. (66) and using Eq. (23), we obtain (for sufficiently small L_1, L_2, L_3 and in the neighborhood of criticality, $m^2 \approx 0$)

$$\begin{aligned} m^2(L_1, L_2, L_3, T) \;\approx\; & \bar{m}_0^2(L_1, L_2, L_3, T) + \frac{6\lambda}{\pi^{D/2}} \Gamma\left(\frac{D-2}{2}\right) \\ & \times \left[\sum_{i=1}^{3} \frac{\zeta(D-2)}{L_i^{D-2}} + 2 \sum_{i<j=1}^{3} E_2\left(\frac{D-2}{2}; L_i, L_j\right) \right. \\ & \left. + 4 E_3\left(\frac{D-2}{2}; L_1, L_2, L_3\right) \right], \end{aligned} \tag{83}$$

where

$$E_3(\nu; L_1, L_2, L_3) = \sum_{n_1, n_2, n_3 = 1}^{\infty} \left[L_1^2 n_1^2 + L_2^2 n_2^2 + L_3^2 n_3^2 \right]^{-\nu} \tag{84}$$

and the function E_2 is given by Eq. (69).

The analytical structure of the function $E_3\left(\frac{D-2}{2}; L_1, L_2, L_3\right)$ can be obtained from the general symmetrized recurrence relation given by Eqs. (72) and (73); explicitly, one has

$$\begin{aligned} E_3\left(\frac{D-2}{2}; L_1, L_2, L_3\right) \;=\; & -\frac{1}{6} \sum_{i<j=1}^{3} E_2\left(\frac{D-2}{2}; L_i, L_j\right) + \frac{\sqrt{\pi}\,\Gamma\left(\frac{D-3}{2}\right)}{6\Gamma\left(\frac{D-2}{2}\right)} \\ & \times \sum_{i,j,k=1}^{3} \frac{(1 + \varepsilon_{ijk})}{2} \frac{1}{L_i} E_2\left(\frac{D-2}{2}; L_j, L_k\right) \\ & + \frac{2\sqrt{\pi}}{3\Gamma\left(\frac{D-2}{2}\right)} W_3\left(\frac{D-3}{2}; L_1, L_2, L_3\right), \end{aligned} \tag{85}$$

where ε_{ijk} is the totally antisymmetric symbol and the function W_3 is a particular case of

Eq. (73). Using Eqs. (74) and (85), the boundary dependent mass can be written as

$$\begin{aligned} m^2(\{L_i\},T) \approx\ & \bar{m}_0^2(L_1,L_2,L_3,T) + \frac{6\lambda}{\pi^{D/2}} \\ & \times \left[\frac{1}{3}\Gamma\left(\frac{D-2}{2}\right) \sum_{i=1}^{3} \frac{1}{L_i^{D-2}} \zeta(D-2) + \frac{\sqrt{\pi}}{6}\zeta(D-3) \right. \\ & \times \Gamma\left(\frac{D-3}{2}\right) \sum_{i<j=1}^{3} \left(\frac{1}{L_i^{D-3} L_j} + \frac{1}{L_j^{D-3} L_i} \right) + \frac{4\sqrt{\pi}}{3} \\ & \times \sum_{i<j=1}^{3} W_2\left(\frac{D-3}{2}; L_i, L_j\right) + \frac{\pi}{6}\zeta(D-4)\Gamma\left(\frac{D-4}{2}\right) \\ & \times \sum_{i,j,k=1}^{3} \frac{(1+\varepsilon_{ijk})}{2} \frac{1}{L_i} \left(\frac{1}{L_j^{D-4} L_k} + \frac{1}{L_k^{D-4} L_j} \right) \\ & + \frac{2\pi}{3} \sum_{i,j,k=1}^{3} \frac{(1+\varepsilon_{ijk})}{2} \frac{1}{L_i} W_2\left(\frac{D-4}{2}; L_j, L_k\right) \\ & \left. + \frac{8\sqrt{\pi}}{3} W_3\left(\frac{D-3}{2}; L_1, L_2, L_3\right) \right] \end{aligned} \tag{86}$$

The first two terms in the square bracket of Eq. (86) diverge as $D \to 3$ due to the poles of the Γ and ζ-functions. However, as it happens in the case of wires, using Eqs. (34) and (76) it can be shown that these divergences cancel exactly one another. After some simplifications, for $D = 3$, the boundary dependent mass (86) becomes

$$\begin{aligned} m^2(\{L_i\},T) \approx\ & \bar{m}_0^2(L_1,L_2,L_3,T) + \frac{6\lambda}{\pi} \left[\frac{\gamma}{2} \sum_{i=1}^{3} \frac{1}{L_i} \right. \\ & + \frac{4}{3} \sum_{i<j=1}^{3} W_2(0; L_i, L_j) + \frac{\pi}{18} \sum_{i,j,k=1}^{3} \frac{(1+\varepsilon_{ijk})}{2} \frac{L_i}{L_j L_k} \\ & + \frac{2\sqrt{\pi}}{3} \sum_{i,j,k=1}^{3} \frac{(1+\varepsilon_{ijk})}{2} \frac{1}{L_i} W_2\left(-\frac{1}{2}; L_j, L_k\right) \\ & \left. + \frac{8}{3} W_3(0; L_1, L_2, L_3) \right] . \end{aligned} \tag{87}$$

As before, since no divergences need to be suppressed, we can take the bare mass given by $\bar{m}_0^2(L_1,L_2,L_3,T) = \alpha(T - T_0)$ and rewrite the renormalized mass as $m^2(L_1,L_2,L_3,T) \approx \alpha\,(T - T_c(L_1,L_2,L_3))$. The expression of $T_c(L_1,L_2,L_3)$ can be easily obtained from Eq. (87), but it is a very complicated formula, involving multiple sums, which makes almost impossible a general analytical study for arbitrary parameters L_1, L_2, L_3; thus, we restrict ourselves to the situation where $L_1 = L_2 = L_3 = L$, corresponding to a cubic box of volume $V = L^3$. In this case, the boundary dependent critical temperature reduces to

$$T_c(V) = T_0 - C_3 \frac{\lambda}{\alpha V^{1/3}} , \tag{88}$$

where the constant C_3 is given by (using $K_{-\frac{1}{2}}(z) = \sqrt{\frac{\pi}{2z}}e^{-z}$)

$$C_3 = 1 + \frac{9\gamma}{\pi} + \frac{12}{\pi} \sum_{n_1,n_2=1}^{\infty} \frac{e^{-2\pi n_1 n_2}}{n_1} + \frac{48}{\pi} \sum_{n_1,n_2=1}^{\infty} K_0(2\pi n_1 n_2)$$
$$+ \frac{48}{\pi} \sum_{n_1,n_2,n_3=1}^{\infty} K_0\left(2\pi n_1 \sqrt{n_2^2 + n_3^2}\right) \approx 2.7657 \,. \tag{89}$$

Thus, the minimal volume of the cubic grain sustaining the transition is

$$V_{\min} = \left(C_3 \frac{\lambda}{\alpha T_0}\right)^3 . \tag{90}$$

Concluding, in this section we focused on the situations with $D = 3$ and $d = 2, 3$, corresponding (in the context of condensed matter systems) to wires and grains, respectively, undergoing phase transitions which may be described by (mean-field) Ginzburg-Landau models. This generalizes to more compactified dimensions the investigation on the superconducting transition in films, described in the subsection **1.2.2.** In all cases studied so far, in the absence of gauge fluctuations, we found that the boundary-dependent critical temperature decreases linearly with the inverse of the linear dimension L:

$$T_c(L) = T_0 - C_d \frac{\lambda}{\alpha L} , \tag{91}$$

where α and λ are the Ginzgurg-Landau parameters, T_0 is the bulk transition temperature and C_d is a constant equal to 1.1024, 1.6571 and 2.6757 for $d = 1$ (film), $d = 2$ (square wire) and $d = 3$ (cubic grain), respectively. Such behavior suggests the existence of a minimal size of the system below which the transition is suppressed.

These findings seems to be in *qualitative* agreement with results for the existence of a minimal thickness for disappearance of superconductivity in films [28, 29, 30, 31, 32, 33, 34]. Experimental investigations in nanowires searching to establish whether there is a limit to how thin a superconducting wire can be, while retaining its superconducting character, have also drawn the attention of researchers; for example, in Ref. [35] the behavior of nanowires has been studied. Similar questions have also been raised concerning the behavior of superconducting nanograins [36, 37]. Nevertheless, an important point to be emphasized is that our results are obtained in a field-theoretical framework and do not depend on microscopic details of the material involved nor account for the influence of manufacturing aspects of the sample; in other words, our results emerge solely as a topological effect of the compactification of the Ginzburg-Landau model in a subspace. Detailed microscopic analysis is required if one attempts to account quantitatively for experimental observations which might deviate from our mean field results.

4. Critical Behavior of Type-II Superconducting Films in a Magnetic Field

Until now, we have considered second-order phase transitions in confined systems not taking into account the possible existence of an external field interacting with the order parameter. In this respect, for the case of superconductors, we neglect the minimal coupling with the vector potential when an external magnetic field is applied and, in its absence, the intrinsic gauge fluctuations. In this section, we investigate how the transition temperature for a film behaves as a function of its thickness and of the intensity of an applied magnetic field.

4.1. Coupling-Constant Correction in the Presence of an External Magnetic Field

We shall take the uniform external magnetic field, $\mathbf{H} = H\hat{e}_3$, normal to the film. To describe the critical curve of a type-II superconductor close to the upper critical field, we will neglect gauge fluctuations of the magnetic vector potential, that is, we take $\nabla \times \mathbf{A} = \mathbf{H}$. Therefore, the hamiltonian density of interest is given by

$$\mathcal{H} = |(\nabla - ie\mathbf{A})\phi|^2 + \bar{m}_0^2 |\phi|^2 + u \left(|\phi|^2\right)^2, \tag{92}$$

where, as before, ϕ is an N-component vector and $\bar{m}_0^2$ is the mass parameter which depends on L, H and T, in such way that

$$\lim_{L\to\infty} \lim_{H\to 0} \bar{m}_0^2(L,H,T) = m_0^2(T) \equiv \alpha\,(T - T_0)\,. \tag{93}$$

Let us initially consider the hamiltonian (92) refering to a bulk superconductor. We choose a gauge such that $\mathbf{A} = (0, xH, \mathbf{0})$. In this case, the part of hamiltonian $H = \int d^D r\, \mathcal{H}$ quadratic in ϕ becomes, after integration by parts, $-\int d^D r\, \phi \mathcal{D} \phi$ where the differential operator $\mathcal{D}$ is

$$\mathcal{D} = \nabla^2 - 2i\omega x \partial_y - \omega^2 x^2 - m_0^2, \tag{94}$$

with $\omega = eH$ being the cyclotron frequency. So, the natural basis to expand the field operators is the set of the normalized eigenfunctions of the operator $\mathcal{D}$ (the Landau basis),

$$\chi_{l,p_y,\mathbf{p}}(\mathbf{r}) = \frac{1}{\sqrt{2^l l!}} \left(\frac{\omega}{\pi}\right)^{\frac{1}{4}} e^{i\mathbf{p}\cdot\mathbf{z}} e^{ip_y y} e^{-\omega(x - p_y/\omega)^2/2} H_l\left[\sqrt{\omega}(x - \frac{p_y}{\omega})\right], \tag{95}$$

where H_l are the Hermite polynomials; the corresponding energy eigenvalues are

$$E_l(|\mathbf{p}|) = |\mathbf{p}|^2 + (2l+1)\omega + m_0^2 \tag{96}$$

and, in the above equations, $\mathbf{p}$ and $\mathbf{z}$ are $(D-2)$-dimensional vectors.

Therefore, the propagator (incorporating the effects of the magnetic field) can be written as [4, 5]

$$G(\mathbf{r}, \mathbf{r}') = \int \frac{d^{D-2}p}{(2\pi)^{D-1}} \int dp_y \sum_{l=0}^{\infty} \frac{\chi_{l,p_y,\mathbf{p}}(\mathbf{r}) \chi^{\star}_{l,p_y,\mathbf{p}}(\mathbf{r}')}{|\mathbf{p}|^2 + (2l+1)\omega + m_0^2}. \tag{97}$$

To calculate the coupling-constant correction, we have to find the appropriated expression for the single one-loop subdiagram (the single bubble in Fig. 1). This is a complicated expression that will be treated here with the approximation of neglecting all Landau levels except the lowest one. Such an approximation can only be used when the magnetic field is high and, so, the Landau levels are well separated. Thus, the use of this approximation will lead to results that are only valid at low temperatures. Notice that, this observation is not restricted to the bulk case but it also applies to a film if the magnetic field is perpendicular to the film planes.

Now, taking into account the generalized Matsubara prescription (10) for a film, we obtain the expression of the single one-loop subdiagram in the form

$$\Pi(D,L,\omega) = \frac{1}{L}\sum_{n=-\infty}^{\infty}\frac{\omega}{2\pi}\int\frac{d^{D-3}k}{(2\pi)^{D-3}}\frac{1}{[\mathbf{k}^2+\omega_n^2+m^2+\omega]^2}, \tag{98}$$

$\mathbf{k}$ being a $(D-3)$-dimensional vector and $\omega_n = 2\pi n/L$.

The sum over n and the integral over k can be treated using the formalism developed in the preceding sections, adapted to the situation under study. Again, using the dimensional regularization formula (14) and the representations (18) and (23) for the gamma and the Bessel functions, respectively, we obtain

$$\begin{aligned}\Pi(D,L,\omega) &= \frac{\omega}{2(2\pi)^{D/2}}\left[\frac{1}{4}\Gamma\left(3-\frac{D}{2}\right)\left(\frac{m^2+\omega}{2}\right)^{\frac{D}{2}-3}\right. \\ &\left.+\sum_{n=1}^{\infty}\left(\frac{\sqrt{m^2+\omega}}{nL}\right)^{\frac{D}{2}-3}K_{\frac{D}{2}-3}(nL\sqrt{m^2+\omega})\right]. \end{aligned}\tag{99}$$

As before, we suppress the polar term proportional to $\Gamma(3-\frac{D}{2})$ in the above equation, leading to the renormalized one-loop subdiagram

$$\Pi_R(D,L,\omega) = \frac{\omega}{2(2\pi)^{D/2}}\sum_{n=1}^{\infty}\left(\frac{\sqrt{m^2+\omega}}{nL}\right)^{\frac{D}{2}-3}K_{\frac{D}{2}-3}(nL\sqrt{m^2+\omega}). \tag{100}$$

The renormalized zero-external momenta four-point function at leading order in $\frac{1}{N}$ is given by the sum of all chains of single one-loop diagrams of the kind shown in Fig. 1. This sum gives for the L and ω -dependent four-point function, at the lowest Landau level approximation, the expression

$$\Gamma^{(4)}_{D,R}(0,L,\omega) \equiv u_R(D,L,\omega) = \frac{u}{1+Nu\,\Pi_R(D,L,\omega)}, \tag{101}$$

from which, in an analogous way as done in subsection **1.2.3**, we define the renormalized coupling constant

$$Nu_R(D,L,\omega) \equiv \lambda_R(D,L,\omega) = \frac{\lambda}{1+\lambda\,\Pi_R(D,L,\omega)}, \tag{102}$$

where $\lambda = Nu$ corresponds to the renormalized coupling constant in absence of boundaries and of external field.

4.2. The Gap Equation and the Critical Curve

In order to study the critical behavior of a type-II film of thickness L, in an external magnetic field perpendicular to it, we have to consider the large-N gap equation for the correlation length (30), properly adapted to the present situation. In the disordered phase, $\varphi_0 = 0$, and we have

$$\xi^{-2} = \bar{m}_0^2 + \omega + \frac{24\,\lambda_R}{L} \sum_{n=-\infty}^{\infty} \frac{\omega}{2\pi} \int \frac{d^{D-3}k}{(2\pi)^{D-3}} \frac{1}{\mathbf{k}^2 + \omega_n^2 + \xi^{-2}}. \tag{103}$$

It is important to observe that, in the presence of a magnetic field, the correlation length is related to the renormalized mass by $\xi^{-2} = m^2(L,\omega) + \omega$, since the pole of the propagator occurs at $m^2(L,\omega) = -\omega$ [4, 5]. Notice that $\lambda_R(D, L, \omega)$, the renormalized (L,ω)-dependent coupling constant, is itself a function of ξ^{-2}, via the mass $m(L,\omega)$ present in Π_R. Criticality, as usual, is reached making $\xi^{-2} \to 0$.

Calculating the sum and the integral in the above expression, in a similar manner as that used to get Eq. (29), Eq.(103) becomes

$$\xi^{-2} = \bar{m}_0^2 + \omega + \frac{24\omega\,\lambda_R}{(2\pi)^{D/2}} \sum_{n=1}^{\infty} \left[\frac{\xi^{-1}}{nL}\right]^{\frac{D-4}{2}} K_{\frac{D-4}{2}}(nL\xi^{-1})\,. \tag{104}$$

Close to criticality $\xi^{-2} \approx 0$ and thus, using (31), we obtain

$$\xi^{-2} \approx \bar{m}_0^2 + \omega_c + \frac{3\,\omega_c \lambda_R(D, L, \omega_c)}{\pi^{D/2}} \Gamma(\frac{D}{2} - 2) L^{4-D} \zeta(D-4), \tag{105}$$

where

$$\lambda_R(D, L, \omega_c) \approx \frac{\lambda}{1 + \lambda\,\omega_c\, A(D) L^{6-D} \zeta(D-6)}, \tag{106}$$

with $A(D) = \frac{1}{32\pi^{D/2}}\Gamma(\frac{D}{2} - 3)$. Notice that we have used the subscript c in the cyclotron frequency (actually, the external field) to indicate that we are in the region of criticality.

We find immediately that there are no divergences in the parcels appearing in Eq. (105) and (106) as $D \to 3$. Therefore, no renormalization procedure is needed and we are allowed to make the simplest choice for the mass parameter: $\bar{m}_0^2 = \alpha(T - T_0)$. In this way, making $D = 3$ and $\xi^{-2} = 0$ in Eqs. (105) and (106), we obtain the critical curve for a film in a perpendicular magnetic field as

$$\alpha(T_c - T_0) + \omega_c + \frac{1440\,\lambda\,\omega_c\,L}{2880\,\pi + \lambda\,\omega_c\,L^3} = 0\,, \tag{107}$$

where we have used that $\Gamma(-3/2) = 4\sqrt{\pi}/3$ and $\zeta(-3) = 1/120$. Notice that, as $L \to \infty$, the critical curve (107) reduces to $\alpha(T_c - T_0) + \omega_c = 0$, reproducing the known result for the upper critical field in bulk superconductors $H_{c2} = \eta^{-2}$.

We can rewrite Eq. (107) in a way such that the upper critical field is expressed as a function of the critical temperature and the film thickness. Using the zero-temperature Ginzburg-Landau coherence length, $\eta_0 = (\alpha T_o)^{-1/2}$, we can introduce the following dimensionless quantities,

$$h = \omega_c\,\eta_0^2\,, \qquad t = \frac{T_c}{T_0}\,, \qquad l = \frac{L}{\eta_0}\,, \tag{108}$$

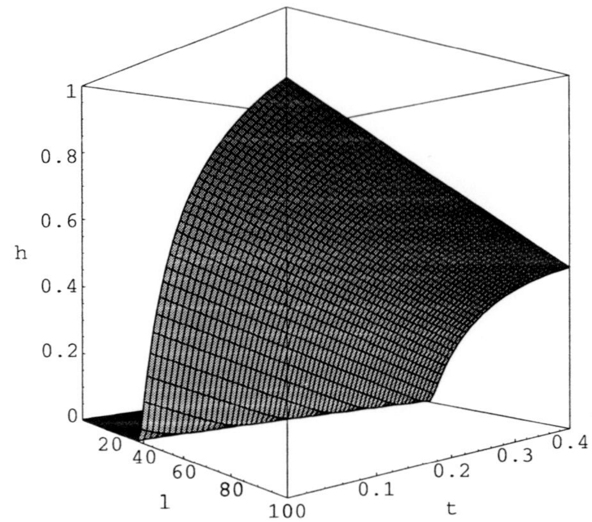

Figure 3. Plot of the surface $h = h(l,t)$, fixing $\beta = 10^3$.

corresponding respectively to reduced critical field, critical temperature and film thickness. With these parameters, Eq. (107) becomes

$$\begin{aligned} h(l,t) &= \frac{1}{2\beta l^3}\left\{-1440\beta l + \beta l^3(1-t) - 2880\pi \right. \\ &\left. + \left[(1440\beta l - \beta l^3(1-t) + 2880\pi)^2 + 11520\pi\beta l^3(1-t)\right]^{1/2}\right\}, \end{aligned} \tag{109}$$

where $\beta = \lambda\eta_0$. The surface $h = h(l,t)$ is illustrated in Fig. 3. We recall that, since we have used the lowest Landau level approximation in our calculations, this surface is only meaningful for high values of the external field, that is for low temperatures and thick films.

We can see from Fig. 3 that each value of l defines a critical line on the $h \times t$ plane, corresponding to a film of thickness L. This set of critical lines also suggests the existence of a minimal value for the thickness L below which superconductivity is suppressed. Indeed, this can be seen from the plot of the reduced critical field at zero temperature, h_0, as a function of the inverse of the reduced film thickness, shown in Fig. 4. As before, we emphasize that our results are obtained within the framework of field theory, emerging from the topological nature of the spatial compactification, and so do not depend on specific characteristics of the material and of the sample.

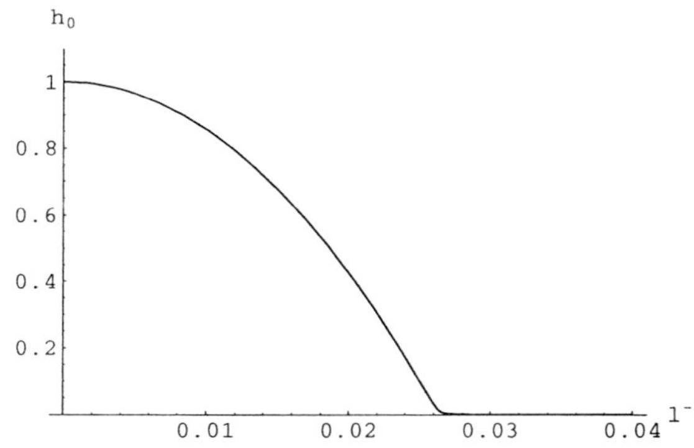

Figure 4. Zero-temperature reduced critical field as a function of the inverse of the reduced film thickness, for $\beta = 10^3$.

5. Concluding Remarks

In this chapter we have reviewed the application of methods of the compactified quantum field theory to the Ginzburg-Landau model for second-order phase transitions. In particular, we have addressed the problem of superconducting transitions in films, wires and grains. Our motivation is the question of whether there is a minimal size of the system (film thickness, transverse-section area of wires and volume of grains) for the superconductivity to be sustained, which has received great attention from experimentalists [28]-[37].

Here, using a generalization of the Matsubara prescription to account for spatial confinement, we have shown that a minimal size of the system (below which superconductivity is suppressed) does exist. The critical temperature decreases linearly with the inverse of the relevant linear dimension characterizing the confined system, Eq. (91). This agrees generally with scaling arguments [9, 10]. Such behavior reproduces, at least qualitatively, the mentioned experimental findings for thin films, nanowires and nanograins. From a microscopic point of view, this has be discussed as due to proximity, localization and Coulomb-interaction effects. Our theoretical investigation of the critical behavior of type-II superconducting films in a magnetic field also supports the existence of such a minimal thickness.

It is worthy to emphasize that our results, obtained in the context of the quantum field theory, are independent of microscopic details of the superconducting material and are not influenced by manufacturing aspects of the sample, as coating or the nature of the substrate. In fact, our results arise as a topological effect of the compactification of the Ginzburg-Landau model in one, two or three spatial dimensions.

Acknowledgements

We are grateful to our colleagues L. M. Abreu, C. de Calan, F. C. Khanna and I. Roditi for interesting discussions and their interest in this work. We also thanks CNPq (Brazil) for financial support.

References

[1] L. D. Landau, *Phys. Z. Sowjet* **11**, 545 (1937).

[2] V. L. Ginzburg and L. D. Landau, *Zh. Eksp. Teor. Fiz.* **20**, 1064 (1950).

[3] I. Afleck and E. Brézin, *Nucl. Phys.* **257**, 451 (1985).

[4] I. D. Lawrie, *Phys. Rev. B* **50**, 9456 (1994).

[5] I. D. Lawrie, *Phys. Rev. Lett.* **79**, 131 (1997).

[6] E. Brézin, D. R. Nelson and A. Thiaville, *Phys. Rev. B* **31**, 7124 (1985).

[7] L. Radzihovsky, *Phys. Rev. Lett.* **74**, 4722 (1995).

[8] C. de Calan, A. P. C. Malbouisson and F. S. Nogueira, *Phys. Rev. B* **64**, 212502 (2001).

[9] J. Zinn-Justin, *Quantum Field Theory and Critical Phenomena* (Clarendon Press, Oxford, 1996).

[10] J. L. Cardy ed., *Finite Size Scaling* (North Holland, Amsterdam, 1988).

[11] J. Polchinski, *Commun. Math. Phys.* **104**, 37 (1986).

[12] J. J. Atick and E. Witten, *Nucl. Phys.* **B310**, 291 (1988).

[13] L. H. Ford and N. F. Svaiter, *Phys. Rev. D* **51**, 6981 (1995).

[14] A. E. Santana, J. M. C. Malbouisson, A. P. C. Malbouisson and F. C. Khanna, "Thermal Field Theory: Algebraic Aspects and Applications to Confined Systems", in *Non-Linear Dynamics and Fundamental Interactions* , eds. F. Khanna and D. Matrasulov (Springer, New York, 2005).

[15] A. P. C. Malbouisson and J. M. C. Malbouisson, *J. Phys. A: Math. Gen.* **35**, 2263 (2002).

[16] A. P. C. Malbouisson, J. M. C. Malbouisson and A. E. Santana, *Nucl. Phys. B* **631**, 83 (2002).

[17] A. P. C. Malbouisson, *Phys. Rev. B* **66**, 092502 (2002)

[18] A. P. C. Malbouisson, J. M. C. Malbouisson and A. E. Santana, *Phys. Lett. A* **318**, 406 (2003).

[19] L. M. Abreu, A. P. C. Malbouisson, J. M. C. Malbouisson and A. E. Santana, *Phys. Rev. B* **67**, 212502 (2003).

[20] L. M. Abreu, A. P. C. Malbouisson and I. Roditi, Physica A **331**, 99 (2004); *Eur. Phys. J. B* **37**, 515 (2004).

[21] L. M. Abreu, C. de Calan and A. P. C. Malbouisson, *Phys. Lett. A* **333**, 316 (2004).

[22] A. P. C. Malbouisson, J. M. C. Malbouisson, A. E. Santana and F. C. Khanna, *Mod. Phys. Lett. A* **20**, 965 (2005).

[23] L. M. Abreu, C. de Calan, A. P. C. Malbouisson, J. M. C. Malbouisson and A. E. Santana, *J. Math. Phys.* **46**, 012304 (2005).

[24] C. Itzykson, J-B. Zuber, *Quantum Field Theory* (McGraw-Hill, New York, 1980), p. 451.

[25] E. Elizalde and E. Romeo, *J. Math. Phys.* **30**, 1133, (1989); E. Elizalde, "Ten physical applications of spectral *zeta* functions", *Lecture Notes in Physics*, Springer-Verlag, Berlin, (1995).

[26] M. Abramowitz, I.A. Stegun, eds., *Handbook of Mathematical Functions* (Dover, New York, 1965).

[27] K. Kirsten, *J. Math. Phys.* **35**, 459 (1994).

[28] D. G. Naugle and R. E. Glover, *Phys. Lett. A* **28**, 611 (1969).

[29] H. Raffy, R. B. Laibowitz, P. Chaudhari and S. Maekawa, *Phys. Rev. B* **28**, R6607 (1983).

[30] J. H. Quateman, *Phys. Rev. B* **34**, 1948 (1986).

[31] J. Simonin, *Phys. Rev. B* **33**, 7830 (1986).

[32] A. Asamitsu, M. Iguchi, A. Ichikawa and N. Nishida, *Physica B* **194**, 1649 (1994).

[33] M. S. M. Minhaj, S. Meepagala, J. T. Chen and L. E. Wenger, *Phys. Rev. B* **49**,15235 (1994).

[34] A. V. Pogrebnyakov, et al, *Appl. Phys. Lett.* **82**, 4319 (2003).

[35] A. Bezryadin, C. N. Lau and M. Tinkham, *Nature* **404**, 971 (2000);
N. Markovic, C. N. Lau and M. Tinkham, *Physica C* **387**, 44 (2003).

[36] D. C. Ralph, C. T. Black and M. Tinkham, *Phys. Rev. Lett.* **76**, 688 (1996); *Phys. Rev. Lett.* **78**, 4087 (1997).

[37] J. von Delft and D. C. Ralph, *Phys. Rep.* **345**, 61 (2001);
J. von Delft, *Ann. Phys.* (Leipzig) **10**, 219 (2001).

INDEX

D

E

L

M

N

O

P

Q

R

S

T

U

V

W

X

Y